T0236029

FIBER LASERS

MATLAB® is a trademark of The MathWorks, Inc. and is used with permission. The MathWorks does not warrant the accuracy of the text or exercises in this book. This book's use or discussion of MATLAB® software or related products does not constitute endorsement or sponsorship by The MathWorks of a particular pedagogical approach or particular use of the MATLAB® software.

First edition published 2022
by CRC Press
6000 Broken Sound Parkway NW, Suite 300, Boca Raton, FL 33487-2742

and by CRC Press
2 Park Square, Milton Park, Abingdon, Oxon, OX14 4RN

© 2022 selection and editorial matter, Johan Meyer, Justice Sompo, Suné von Solms; individual chapters, the contributors

CRC Press is an imprint of Taylor & Francis Group, LLC

Reasonable efforts have been made to publish reliable data and information, but the author and publisher cannot assume responsibility for the validity of all materials or the consequences of their use. The authors and publishers have attempted to trace the copyright holders of all material reproduced in this publication and apologize to copyright holders if permission to publish in this form has not been obtained. If any copyright material has not been acknowledged please write and let us know so we may rectify in any future reprint.

Except as permitted under U.S. Copyright Law, no part of this book may be reprinted, reproduced, transmitted, or utilized in any form by any electronic, mechanical, or other means, now known or hereafter invented, including photocopying, microfilming, and recording, or in any information storage or retrieval system, without written permission from the publishers.

For permission to photocopy or use material electronically from this work, access www.copyright.com or contact the Copyright Clearance Center, Inc. (CCC), 222 Rosewood Drive, Danvers, MA 01923, 978-750-8400. For works that are not available on CCC please contact mpkbookspermissions@tandf.co.uk

Trademark notice: Product or corporate names may be trademarks or registered trademarks and are used only for identification and explanation without intent to infringe.

ISBN: 978-0-367-54348-8 (hbk)
ISBN: 978-1-032-18816-4 (pbk)
ISBN: 978-1-003-25638-0 (ebk)

DOI: 10.1201/9781003256380

Typset in Times
by MPS Limited, Dehradun

FIBER LASERS

Fundamentals with MATLAB®
Modelling

Johan Meyer, Justice Sompo, and
Suné von Solms

CRC Press
Taylor & Francis Group
Boca Raton London New York

CRC Press is an imprint of the
Taylor & Francis Group, an **Informa** business

Dedication

This book contains the authors' original research. It thus represents a new contribution to the development and modelling of optical fiber lasers.

Each chapter of the book was scrutinised and reviewed before inclusion in the book. The book as a whole was subjected to a double-blind peer review by three independent experts in the field of optical systems and fiber lasers. At least one of the peer reviewers were of international origin. All the review reports as well as the authors' responses thereto, have been kept. After the round of critical peer review, changes were suggested based on the comments received from the peer reviewers.

The authors revised the book contents taking into consideration the comments and suggestions from the peer reviewers.

This book contributes to the scholarly knowledge of the development and understanding of optical fiber lasers. The intended audience of the book include scholars, designers, mathematical modellers, and manufacturers of optical systems and fiber lasers.

Contents

Preface

It is over 60 years since the first fiber laser was experimentally demonstrated. Fiber laser is a technology for the creation of light by the process of simulated emission using an active medium made of optical fiber that has been doped with rare-earth material such as Erbium, Ytterbium, Neodymium, Thulium, Praseodymium, and Holmium. Since its invention, we have witnessed its presence in various applications in the field of science and engineering. The scientific and technological improvements have allowed fiber laser light beams to be produced from continuous-wave operation up to as short as a femtosecond and below. The wavelengths covered spanning almost the entire optical spectrum with output powers of magnitude up to the order of terawatt. The possible wavelength range, power scale, pulse width, and light coherence length make fiber lasers suitable for a variety of industrial and scientific applications. The creation of the Internet and World Wide Web was possible due to tremendous scientific advancement created by photonics devices including light sources, optical fibers, and optical detectors. Fiber lasers can be engineered for industrial machining for materials processing and marking.

This book serves to introduce professionals to the basic concepts of light propagation in optical fiber and the working principles of fiber lasers. The book provides a coherent presentation of the issues of computational photonics applied to fiber lasers. The main motivation for developing an approach described in the book was to establish the foundations needed to understand the working principles of fiber lasers and their different component through step-by-step mathematical modelling and simulation. In this book authors propose a simulation-type approach to explain the fundamentals of fiber lasers. A self-contained development that includes theoretical foundations of fiber lasers and the MATLAB® code aimed at detailed the simulation of different fiber laser cavity configurations both in continuous-wave and pulsing operating regime based on the Q-switching technique are presented. The modelling coverage of fiber lasers given in this book is unique and should serve as a very good research guideline and design understanding to the subject of fiber lasers.

The book has evolved from the research and lectures given by the authors to scholars and professionals in Photonics. The authors believe that one can master the concepts of fiber lasers by studying not only the theory but also by developing numerical models and performing numerical experiments via computer simulations using commercial software or those developed and customized in-house. With this methodology of presentation approach, the authors are of the view that readers should have some sense of perspective of the design and implementation of fiber laser cavity configuration which, unfortunately, current literature does make not make it easy to read due to considerable mathematical complexity.

The book contains 8 chapters. An important part of the book is the MATLAB code provided to solve the mathematical models and simulate key concepts elaborated in each chapter. The first chapter introduces the history, technological progress milestones in the field of fiber laser. Chapter 2 discusses the important concepts of fiber optics and doped optical fiber which are key components in the fiber laser system. Chapter 3 discusses the physics of rare-earth ions and introduce general concepts of lasers. Numerical methods used throughout the book are presented in Chapter 4. Chapter 5 is about modelling continuous wave fiber laser systems. In Chapter 6, Q-switched fiber laser modelling and simulation are discussed. Chapter 7 and 8 are chronological presentations of state of art of narrow linewidth and high-power fiber lasers, respectively. Finally, I would like to wish the potential readers the similar joy of reading the book and experimenting with programs as I had reading it.

Dr Kaboko Jean-Jacques Monga

1 Fundamentals of Fiber Lasers

1.1 INTRODUCTION

Fiber lasers were proposed right after the first demonstration of a working laser by Theodore Maiman in the 1960s (Maiman 1960). The possibility of using silica fiber as a gain medium for lasers was introduced by Elias Snitzer et al. (1989). Soon after, they made the first demonstration of a working silica fiber laser (Snitzer and Koester 1963). Since then, fiber lasers have evolved to be one of the most successful technologies in photonics.

Fiber lasers are low cost and relatively easy to produce sources of excellent beam quality light in the infrared, visible, and ultraviolet ranges of the optical spectrum. For this reason, they are desirable for applications in telecommunication, medicine, manufacturing industry, and metrology. Several types of fiber lasers have been developed to meet an always-increasing demand from various industries. These include continuous wave, pulsed, narrow-linewidth, single frequency, tuneable and high-power fiber lasers. The fiber laser basic structure consists of a gain medium made of an optical fiber doped with rare-earth ions and a feedback mechanism brought about by a resonant cavity formed by either physical mirrors or the frequency selective fiber Bragg grating (FBG), printed into the core of the fiber.

In this introductory chapter, we will provide a short overview of the important properties of fiber lasers as well as the milestones in their development. We will highlight why fiber lasers are such a desirable light source. We will briefly discuss the different types of fiber lasers. Finally, we will mention some of the most important applications of fiber lasers.

1.2 INTEREST OF LASERS IN FIBER FORM

The telecommunications industry became first interested in Erbium-doped fiber because of their excellent amplifying characteristics in the 1550 nm region. With the maturation of fabrication techniques of fiber optics such as Modified Chemical Vapour Deposition (MCVD) (Cognolato 1995) and the availability of powerful and reliable semiconductor lasers diodes emitting in the absorption band of most rare-earth ions, it became possible to manufacture Erbium-Doped Fiber Amplifiers (EDFA) (W Naji et al. 2011), useful for long haul optical communication systems.

Obtaining a laser from these amplifiers only required adding a feedback mechanism to them. This can be achieved using mirrors in a Fabry-Perot configuration. These mirrors can be of various types, the simplest being the coating of the end faces of cleaved optical fiber with reflective material. The reflection at the air-glass interface is enough to trigger oscillation because of the particularly high gain

DOI: 10.1201/9781003256380-1

available in the rare-earth-doped fiber gain medium. A more effective way of obtaining feedback in fiber lasers is by using fiber Bragg gratings (Hill and Meltz 1997). Fiber Bragg gratings have the advantage of being frequency-selective allowing the realization of single longitudinal mode fiber lasers (Jauncey et al. 1988).

The core diameter of a standard single-mode fiber can vary from 3 to 10 micron; therefore, significant light intensity can develop with relatively low propagating power. The fiber geometry results in high confinement of the pump and laser fields over long distances thereby providing long interaction lengths between the rare-earth dopant and the pump field. The consequence of the foregoing is the high population inversion which means high optical gains at relatively low pump powers. The high gain also allows the use of high loss elements such as etalons and diffraction gratings without a significant increase of threshold or reduction of output power. Such components are often essential to achieve single longitudinal mode operation of the fiber laser (Huang et al. 2005).

The broadening of the absorption bands of rare-earth ions provides an incomparable opportunity to absorb a large band of optical frequencies. Therefore, fiber lasers possess the ability to convert the low-quality output radiation from low-cost laser diodes, transmitting at several wavelengths, into a high-brightness coherent source, desirable for applications such as remote sensing (Fu et al. 2012) and fiber-based communications systems (Gangwar and Sharma 2012). On the other hand, the emission spectrum is also broad, allowing the design of lasers emitting over several wavelengths as well as tuneable lasers. Tuneable fiber lasers are designed by incorporating a wavelength selective element into the cavity making it possible to achieve tunability over wavelengths of 50 nm or more (Chen et al. 2005). Furthermore, the flexibility of the fiber enables long cavity lengths to be established while taking up a small volume of space. Long cavities result in narrow-linewidth fiber lasers which are preferred in several applications (Yarutkina et al. 2013). The compatibility of the fiber with several optical components such as couplers, isolators, and wavelength division multiplexers (WDMs) makes it possible to have a compact and robust design because the laser can be made in an all fiber configuration without light leaving the fiber, avoiding tedious alignment of bulk optics. Fiber lasers also offer several design possibilities that allow the control of optical properties such as dispersion and polarization, resulting in a large performance improvement possibility.

The small surface to volume ratio of optical fibers, which is 10 to 50 times larger than that of other types of solid-state lasers, is at the origin of good heat dissipation making fiber lasers excellent candidates for high power lasers. This surface to volume ratio prevents detrimental phenomena resulting from heat build-up such as thermal aberrations (Paun et al. 2009).

The doping of the glass matrices with various rare-earth ions opens the possibility for multiple operation wavelengths in the near and far-infrared spectrum. The most used dopants are Neodymium and Ytterbium (1 μm) (Zervas 2014), Erbium, Erbium, and Ytterbium (1.5 μm) (Song et al. 2009), Thulium and Holmium (around 2 μm) (Hanna et al. 1988). Fiber lasers built using these dopants offer a variety of output wavelengths, some of which are of great interest in telecommunications and other applications. For example, the output in the 1.3 to 1.5-micron spectral region which corresponds to low-loss transmission windows for silica glass has been extremely useful in the past decade (Agrawal 2012). Nowadays, the 2 to 3-micron spectral region is subject to intense research because lasers at these wavelengths have potential application as light sources for future generations of low-loss mid-infrared telecommunications fiber systems (Pollnan and Jackson 2001). Another benefit of fiber lasers results from their inherent wave guiding property which allows easy coupling of fibers and various optical components.

1.3 CHRONOLOGICAL REVIEW OF FIBER LASERS

The origin of fiber lasers can be traced back to early years of laser technology when Snitzer and Koester published results for a multi-component fiber laser in 1964 (Snitzer and Koester 1963). Soon after, intensive research was conducted to investigate the possibility of using fiber lasers in optical information processing (Luo et al. 2017) as well as optical amplification (Giles and Desurvire 1991). It became obvious that fiber lasers have huge potential after Kao and Sham speculated on the possibility of using fiber optics in telecommunication . However, due to the weak output power of earlier devices, the concept did not find practical application and was regarded as a mere laboratory interest. As a result, from 1975 to 1985 little research was published. The regain of interest in fiber lasers was stimulated by a conjunction of factors, amongst them the improvement in the fiber manufacturing techniques, availability of high-quality laser diodes emitting at a wavelength corresponding to the absorption wavelength of rare-earth ions, new host materials and the improvement in passive optical components including isolators, couplers, and fiber Bragg gratings. The research group at Southampton University contributed significantly to the field during that period. Using an extension of MCVD technologies, they were able to manufacture doped fibers allowing the demonstrating of Q-switching, mode-locking, and single longitudinal mode operation (Nagel et al. 1982). Soon afterwards, the British Telecom Research Laboratory made significant advances in understanding the core concepts of fiber laser technology such as gain, excited-state absorption, and the relation between host material and the range of lasing wavelength. Specifically, they pioneered the use of Fluorozirconate (Kaczmarek and Karolczak 2007) glass as host to increase the range of emission wavelength.

1.3.1 ERBIUM AND THE YTTERBIUM CO-DOPING

Later, Erbium-doped fiber lasers were reported. Erbium has an important property which is lasing in the 1530 to approximately 1580 nm region corresponding to the second telecommunication window. However, because of luminosity quenching resulting from Erbium ion clustering (Auzel and Goldner 2001), the output power of Erbium-doped fiber lasers was limited to a few milliwatts. In fact, the Erbium ions in glass result in detrimental effects at high concentrations. At doping concentrations around 24×10^{25} ions/m^3, because of the low solubility of the ions in the amorphous glass, Erbium ions tend to form clusters and exchange energy among themselves. This form of energy exchange tends to reduce the conversion efficiency of the laser. The energy efficiency reduction is commonly known in technical terms as luminosity quenching. Such luminosity quenching phenomena include mainly cooperative up-conversion (CUC) (Hwang et al. 2000) and pair induced quenching (PIQ) (Federighi and Di Pasquale 1995). These two phenomena along with excited-state absorption (ESA) (Barmenkov et al. 2009) contribute significantly to the degradation of the overall efficiency of the fiber laser because the energy supposed to contribute to stimulated emission and participate in lasing action is now spent in energy exchange phenomena.

To prevent these up-conversion phenomena from taking place, several techniques have been used when designing the glass of the fiber. Such techniques involve using phosphate or aluminosilicate glass with higher solubility for Erbium ions rather than silica glass (Taccheo et al. 1999). An experiment conducted and reported in a 2001 paper confirmed efficiency of up to 20% when using phosphate glass compared to the case of silica glass (Moghaddam et al. 2011). In addition to using phosphate glasses, co-doping Erbium with Ytterbium (Federighi and Di Pasquale 1995) ions proved to be very beneficial in terms of reducing cluster centres, therefore, luminosity quenching. Ytterbium ions present a simple structure of only two energy levels making any possibility of excited-state absorption impossible. In addition, the emitting spectra of Ytterbium overlap perfectly with that of Erbium at a wavelength around 1550 nm, therefore, when pumping such a structure with appropriate pump wavelength, the energy is first captured by Ytterbium ions and later resonantly transferred to Erbium ions. The Erbium ions decay radiatively to their ground level, releasing photons at the appropriate wavelength. One must notice that for this combination to work perfectly, the concentration of Ytterbium must be higher than that of Erbium. Ytterbium concentrations of 10 to 20 times that of Erbium were reported for optimum results (Yelen et al. 2005).

1.3.2 CONTINUOUS-WAVE AND PULSED FIBER LASERS

Depending on the application of interest, continuous-wave or pulsed fiber lasers may be required. Continuous-wave fiber lasers find applications mainly in low power applications for telecommunication and sensing (Tokita et al. 2010). These types of fiber lasers comprise, the short cavity fiber laser such as the distributed Bragg reflectors fiber laser (DBR) and Distributed Feedback fiber (DFB) laser which most interesting property is single longitudinal mode operation (Ronnekleiv et al. 2003). Single longitudinal mode fiber lasers are described with much detail in Chapter 7 of this book. Other applications of continuous wave fiber lasers include light detection and ranging (LIDAR) and spectroscopy. Applications like tomography in medicine also use continuous-wave fiber lasers. Pulsed fiber lasers, on the other hand, are mostly used in manufacturing and rangefinders. The two main techniques of obtaining pulsed fiber lasers are Q-switching (Russo et al. 2002) and mode-locking. The Q-switched fiber lasers can, in turn, be divided into active Q-switching and passive Q-switching. The active Q-switching involves using an external Q-switcher to open and close the laser cavity periodically changing its Q-factor thus releasing laser pulses periodically. The Q-switcher is in most of the cases an acousto-optic modulator (Cuadrado-Laborde et al. 2007). Most recently, other devices were also used as Q-switchers, including electro-optic modulators, tunable fiber Bragg grating using a piezoelectric stretcher (Kaneda et al. 2004) as well as a combination of a Fabry-Perrot tunable filter and a fiber Bragg grating (Manuel et al. 2016). All these techniques were available for linear and ring cavity laser configurations (Cheng et al. 2007).

1.3.3 SINGLE LONGITUDINAL MODE FIBER LASERS

Applications in spectroscopy, sensing, and medicine require single longitudinal mode operation of the fiber laser. To understand the engineering of single longitudinal mode fiber lasers, one has to understand the cavity longitudinal modes. In a long linear cavity fiber laser, thousands of longitudinal modes can oscillate simultaneously. The number of such modes is proportional to the gain profile of the amplifying medium as well as the length of the cavity. To obtain a single longitudinal mode operation the first method one can think of is to use an external bulk pass-band filter which can filter out undesirable modes and leave only the mode of interest. However, using such a bulk filter in the laser configuration comes with a cost. Filters, most of the time, have insertion losses up to 3 dB, increasing losses in the cavity. The threshold of such lasers is often high and overall efficiency is reduced significantly. In addition, the filter introduces noise which has a negative impact on the noise figure of the laser. Filters also yield cumbersome, less portable devices that offer little interest outside of the laboratory. Such devices have been widely reported combined with different sophisticated methods of reducing the disadvantages (Chen et al. 2005). The most successful techniques involve using fiber Bragg gratings which are wavelength-selective mirrors. This technique leads to robust and more compact devices. Linewidths of up to 47 kHz have been reported for such devices in recent years (Ball et al. 1991). The other method to obtain single longitudinal mode fiber lasing consist of reducing the length of the gain medium in such a way that only one longitudinal mode can oscillate. Such designs have the advantage of overcoming the high threshold limitation inherent to fiber lasers with external filters. Most importantly they offer lightweight and very compact devices. Amongst these devices, Distributed Bragg Reflector fiber lasers (Lin et al. 2012) and Distributed Feedback (Lauridsen et al. 1997) fiber lasers have been sufficiently investigated in the past few decades and have successfully been used as a mature technology. However, the short cavity fiber lasers as they are known in the current literature, are plagued by a severe drawback, which is their low output power. In order to compensate for their short length and still be able to achieve lasing, a high dopant concentration is required. This leads to poor fiber laser performance, therefore solution like using phosphate matrix glass and Ytterbium co-doping has been used. The first attempt to reduce cavity length and achieve single longitudinal mode operation of fiber lasers was through DBR configurations. A DBR fiber laser is a Fabry-Perot cavity that uses fiber Bragg gratings as mirrors and with a cavity length reduced to the minimum acceptable. DBR is a popular type of laser because of their simplicity. However, they are not totally immune to mode hopping (Zhang et al. 2009). Mode hopping is the abrupt change in lasing frequency observed in most lasers and fiber lasers, due to external, uncontrollable factors such as temperature changes or vibrations. The DFB configuration came into play as an elegant solution to the problem of mode hopping. In such a configuration, instead of having a piece of highly doped rare-earth fiber sandwiched between two perfectly wave-length matching fiber Bragg gratings, a fiber Bragg grating is printed throughout the entire length of the doped fiber gain medium with a pi-phase shift in the middle of the Bragg grating. In this way, a very stable single longitudinal mode operation can be achieved. From their modest beginning with a phase shift obtained with a heated wire reported in the Kribledotn paper of 1994 (Asseh et al. 1995), distributed feedback fiber lasers have come a long way to become a mature and reliable technology. This progress was mainly due to less reliable technologies for printing Bragg gratings including processor-controlled translation stages and rare-earth-doped photosensitive fibers.

1.3.4 Power scalability and High Power Fiber Lasers

Applications in manufacturing, defence industries, and medicine required the use of high-power lasers. The gas lasers, namely the CO_2 laser, have been used success-fully since the early days of the lasers industry. The main drawbacks of these gas lasers include poor beam quality, high maintenance cost and low reliability. The need to find more reliable, excellent beam quality high power lasers trigged research in high power fiber lasers. Compared to their gas counterpart, fiber lasers offer desirable characteristics such as robustness, low cost of ownership, low main-tenance cost as well as excellent beam quality. Research on high power fiber lasers can be divided into research in new rear-earth materials and research in the new fiber geometries to optimize laser beam and doped area interaction inside the gain medium. The first high power fiber lasers were Ytterbium-doped fiber lasers (E Snitzer et al. 1988) soon after Thulium and Praseodymium fiber lasers were developed. The most power scaling improvement technique is the cladding pumping introduced by E. Snitzer et al. in 1988 (Snitzer et al. 1988). In this scheme, instead of launching power into the small size and numerical aperture of the fiber core, the pump power is launched into the cladding which has a much larger nu-merical aperture. This improves dramatically the light coupling efficiency. Because of the bigger transverse section, the optical field in the cladding propagates in multimode fashion whereas in the core single-mode propagation is obtained for its diameter is much smaller at the pump wavelength. The generated laser light, however, is perfectly trapped in the small section of the core resulting in a much brighter and intense laser output. For this reason, cladding-pumped fiber lasers are regarded as efficient light brightness converters. The probability of interaction between the pump beam and rare-earth ions is directly proportional to the number of modes supported by the cladding. In other words, the higher the modes supported, the higher the absorption. Therefore, power scaling for the fiber lasers relies on developing fibers with large cladding transverse section sizes and numerical aper-tures. It was shown that the best absorption efficiency was achieved by designing a cladding shape in such a way that it scrambles the propagating modes (Mortensen 2007). In the case of a normal circular core fiber, all the meridional mode, LP_{0n} and a small fraction of skewed, LP_{m0}, modes satisfy this criterion leaving the majority of LP_{mn} mode totally out of contact with the doped core. Consequently, only a tiny fraction of the propagating modes interact with the doped core (Liu et al. 1997).

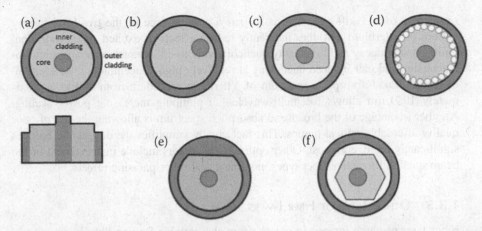

FIGURE 1.1 Most used double-clad-pumped fiber design: (a) centred core (b) off-centred (c) rectangular inner cladding (d) cylindrically arranged air holes inner cladding (e) D-shape inner cladding (f) hexagonal inner cladding.

A straightforward solution to increase the overall absorption is to break the cladding's circular symmetry and increase the fraction of the cladding modes overlapping with the doped core. This can be effectively achieved by designing the fiber transverse section in various ways (Muendel 1996; Liu et al. 1997; Leproux et al. 2001; Doya et al. 2001; Kouznetsov et al. 2001; Morchead and Muendel 2011) as shown in Figure 1.1.

There exist several cladding pumping schemes that have been proposed and used over the years to enhance the brightness of cladding-pumped fiber lasers. These schemes can be classified into two categories, namely side-pumping and end-pumping. In his 1991 paper, T.Y. Fan demonstrated a side pumping scheme, where he successfully used three free-space geometric combinations of pump modules to pump the fiber laser. Additionally, he demonstrated that coupling for up to 100 of such sources could be possible (Fan 1991). Other side pumping techniques involve using a tapered fiber bundle (TFB) (Kosterin et al. 2004; DiGiovanni and Stentz 1999). In each one of these cases, the number of combined pump modules is directly proportional to the cladding diameter as well as the numerical aperture. Additional techniques include using wavelength-multiplexed pumps. In this case, a bulk WDM coupler is often required (Luo et al. 2017). This last technique has the advantage of increasing the brightness of the resulting pump as opposed to the previous technique where the resulting brightness is lower than that of the individual laser's diodes. Side pumping has been investigated and reported with one scheme using total internal reflection occurring in a V-groove made on the cladding (Elias Snitzer et al. 1989). However, this approach has proved to be very difficult to realize in practice for power values in the range of kilowatt. Other side pumping techniques use tapered or angle-polished pump fiber fused to the cladding (Elias Snitzer et al. 1989; Xiao et al. 2010; Valentin and Igor 1999).

The majority of high-power fiber lasers use Thulium and Ytterbium rare-earth doping. This is because the two, ions have shown the best power scaling with power range well over 1 kilowatt. This is in part a consequence of the two-level energy system of Ytterbium that does inherently reduce effects of excited state absorption, multi-phonon decay and luminosity quenching due to effects such as cooperation up-conversion and pair induced quenching at relatively high concentrations. In addition, the rather broad absorption spectrum of Ytterbium extending from 850 to approximately 1080 nm allows for multi-wavelength pumping increasing power scaling. Another advantage of the broadband absorption spectrum is allowing the use of poor quality affordable optical pumps. This fact greatly simplifies the design and reduces significantly the overall cost. Other optimization factors include improvement of the beam quality, the active fiber types, nonlinear and other parasitic effects.

1.3.5 OTHER TYPES OF FIBER LASERS

Fiber laser research interest in recent years also involve Raman fiber lasers (Svane and Rottwitt 2012) and application in fiber sensing using non-linear effects such as stimulated Raman and Brillouin scattering (Frazão et al. 2009). The Raman amplifier is particularly interesting compared to normal rare-earth-doped fiber amplifiers such as Erbium-Doped Fiber Amplifiers (EDFAs) because of its low threshold and therefore high-efficiency conversion. It has also shown excellent thermal characteristics. Raman doped fiber laser has a stable frequency emission (Yao et al. 2015). However, they are more difficult to design and build than normal rare-earth-doped fiber lasers, yet easy to build compared to their Brillouin counterparts.

1.4 FIBER LASER APPLICATIONS

1.4.1 MANUFACTURING

Lasers are widely used in manufacturing for cutting, drilling, welding, cladding, soldering (brazing), hardening, ablating, surface treatment, marking, engraving, micromachining, pulsed lasers deposition, lithography, and alignment applications.

Advantages of laser processing methods compared to mechanical approaches are:

- The fabrication of very fine structures with high quality avoiding mechanical stresses caused by mechanical drills and blades. A laser beam with high beam quality can be used to drill very fine and deep holes (e.g. injection nozzles);
- High processing speed is often achieved (e.g. fabrication of filter sieves);
- The lifetime limitation of mechanical tools is removed and;
- It is also advantageous to process materials with non-contact methods.

1.4.2 MEDICAL APPLICATIONS

Lasers are used for surgery exploiting the possibility to cut tissues while causing minimal bleeding. Lasers are applied in eye surgery and vision correction, dentistry, dermatology (e.g. photodynamic therapy), various kinds of cosmetic treatments such as tattoo removal and hair removal.

Different types of lasers are required for medical applications depending on the optical wavelength, output power, pulse format, etc. In many cases, the laser wavelength is chosen such that certain substances (e.g. pigments in tattoos or cavities in teeth) absorb light more strongly than the surrounding tissue so that they can be more precisely targeted.

1.4.3 SPECTROSCOPY

Laser spectroscopy is used in many different forms and in a wide range of applications. For example, atmospheric physics and pollution monitoring profits from trace gas sensing with differential absorption Light Detection and Ranging (LIDAR) technologies. Solid materials can be analysed with laser-induced breakdown spectroscopy. Laser spectroscopy also plays a role in medicine (e.g. cancer detection), biology, and various types of fundamental research, partly related to metrology.

1.4.4 VARIOUS SCIENTIFIC APPLICATIONS

Laser cooling makes it possible to bring clouds of atoms or ions to extremely low temperatures. This has applications in fundamental research and for industrial purposes. Particularly, in biological and medical research, optical tweezers can be used for trapping and manipulating small particles, such as bacteria or parts of living cells.

1.5 CONCLUSION

In this overview chapter, we briefly discussed the fundamentals of fiber lasers. The technology is as old as the laser itself, produced little interest until the development of reliable optical components starting from the mid-1970s. The interest of lasers in fiber form was discussed highlighting the unique properties that make the fiber laser a desirable device for applications ranging from, telecommunication, to manufacturing including medicine and metrology. Different configurations of fibers lasers were also briefly discussed which included continuous wave fiber lasers, pulsed fiber laser. Other lasers including single longitudinal mode fiber lasers as well as high power fiber lasers were also discussed.

REFERENCES

A. Liu and K. Ueda. 1996. "The Absorption Characteristics of Circular, Offset, and Rectangular Double-Clad Fibers." *Optical Fiber Communication Conference (OFC)* 132: 511–518.

Agrawal, Govind P. 2012. *Fiber-Optic Communication Systems*. Vol. 222. John Wiley & Sons.

Asseh, A., H. Storoy, Jon Thomas Kringlebotn, Walter Margulis, B. Sahlgren, S. Sandgren, R. Stubbe, and G. Edwall. 1995. "10cm Yb3+ DFB Fiber Laser with Permanent Phase Shifted Grating." *Electronics Letters* 31 (July): 969–970. 10.1049/el:19950672.

Auzel, F. and P. Goldner. 2001. "Towards Rare-Earth Clustering Control in Doped Glasses." *Optical Materials* 16 (1–2): 93–103.

Ball, G.A., W.W. Morey, and W.H. Glenn. 1991. "Standing-Wave Monomode Erbium Fiber Laser." *IEEE Photonics Technology Letters* 3 (7): 613–615.

Barmenkov, Yu O, A.V. Kir'yanov, A.D. Guzmán-Chávez, José-Luis Cruz, and Miguel V. Andrés. 2009. "Excited-State Absorption in Erbium-Doped Silica Fiber with Simultaneous Excitation at 977 and 1531 Nm." *Journal of Applied Physics* 106 (8): 83108.

Chen, Xiangfei, Jianping Yao, Fei Zeng, and Zhichao Deng. 2005. "Single-Longitudinal-Mode Fiber Ring Laser Employing an Equivalent Phase-Shifted Fiber Bragg Grating." *IEEE Photonics Technology Letters* 17 (7): 1390–1392.

Cheng, X.P., J. Zhang, P. Shum, M. Tang, and R.F. Wu. 2007. "Influence of Sidelobes on Fiber-Bragg-Grating-Based $ Q $-Switched Fiber Laser." *IEEE Photonics Technology Letters* 19 (20): 1646–1648.

Cognolato, L. 1995. "Chemical Vapour Deposition for Optical Fiber Technology." *Le Journal de Physique IV* 5 (C5): C5–975.

Cuadrado-Laborde, C., M. Delgado-Pinar, S. Torres-Peiró, A. Díez, and M.V. Andrés. 2007. "Q-switched All-Fiber Laser Using a Fiber-Optic Resonant Acousto-Optic Modulator." *Optics Communications* 274 (2): 407–411. 10.1016/j.optcom.2007.02.032.

DiGiovanni, David John and Andrew John Stentz. 1999. "Tapered Fiber Bundles for Coupling Light into and out of Cladding-Pumped Fiber Devices." Google Patents.

Doya, Valérie, Olivier Legrand, and Fabrice Mortessagne. 2001. "Optimized Absorption in a Chaotic Double-Clad Fiber Amplifier." *Optics Letters* 26 (12): 872–874.

Fan, Tso Yee. 1991. "Efficient Coupling of Multiple Diode Laser Arrays to an Optical Fiber by Geometric Multiplexing." *Applied Optics* 30 (6): 630–632.

Federighi, M. and F. Di Pasquale. 1995. "The Effect of Pair-Induced Energy Transfer on the Performance of Silica Waveguide Amplifiers with High Er/Sup 3+//Yb/Sup 3+/ Concentrations." *IEEE Photonics Technology Letters* 7 (3): 303–305.

Frazão, O., C. Correia, M.T.M. Rocco Giraldi, M.B. Marques, H.M. Salgado, M.A.G. Martinez, J.C.W.A. Costa, A.P. Barbero, and J.M. Baptista. 2009. "Stimulated Raman Scattering and Its Applications in Optical Communications and Optical Sensors." *The Open Optics Journal* 3 (1): 1–11.

Fu, Hongyan, Daru Chen, and Zhiping Cai. 2012. "Fiber Sensor Systems Based on Fiber Laser and Microwave Photonic Technologies." *Sensors* 12 (5): 5395–5419.

Gangwar, Arun and Bhawana Sharma. 2012. "Optical Fiber: The New Era of High Speed Communication (Technology, Advantages and Future Aspects)." *International Journal of Engineering Research and Development* 4 (2): 19–23.

Giles, C. Randy and Emmanuel Desurvire. 1991. "Modeling Erbium-Doped Fiber Amplifiers." *Journal of Lightwave Technology* 9 (2): 271–283.

Hanna, D.C., I.M. Jauncey, R.M. Percival, I.R. Perry, R.G. Smart, P.J. Suni, J.E. Townsend, and A.C. Tropper. 1988. "Continuous-Wave Oscillation of a Monomode Thulium-Doped Fiber Laser." *Electronics Letters* 24 (19): 1222–1223.

Hill, Kenneth O. and Gerald Meltz. 1997. "Fiber Bragg Grating Technology Fundamentals and Overview." *Journal of Lightwave Technology* 15 (8): 1263–1276.

Huang, Shenghong, Weiping Qin, Yan Feng, Akira Shirakawa, Mitsuru Musha, and Ken-ichi Ueda. 2005. "Single-Frequency Fiber Laser from Linear Cavity with Loop Mirror Filter and Dual-Cascaded FBGs." *Photonics Technology Letters, IEEE* 17 (July): 1169–1171. 10.1109/LPT.2005.846469.

Hwang, Bor-Chyuan, Shibin Jiang, Tao Luo, Jason Watson, Gino Sorbello, and Nasser Peyghambarian. 2000. "Cooperative Upconversion and Energy Transfer of New High Er 3+-and Yb 3+-Er 3+-Doped Phosphate Glasses." *JOSA B* 17 (5): 833–839.

Jauncey, I.M., L. Reekie, J.E. Townsend, C.J. Rowe, and D.N. Payne. 1988. "Single Longitudinal Mode Operation of a Nd3+-Doped Fiber Laser." *Electronics Letters* 24 (February): 24–26. 10.1049/el:19880017.

Kaczmarek, Franciszek and Jerzy Karolczak. 2007. "Infrared-to-Visible Upconversion: Spontaneous Emission and Amplified Spontaneous Emission in a ZBLAN: Er^3 + Optical Fiber." *Optica Applicata* 37 (1/2): 101.

Kaneda, Yushi, Yongdan Hu, Christine Spiegelberg, Jihong Geng, and Shibin Jiang. 2004. "Single-Frequency, All-Fiber Q-switched Laser at 1550-Nm." *in Advanced Solid-State Photonics (TOPS), G. Quarles, ed., Vol. 94 of OSA Trends in Optics and Photonics (Optical Society of America, 2004), paper 126.*

Koontz, Warren L.G. 2005. "Fiber Optic Telecommunications Technology and Systems–A Two-Course Sequence for a Telecommunications Engineering Technology MS Program."

Kosterin, Andrey, Valery Temyanko, Mahmoud Fallahi, and Masud Mansuripur. 2004. "Tapered Fiber Bundles for Combining High-Power Diode Lasers." *Applied Optics* 43 (19): 3893–3900.

Kouznetsov, Dmitrii, Jerome V. Moloney, and Ewan M. Wright. 2001. "Efficiency of Pump Absorption in Double-Clad Fiber Amplifiers. I. Fiber with Circular Symmetry." *JOSA B* 18 (6): 743–749.

Lauridsen, Vibeke Claudia, Thomas Sondergaard, Poul Varming, and Jørn Hedegaard Povlsen. 1997. "Design of Distributed Feedback Fiber Lasers." In Integrated Optics and Optical Fiber Communications, 11th International Conference on, and 23rd European Conference on Optical Communications (Conf. Publ. No.: 448), 3:39–42. IET.

Leproux, Philippe, Sébastien Février, Valerie Doya, Philippe Roy, and Dominique Pagnoux. 2001. "Modeling and Optimization of Double-Clad Fiber Amplifiers Using Chaotic Propagation of the Pump." *Optical Fiber Technology* 7 (4): 324–339.

Lin, Qian, Mackenzie A. Van Camp, Hao Zhang, Branislav Jelenković, and Vladan Vuletić. 2012. "Long-External-Cavity Distributed Bragg Reflector Laser with Subkilohertz Intrinsic Linewidth." *Optics Letters* 37 (11): 1989–1991.

Liu, Anping, Jie Song, Kouichi Kamatani, and Ken-ichi Ueda. 1997. "Effective Absorption and Pump Loss of Double-Clad Fiber Lasers." In *Solid State Lasers VI*, 2986:30–38. International Society for Optics and Photonics.

Luo, Ming-Xing, Hui-Ran Li, and Xiaojun Wang. 2017. "Distributed Atomic Quantum Information Processing via Optical Fibers." *Scientific Reports* 7 (1): 1234.

Maiman T.H. 1960. "Stimulated Optical Radiation in Ruby." *Nature* 187 (12): 493–494.

Manuel, Rodolfo Martínez, J.J.M. Kaboko, and M.G. Shlyagin. 2016. "Active Q-switching of a Fiber Laser Using a Modulated Fiber Fabry–Perot Filter and a Fiber Bragg Grating." *Laser Physics* 26 (2): 025105. 10.1088/1054-660X/26/2/025105.

Moghaddam, M.R.A., Sulaiman Wadi Harun, Roghaieh Parvizi, Z.S. Salleh, Hamzah Arof, Asiah Lokman, and Harith Ahmad. 2011. "Experimental and Theoretical Studies on Ytterbium Sensitized Erbium-Doped Fiber Amplifier." *Optik-International Journal for Light and Electron Optics* 122 (20): 1783–1786.

Morehead, James J. and Martin H. Muendel. 2011. "Nearly Circular Pump Guides." In *Fiber Lasers VIII: Technology, Systems, and Applications*, 7914:79142Y. International Society for Optics and Photonics.

Mortensen, Niels Asger. 2007. "Air-Clad Fibers: Pump Absorption Assisted by Chaotic Wave Dynamics?" *Optics Express* 15 (14): 8988–8996.

Muendel, Martin H. 1996. "Optimal Inner Cladding Shapes for Double-Clad Fiber Lasers." In Summaries of Papers Presented at the *Conference on Lasers and Electro-Optics*, 209. IEEE.

Nagel, Suzanne R., John B. MacChesney, and Kenneth L. Walker. 1982. "An Overview of the Modified Chemical Vapor Deposition (MCVD) Process and Performance." *IEEE Transactions on Microwave Theory and Techniques* 30 (4): 305–322.

Paun, M.A., Mohammad Avanaki, George Dobre, Ali Hojjatoleslami, and Adrian G.H. Podoleanu. 2009. "Wavefront Aberration Correction in Single Mode Fiber Systems." *Journal of Optoelectronics and Advanced Materials* 11 (11): 1681–1685.

Pollnan, M. and Stuart D. Jackson. 2001. "Erbium 3/Spl Mu/m Fiber Lasers." *IEEE Journal of Selected Topics in Quantum Electronics* 7 (1): 30–40.

Ronnekleiv, Erlend, Sigurd Weidemann Lovseth, and Jon Thomas Kringlebotn. 2003. "Er-Doped Fiber Distributed Feedback Lasers: Properties, Applications, and Design Considerations." In Fiber-Based Component Fabrication, *Testing, and Connectorization*, 4943:69–80. International Society for Optics and Photonics.

Russo, N.A., R. Duchowicz, J. Mora, J.L. Cruz, and M.V. Andrés. 2002. "High-Efficiency Q-switched Erbium Fiber Laser Using a Bragg Grating-Based Modulator." *Optics Communications* 210 (3–6): 361–366. 10.1016/S0030-4018(02)01815-1.

Snitzer, Elias, Koester. C.J. 1963. "Amplification in a Fiber Laser." *Applied Optics* 3: 1182–1186.

Snitzer, E., H. Po, F. Hakimi, R. Tumminelli, and B.C. McCollum. 1988. "Double Clad, Offset Core Nd Fiber Laser." In *Optical Fiber Sensors*, PD5. Optical Society of America.

Snitzer, Elias, Hong Po, Richard P. Tumminelli, and Farhad Hakimi. 1989. "Optical Fiber Lasers and Amplifiers." Google Patents.

Song, Feng, Zhenzhou Cheng, Changguang Zou, Lanjun Luo, Yingying Cai, Wenyi Piao, Chen Wei, and Jianguo Tian. 2009. "Experimental Study and Theoretical Simulation for High Gain Compact Er3+-Yb3+-Codoped Fiber Laser." In 2009 International Conference on Optical Instruments and Technology: Optoelectronic Devices and Integration, 7509:75090N. International Society for Optics and Photonics.

Svane, Ask Sebastian and Karsten Rottwitt. 2012. "PM Raman Fiber Laser at 1679 Nm." In *Integrated Photonics Research, Silicon and Nanophotonics*, JTu5A-28. Optical Society of America.

Taccheo, Stefano, Gino Sorbello, Stefano Longhi, and Paolo Laporta. 1999. "Measurement of the Energy Transfer and Upconversion Constants in Er–Yb-Doped Phosphate Glass." *Optical and Quantum Electronics* 31 (3): 249–262.

Tokita, Shigeki, Mayu Hirokane, Masanao Murakami, Seiji Shimizu, Masaki Hashida, and Shuji Sakabe. 2010. "Stable 10 W Er: ZBLAN Fiber Laser Operating at 2.71–2.88 Mm." *Optics Letters* 35 (23): 3943–3945.

Valentin, Gapontsev P and Samartsev Igor. 1999. "Coupling Arrangement between a Multi-Mode Light Source and an Optical Fiber through an Intermediate Optical Fiber Length." Google Patents.

W Naji, A., Belal Hamida, Cheng San, Mohd Adzir Mahdi, Sulaiman Wadi Harun, Sadrudin Khan, W.F. Al-Khateeb, A. Zaidan, Bilal Bahaa, and H. Ahmad. 2011. "Review of Erbium-Doped Fiber Amplifier." *International Journal of Physical Sciences* 6 (September).

Xiao, Qi-rong, Ping Yan, S. Yin, J. Hao, and M. Gong. 2010. "100 W Ytterbium-Doped Monolithic Fiber Laser with Fused Angle-Polished Side-Pumping Configuration." *Laser Physics Letters* 8 (2): 125.

Yao, Tianfu, Achar Harish, Jayanta Sahu, and Johan Nilsson. 2015. "High-Power Continuous-Wave Directly-Diode-Pumped Fiber Raman Lasers." *Applied Sciences* 5 (4): 1323–1336.

Yarutkina, Irina, O. Shtyrina, M. Fedoruk, and S.k. Turitsyn. 2013. "Numerical Modeling of Fiber Lasers with Long and Ultra-Long Ring Cavity." *Optics Express* 21 (May): 12942–12950. 10.1364/OE.21.012942.

Yelen, Kuthan, Louise M.B. Hickey, and Mikhail N. Zervas. 2005. "Experimentally Verified Modeling of Erbium-Ytterbium Co-Doped DFB Fiber Lasers." *Journal of Lightwave Technology* 23 (3): 1380.

Zervas, Michalis. 2014. "High Power Ytterbium-Doped Fiber Lasers - Fundamentals and Applications." *International Journal of Modern Physics B* 28 (April). 10.1142/S021 7979214420090.

Zhang, Yang, Bai-Ou Guan, and Hwa-Yaw Tam. 2009. "Ultra-Short Distributed Bragg Reflector Fiber Laser for Sensing Applications." *Optics Express* 17 (12): 10050–10055.

2 Optical Fibers

2.1 INTRODUCTION

An optical fiber can be considered as a cylindrical waveguide. Its working principle is based on the total internal reflection, where a beam of light is trapped inside the cylindrical core of the fiber, which is surrounded by a cladding of lower refractive index. Although the phenomenon of total internal reflection is known since 1854 (Tyndall 1854), the first attempt to make optical fiber with glass dates from the 1920s (Baird 1927; Hansell 1930; Lamm 1930). These first optical fibers were unpractical until a significant improvement was made to the guiding characteristics by using a cladding layer surrounding the core layer (Van Heel 1954; Kapany 1959). Because of their high loss, the obtained fibers were not suitable for optical communication, thus they were used mostly for medical imaging over short distances (Kapany 1967). The use of glass fiber became possible in 1970 when the losses of silica glass were reduced to below the mark of 20 dB/km (Kao and Hockham 1966) (Kapron et al. 1970). Further series of improvements resulted in a loss of around 0.2 dB/km in the 1.55 μm spectral region. The availability of low-loss fiber at an affordable price combined with the discovery of technologies such as Erbium-doped fiber amplifiers (EDFA) (Miya et al. 1979) triggered a revolution in optical communication. The full description of optical fiber is outside of the scope of this book and several books entirely devoted to the subject have been published (Adams 1981; Okoshi 1982; A.W. Snyder and Love 1983; Jeunhomme 1990; Ed 1985; T. Izawa and Sudo 1987; E. G. Neumann 1988; Marcuse 1991; Cancellieri 1991; Buck 1995). In this chapter, the authors are going to provide a brief overview of optical fiber concepts focusing on characteristics relevant to the design and understanding of fiber lasers.

In section 2.2, a geometrical-optics description is used to explain the guiding mechanism of light in the fiber. This approach provides an intuitive understanding of trapping light inside the core of an optical fiber by total internal reflection. However, the approach cannot be used to describe transverse mode energy distribution. Maxwell's equations are used in section 2.3 to describe wave propagation in optical fibers. Section 2.4 discusses the loss mechanisms in optical fibers, and section 2.5 is all about dispersion. Section 2.6 is devoted to nonlinear effects. Glass materials for optical fibers are discussed in section 2.7 and the last section is about different fabrication techniques of both standard and rare-earth-doped fibers.

DOI: 10.1201/9781003256380-2

2.1.1   Ray Optic Description

2.1.1.1   Total Internal Reflection (TIR)

The guidance of a light beam in the optical fiber takes place because of the phenomenon of total internal reflection. Total internal reflection occurs when light
passes from a higher refractive index material to a lower refractive index material.
However, for total internal reflection to take place, specific conditions must be met.
Let us consider Figure 2.1 where a ray of light is incident at the interface of two
media of different refractive indices (i.e. air and glass), experiment shows that the
ray will undergo partial reflection and partial refraction.

In Figure 2.1(a), the vertical line represents the normal to the surface. The angles
ϕ_1, ϕ_2 and, ϕ_r represent the angles that the incident ray, refracted ray, and reflected
ray make with the normal, respectively. The relation between these angles is given
by Snell's law as

$$n_1 sin\phi_1 = n_2 sin\phi_2 \qquad (2.1)$$

where, n_1 and n_2 are the refractive indices of two materials.

The angle of incidence, for which the angle of refraction is equal to 90 degrees,
Figure 2.1(b), is known as the critical angle and is given by,

$$\phi_c = \phi_1 = sin^{-1}\left(\frac{n_2}{n_1}\right) \qquad (2.2)$$

The total internal reflection occurs when the incident angle is greater than the
critical and as shown in Figure 2.1(c).

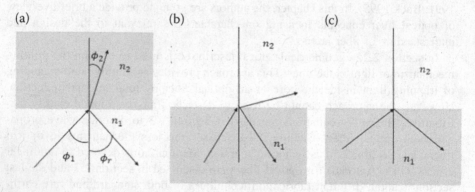

FIGURE 2.1   a) Reflection and total internal refraction. A ray is incident on a lower refractive index medium (n2 < n1). b) If the angle of incidence is small than the critical angle,
the beam will undergo partial refraction. c) If the angle of incidence is greater than the critical
angle, the beam will undergo total internal reflection.

2.1.1.2 Optical Fiber

An optical fiber is a cylindrical glass or plastic waveguide made of different parts, namely a core, with a higher refractive index surrounded by a cladding with lower refractive index, which is surrounded by a buffer and a protective jacket. Figure 2.2 shows a schematic representation of an optical fiber.

The light-guiding element is the core. Light is trapped inside the core of the optical fiber by total internal reflection taking place at the core-cladding boundary. Figure 2.3 represents the principle of light confinement in an optical fiber.

Using equations (2.1) and (2.2), one can find the maximum angle that the incident beam should make with the fiber axis to remain trapped inside the core. That angle is given by $\theta_r = \pi/2 - \phi_c$ and substituting into equation (2.1), we obtain:

$$n_0 sin\theta_i = n_1 cos\phi_c = (n_1^2 - n_2^2)^{1/2} \tag{2.3}$$

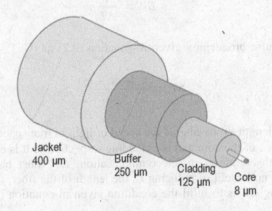

FIGURE 2.2 Schematic representation of an optical showing the core, cladding, primary buffer and external protective jacket.

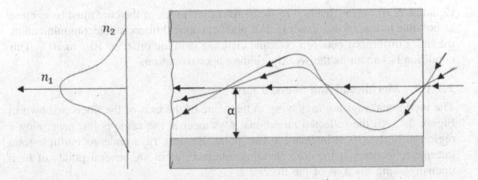

FIGURE 2.3 Light confinement through total internal reflection in step-index fibers. Rays for which $\phi < \phi_c$ are refracted out of the core.

The value $n_0 sin\theta_i$ is known as the numerical aperture (NA) of the fiber. It represents the light-gathering capacity of an optical fiber. For $n_1 \approx n_2$ the NA can be approximated by:

$$NA = n_1 (2\Delta)^{1/2}, \tag{2.4}$$

where Δ is the fractional index change at the core-cladding interface given by:

$$\Delta = \frac{(n_1 - n_2)}{n_1} \tag{2.5}$$

The refractive index change, Δ, should be made as large as possible to couple maximum light into the fiber. However, it can also be shown that the bandwidth of the optical fiber is given by the relation (G. P. Agrawal 2012):

$$BW = \frac{1}{\Delta T} \tag{2.6}$$

where ΔT is pulse broadening given in equation (2.7) as (G. P. Agrawal 2012):

$$\Delta T = \frac{L}{c} \frac{n_1 (n_1 - n_2)}{n_2}, \tag{2.7}$$

where L is the length of the fiber, c the speed of light in free space and n_1 and n_2 the refractive indices of the core and the cladding respectively. It is easy to see that for large bandwidths required in telecommunication, ΔT must have the minimum value. There is no benefit of reducing L, the length of the fiber, therefore, the only way of reducing ΔT is to fulfil the condition given in equation (2.8) as:

$$\Delta = \frac{(n_1 - n_2)}{n_2} \ll 1 \tag{2.8}$$

Equation (2.8) means that $n_1 \approx n_2$ or the refractive index of the core must be as close as possible to that of the cladding. For practical optical fibers in telecommunication, the index difference between core and cladding is in the order of 10^{-3} to 10^{-4}. This condition is known as the weakly guiding approximation.

2.1.1.3 Meridional and Skewed Rays

The meridional rays are rays lying in the plane of the axis of the fiber as shown in Figure 2.4. All the reflected meridional rays meet at the same point, generating a region of high optical intensity at that point. Because rays undergo multiple total internal reflections at the core-cladding interface, there are several points of high intensity along the axis of the fiber.

The other category of propagating rays in the core of an optical fiber is skewed rays. Skewed rays propagate helicoidally around the axis of the optical fiber without

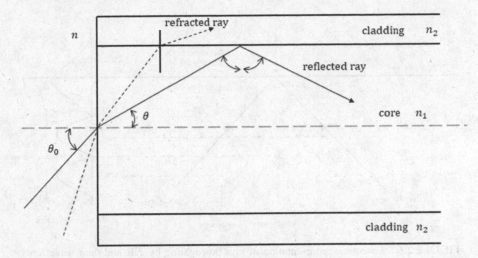

FIGURE 2.4 Launching of meridional rays into the fiber.

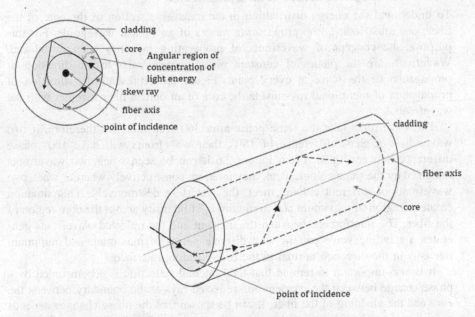

FIGURE 2.5 Launching of skewed rays into the fiber.

crossing it. Their energy is confined in an annular region around the axis as shown in Figure 2.5. As a result, skewed rays have maximum energy concentrated towards the core-cladding interface and gradually reducing to a minimum value towards the axis of the fiber.

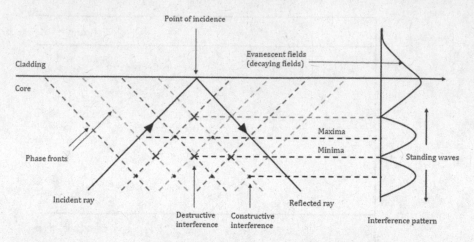

FIGURE 2.6   Schematic representation of rays propagating by TIR and their wavefronts.

2.1.1.4   Propagating Modes of an Optical Fiber

To understand the energy distribution in the transverse section of the core of the fiber, one must look at the propagating modes of an optical waveguide. For this purpose, the concept of wavefronts of propagating rays has been introduced. Wavefronts are the planes of constant phase perpendicular to the direction of propagation of the wave at every point. Figure 2.6 represents the direction of propagation of meridional rays inside the core of an optical fiber together with the wavefronts.

The wavefronts take the same phase after 360° (2π radians), therefore, if two waves have a phase difference of 180°, their wavefronts will have 180° phase difference between them. From Figure 2.6 it can be seen when two wavefronts indicated by the same colour, meet, they interfere constructively whereas when two wavefronts of different colours meet, they interfere destructively. This situation creates a region of maximum and minimum light intensity across the core region of the fiber. The interference between the incident and the reflected wavefronts generates a standing wave pattern with discrete points of maximum and minimum intensity in the direction normal to the core-cladding interface.

It is also important to remind that total internal reflection is accompanied by a phase change between the incident and reflected rays at the boundary between the core and the cladding of the fiber. It can be shown that the phase changes depends on the angle of incidence of the ray at the core-cladding boundary, and the refractive index of the core and cladding (G. Agrawal 2012). This can be expressed by equation (2.9) below:

$$\frac{2\pi n_1 \sin\phi}{\lambda} + \delta = \pi m \tag{2.9}$$

where, n_1 is the refractive index of the core, ϕ the angle of refraction inside the core of the fiber, λ the wavelength of the light δ the phase change undergone by the ray, and m an integer.

Equation (2.5) tells that only rays of light that are incident on the tip of the fiber in such a way that their angle of reflection inside the fiber, ϕ satisfy equation (2.5) will successfully propagate inside the core of the optical fiber. It is easy to see that since m is an integer; the angle ϕ can only take discrete values. The meaning of this is that only some discrete launching angles within the acceptance cone will result in a propagating field inside the core. The discrete propagating rays that fulfil the condition of launching angles are known as modes of the optical fiber.

The number of discrete "m" will have different meaning as it will be explained later with the wave optics model. For example, the ray that is launched along the axis of the fiber propagate without any phase condition requirement and corresponds to the first propagating mode also known as the fundamental or zero-order mode of propagation. Finally, it is worth reminding that depending on the numerical aperture of the fiber (acceptance cone), there can be N possible propagation modes resulting in unique intensity patterns around the axis of the core.

2.1.1.5 Single-mode and Multi-mode Fiber

The number of modes inside the core of an optical fiber depends on the NA which in turn depends on the core radius of the fiber. In general, a fiber with a large core diameter is capable of carrying several modes at a given wavelength. This type of fiber is known as a multi-mode fiber. If the core radius is reduced so that only one mode can propagate, a single-mode fiber is obtained as illustrated in Figure 2.7.

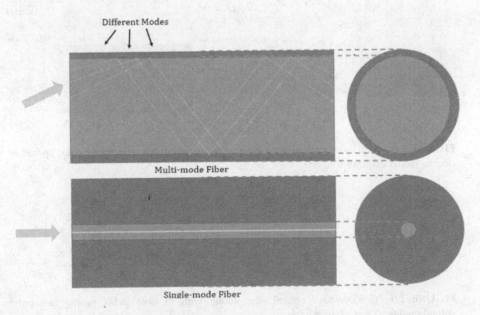

FIGURE 2.7   A schematic representation of multi-mode and single-mode fibers.

In telecommunication applications, common core diameters for multi-mode fibers are 62.5 and 50 μm and 9 μm for single-mode fibers. It is possible to have other diameters, for example, rare-earth-doped fiber used in fiber amplifiers and lasers can have a small core diameter of 3 μm, whereas plastic core optical fibers have core diameters of about 100 μm.

2.1.1.6   Step Index and Graded-index Fiber

Depending on the refractive index profile, the optical fiber can be categorised into two types namely, step-index and graded-index optical fiber. The transverse profile of the refractive index in the step-index fiber is uniform. The modes inside the core of a step-index fiber propagate at the same velocity covering different distances. Consequently, there is a delay between modes travelling the short distance and those travelling the longer distance. This propagation delay is at the origin of modal dispersion leading into pulse spreading as illustrated in Figure 2.8.

The graded index fiber was designed to reduce the modal dispersion problem of step-index fibers. In this fiber, the core refractive index profile varies gradually from highest to lowest as one moves away from the central core towards the cladding interface as shown in Figure 2.9.

2.2   WAVE OPTIC DESCRIPTION

In this section, the propagation of light in step-index fibers is considered by using Maxwell's equations. Assuming that the core of the fiber is a pure dielectric (there is no charge and no conduction current), Maxwell's equations can be written as:

$$\nabla . D = 0 \tag{2.10}$$

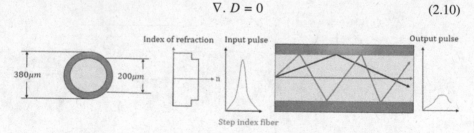

FIGURE 2.8   Step index optical fiber along with its core index profile and pulse spreading due to modal dispersion.

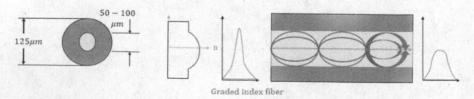

FIGURE 2.9   Graded-index optical fiber along with its core index profile and pulse spreading due to modal dispersion.

$$\nabla . B = 0 \tag{2.11}$$

$$\nabla \times E = -\frac{\partial B}{\partial t} \tag{2.12}$$

$$\nabla \times H = -\frac{\partial D}{\partial t} \tag{2.13}$$

$$D = \varepsilon E \tag{2.14}$$

$$B = \mu H \tag{2.15}$$

where D is the electric displacement vector, B the magnetic flux density, E the electric field, H the magnetic field, ε the relative permittivity of the material, and μ its magnetic permeability.

These equations are coupled and must be separated to be solved. This can be done by taking the curl of equations (2.12) or (2.13) as follows:

$$\nabla \times \nabla \times E = -\nabla \times \frac{\partial B}{\partial t} \tag{2.16}$$

Substituting (2.15) in (2.16) yields:

$$\nabla \times \nabla \times E = -\frac{\partial}{\partial t} \nabla \times (\mu H) \tag{2.17}$$

We assume a homogeneous medium, therefore permeability is not a function of space and can be taken out of the ∇ operator and equation (2.17) can be written as:

$$\nabla \times \nabla \times E = -\mu \frac{\partial}{\partial t} (\nabla \times H) \tag{2.18}$$

Substituting (2.13) in (2.18) we obtain:

$$\nabla \times \nabla \times E = -\mu \frac{\partial}{\partial t} \left(\frac{\partial D}{\partial t} \right) \tag{2.19}$$

Substituting (2.14) in (2.19) we obtain:

$$\nabla \times \nabla \times E = -\mu \varepsilon \frac{\partial^2 E}{\partial t^2} \tag{2.20}$$

Using the curl of curl identity written as:

$$\nabla \times \nabla \times E = \nabla(\nabla. E) - \nabla^2 E \tag{2.21}$$

Replacing (2.21) in (2.20) yields:

$$\nabla(\nabla. E) - \nabla^2 E = -\mu\varepsilon\frac{\partial^2 E}{\partial t^2} \tag{2.22}$$

Since the medium is homogeneous, ε is independent of space, therefore using (2.14) we can say that $\nabla. E = 0$, and (2.22) becomes:

$$\nabla^2 E = \mu\varepsilon\frac{\partial^2 E}{\partial t^2} \tag{2.23}$$

Using the same reasoning we can find the equation for the magnetic field as:

$$\nabla^2 H = \mu\varepsilon\frac{\partial^2 H}{\partial t^2} \tag{2.24}$$

Equations (2.23) and (2.24) constitute the wave equations and it can be shown that in free space the propagation speed $\frac{1}{\sqrt{\mu\varepsilon}}$ is the speed of light c.

The solutions of these equations help to understand the behaviour of electric and magnetic fields inside the optical fiber. In general, the equation for electric or magnetic field is solved and the quantity found is substituted back to Maxwell's equation to obtain the other quantity. Since the equations (2.23) and (2.24) are perfectly similar, they can be replaced by a generic wave equation of magnitude U.

$$\nabla^2 U = \mu\varepsilon\frac{\partial^2 U}{\partial t^2} \tag{2.25}$$

Also, because the optical fiber has a cylindrical shape, adopting the cylindrical coordinate system will simplify the resolution of the equation. Equation (2.25) in cylindrical coordinates can be written as:

$$\frac{1}{r}\frac{\partial}{\partial r}\left(r\frac{\partial U}{\partial r}\right) + \frac{1}{r^2}\frac{\partial^2 U}{\partial \varnothing^2} + \frac{\partial^2 U}{\partial z^2} = \mu\varepsilon\frac{\partial^2 U}{\partial t^2}. \tag{2.26}$$

The electric and magnetic fields are vectors quantities so that three components are needed to define each one of them resulting in six quantities to be found. Maxwell's equations, on the other hand, show that the electric and magnetic fields are only related by four equations. This means that the six components of the electric and magnetic fields are not independent. Therefore, by considering two components of the electric and magnetic fields as the independent components, Maxwell's equations can be used to find the remaining four components.

The propagation takes place takes place along the z-direction. This direction is regarded as a special direction because the net energy flows into it. Any other direction which is perpendicular to the z-direction is referred to as the "transverse" direction. So, the two independent components chosen will be along the z-direction. Because these components are in the axis of the fiber, they are sometimes referred to as longitudinal components of the electric and magnetic fields. If the independent components are written as E_z and H_z, it can be shown that the transverse field components will be written as:

$$E_r = \frac{-j}{q^2}\left(\beta\frac{\partial E_z}{\partial r} + \frac{\mu\omega}{r}\frac{\partial H_z}{\partial \varnothing}\right),\tag{2.27}$$

$$E_\varnothing = \frac{-j}{q^2}\left(\frac{\beta}{r}\frac{\partial E_z}{\partial \varnothing} - \mu\omega\frac{\partial H_z}{\partial r}\right),\tag{2.28}$$

$$H_r = \frac{-j}{q^2}\left(\beta\frac{\partial H_z}{\partial r} + \frac{\omega\varepsilon}{r}\frac{\partial E_z}{\partial \varnothing}\right),\tag{2.29}$$

$$H_\varnothing = \frac{-j}{q^2}\left(\frac{\beta}{r}\frac{\partial H_z}{\partial \varnothing} - \omega\varepsilon\frac{\partial E_z}{\partial r}\right).\tag{2.30}$$

where, E_r is the radial component of the electric field, $E_\varnothing$ is the azimuthal component of the electric field, H_r is the radial component of the magnetic field, $H_\varnothing$ is the azimuthal component of the magnetic field.

In equations (2.27) to (2.30), $q^2 = \mu\varepsilon\omega^2 - \beta$; ω is the angular frequency of the propagating light and β its propagation constant. Equations (2.27) to (2.30) show that E_r, E_ϕ, H_r and H_ϕ are expressed as functions of E_z and H_z only. It can also be seen that when one of the longitudinal components is zero, there is still energy in the corresponding transverse component. However, when the two longitudinal components are equal to zero, the corresponding transverse component vanishes. The physical meaning of this is that transverse electromagnetic modes (TEM) cannot develop in optical fibers.

In general, there exist three possibilities depending on which longitudinal value is equal to zero, resulting in three types of modes. If $E_z = 0$ and $H_z \neq 0$, the resulting mode is said to be transverse electric or TE mode. This mode has a magnetic component and no electric component in the direction of propagation z. If $E_z \neq 0$ and $H_z = 0$, the resulting mode is said to be transverse magnetic or TM mode. This mode has an electric component and no magnetic component in the direction of propagation z. Finally, if $E_z \neq 0$ and $H_z \neq 0$, the resulting mode is said to be "Hybrid" mode, in which there are both electric and magnetic components in the direction of propagation.

In analogy with the ray optic analysis, the TE and TM modes correspond to meridional rays whereas the hybrid modes correspond to the skewed modes. The descriptions made earlier show that the problem of obtaining the six components of the electric and magnetic fields reduces in finding an analytical solution to the wave equation for its longitudinal components. Thus, the wave equations must be solved for E_z and H_z. To simplify the problem, since E_z and H_z are scalar quantities, we can define a generic scalar quantity ψ that can represent any of the two components. In this way the analysis is strongly simplified since the wave equations for both electric and magnetic longitudinal components are identical, therefore the two solutions are also going to be identical.

Following equation (2.26), the wave equation in cylindrical coordinates in terms of the scalar quantity ψ will be written as:

$$\frac{\partial^2 \psi}{\partial r^2} + \frac{1}{r}\frac{\partial \psi}{\partial r} + \frac{1}{r^2}\frac{\partial^2 \psi}{\partial \phi^2} + \frac{\partial^2 \psi}{\partial z^2} = \mu\varepsilon\frac{\partial^2 \psi}{\partial t^2} \tag{2.31}$$

Assuming all the fields are time-harmonic with angular frequency ω, the field can be written as:

$$\psi \sim e^{j\omega t} \tag{2.32}$$

And the second derivative with respect to time of equation (2.32) can be written as:

$$\frac{\partial^2 \psi}{\partial t^2} = -\omega^2 \psi \tag{2.33}$$

Substituting (2.33) into (2.32) yields:

$$\frac{\partial^2 \psi}{\partial r^2} + \frac{1}{r}\frac{\partial \psi}{\partial r} + \frac{1}{r^2}\frac{\partial^2 \psi}{\partial \phi^2} + \frac{\partial^2 \psi}{\partial z^2} = -\omega^2\mu\varepsilon\psi. \tag{2.34}$$

The second-order differential equation (2.34) can be solved by separation of variables. Each of the resulting functions is a function of whether ϕ, r, or z. Thus, the solution for equation (2.34) can be written as:

$$\psi = R(r)\Phi(\phi)Z(z) \tag{2.35}$$

There are three functions to be determined, namely $R(r)$, $\Phi(\phi)$, and $Z(z)$. This can be easily done using the understanding of the wave propagation in the optical fiber. The optical energy is carried by a travelling mode in the z direction; therefore, the longitudinal component of the field can be written as:

$$Z(z) = e^{-j\beta z} \tag{2.36}$$

The first and second derivatives of ψ with respect to z are respectively given by:

$$\frac{\partial \psi}{\partial z} = -j\beta\psi \tag{2.37}$$

and,

$$\frac{\partial^2 \psi}{\partial z^2} = -\beta^2\psi \tag{2.38}$$

In the same way, for a given value of r and z, there going to be a point $P(r, \phi, z)$ moving along a circle in a plane transverse to the axis of the fiber (propagation direction). It can be easily shown that this point takes the same position on its trajectory for each change of ϕ by a value of 2π. Therefore Φ is a periodic function and one can write:

$$P(r, \phi, z) = P(r, \phi + 2m\pi, z), \tag{2.39}$$

where m is an integer. To meet the criteria of equation (2.39), $\Phi(\phi)$ must be of the form:

$$\Phi(\phi) = e^{j\nu\phi} \tag{2.40}$$

where ν is an integer. Similarly, taking the first and second derivatives of ψ with respect to ϕ, yields:

$$\frac{\partial \psi}{\partial \phi} = j\nu\psi; \quad \frac{\partial^2 \psi}{\partial \phi^2} = -\nu^2\psi. \tag{2.41}$$

Out of the three functions of equation (2.35) only $R(r)$ remains to be determined. This can be found by substituting (2.37), (2.38), and (2.41) in (2.34). The result of the substitution is shown in equation (2.42) below.

$$\frac{\partial^2 \psi(r)}{\partial r^2} + \frac{1}{r}\frac{\partial \psi(r)}{\partial r} + \left\{ (\omega^2\mu\varepsilon - \beta^2) - \frac{\nu^2}{r^2} \right\} \psi(r) = 0. \tag{2.42}$$

Because of the equivalence provided in equation (2.35), equation (2.42) will also govern the quantity $R(r)$ and can be written as:

$$\frac{\partial^2 R(r)}{\partial r^2} + \frac{1}{r}\frac{\partial R(r)}{\partial r} + \left\{ (\omega^2\mu\varepsilon - \beta^2) - \frac{\nu^2}{r^2} \right\} R(r) = 0. \tag{2.43}$$

Equation (2.43) is nothing but the well-known Bessel equation, which solution are the Bessel functions of variable r. In other words, the radial variation of the electric and magnetic field distribution is given by Bessel's functions. A more compact form of equation (2.43) can be written as:

$$\frac{\partial^2 R(r)}{\partial r^2} + \frac{1}{r}\frac{\partial R(r)}{\partial r} + \left(q^2 - \frac{v^2}{r^2}\right)R(r) = 0. \qquad (2.44)$$

where q is defined as:

$$q^2 = \omega^2 \mu \varepsilon - \beta^2 \qquad (2.45)$$

It is important to remember that the phase constant β is still unknown and must be determined by applying appropriate boundary conditions. Once this quantity is known, the value of q will also be known and Bessel's equation (2.44) can be solved and the function $R(r)$ obtained. One can than go back to equation (2.35) and substitute the values of $R(r)$, $\Phi(\phi)$ and $Z(z)$. The obtained expression of ψ then fully represent either electric of magnetic field and the problem is solved. Details of the actual solution are given in the next paragraph.

For solving the Bessel's equation (2.44), one assumes a travelling wave in a homogeneous lossless media, thus β, μ and ε are real, therefore, q is to be either real or purely imaginary. Different families of Bessel's functions are obtained depending on whether q is real or imaginary. If q is real ($q^2 > 0$), there exist two families of solutions, namely the Bessel functions also known as the Bessel functions of the first kind and Neumann n functions also known as Bessel functions of the second kind, denoted by $J_v(qr)$ and $N_v(qr)$ respectively. In this notation, v is the "order" of the function and qr its argument. The general solution of equation (2.44) can then be written as linear combinations of the two kinds of Bessel functions like:

$$R(r) = \alpha_1 J_v(qr) + \alpha_2 N_v(qr). \qquad (2.46)$$

where α_1 and α_2 are arbitrary constants. The two types of Bessel functions are plotted in Figure 2.10.

If q is imaginary ($q^2 < 0$), there exist two families of solutions denoted $K_v\left(\frac{qr}{j}\right)$ and $I_v\left(\frac{qr}{j}\right)$ and known as modified Bessel functions of the first and second kind respectively. The modified Bessel functions are shown in Figure 2.11. In this case, also the general solution of equation (2.44) can be written as a linear combination of the two types of modified Bessel functions.

$$R(r) = \eta_1 K_v\left(\frac{qr}{j}\right) + \eta_2 I_v\left(\frac{qr}{j}\right) \qquad (2.47)$$

where η_1 and η_2 are arbitrary constants.

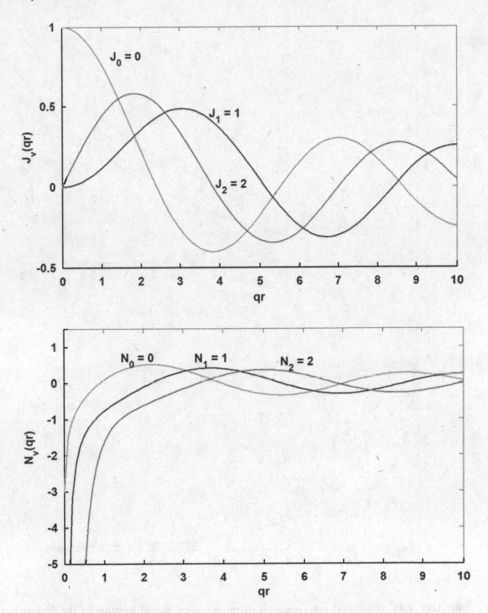

FIGURE 2.10 Graphical representation of Bessel and Neumann functions.

The observations of Figure 2.10 and Figure 2.11 show that:

- Bessel and Neumann functions are oscillatory, whereas both modified Bessel functions of the first and second kind are monotonic functions of the independent variable.
- The function $J_v(qr)$ is finite for all values of qr. Except $J_0(qr)$ which has a value of 1 at $qr = 0$, all other $J_v(qr)$ have values of 0 at $qr = 0$.

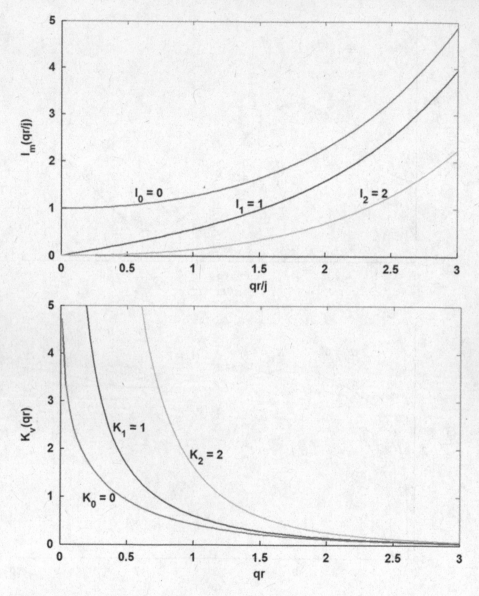

FIGURE 2.11 Graphical representation of the modified Bessel function of the first and second kind.

- The functions $N_v(qr)$ have a singularity at the origin. In other words, they asymptotically approach $-\infty$ when the argument qr approaches zero.
- The functions $I_v(qr/j)$ are monotonically increasing functions of qr/j that asymptotically approach ∞ as qr/j tends to ∞.
- The functions $K_v(qr/j)$ are monotonically decreasing functions of qr/j that asymptotically approach 0 as qr/j approaches ∞.

These considerations are useful when choosing the valid solutions of the wave equation in optical fiber media. From ray optic analysis, it is known that inside the core the interference of phase fronts produces regions of maximum and zero intensities. Therefore, inside the core, the field distribution is oscillatory and the modified Bessel functions cannot be valid solutions. The choice has to be made between Bessel and Neumann functions. Since Neumann functions have a singularity at the origin, which is the axis of the fiber, they are eliminated from valid solutions. Thus, the only possible solution inside the core is the Bessel functions of the first kind $J_\nu(qr)$, where q is real.

Since the energy is confined into the core, the field will be dying down as one move away from the core. The first modified Bessel function shows monotonically increased field amplitude as r increases and cannot be a valid solution. On the other hand, the second modified Bessel function monotonically decreases when r increases and is, therefore, a valid solution for field distribution inside the cladding.

Having two types of solutions in the core and cladding means that the values of β are also comprised between limit values in the core and the cladding. If a is the core radius of the fiber, the solution to the wave equation in the core and cladding regions of the fiber can then be written as,

$$\psi_1(r, \phi, z, t) = \alpha_1 J_\nu(ur) e^{j\nu\phi - j\beta z + j\omega t}, \tag{2.48}$$

for $r < a$,
   and

$$\psi_2(r, \phi, z, t) = \eta_1 K_\nu(wr) e^{j\nu\phi - j\beta z + j\omega t}, \tag{2.49}$$

for $r > a$.

From equations (2.43) and (2.44):

$$u = \sqrt{\omega^2 \mu \epsilon_1 - \beta^2}, \tag{2.50}$$

for $r < a$,
   and

$$jw = \sqrt{\beta^2 - \omega^2 \mu \epsilon_2}, \tag{2.51}$$

for $r > a$.

The equations (2.48) and (2.49) represent the solution to the general wave equation in an optical fiber, leading to the expressions for electric and magnetic fields inside the core and cladding written as:

$$E_{z1} = A J_\nu(ur) e^{j\nu\phi - j\beta z + j\omega t}, \tag{2.52}$$

$$H_{z1} = BJ_\nu(ur)e^{j\nu\phi-j\beta z+j\omega t}. \tag{2.53}$$

and

$$E_{z2} = CK_\nu(wr)e^{j\nu\phi-j\beta z+j\omega t}, \tag{2.54}$$

$$H_{z2} = DK_\nu(wr)e^{j\nu\phi-j\beta z+j\omega t}. \tag{2.55}$$

where A, B, C and D are arbitrary constants and their values can be determined by substituting appropriate boundary conditions.

The value of the propagation constant β also is not known. However, it is easy to show that this value is bounded between specific limits. Inside the core of the optical fiber, q must be real. This translates in having $q^2 > 0$ in equation (2.45) and, thus $\beta^2 < \omega^2\mu\varepsilon_1$. Inside the cladding of the fiber, on the other hand, q is purely imaginary which translate by $q^2 < 0$ in equation (2.45) and $\beta^2 < \omega^2\mu\varepsilon_2$. Combining these two results, the propagation constant can be written in the bounded form as,

$$\omega\sqrt{\mu\varepsilon_2} < \beta < \omega\sqrt{\mu\varepsilon_1}. \tag{2.56}$$

By replacing ω, $\mu\varepsilon_1$, and $\mu\varepsilon_2$ by their respective values, equation (2.56) can be written as:

$$\beta_0 n_2 < \beta < \beta_0 n_1. \tag{2.57}$$

In equation (2.57), β_0 is the phase constant of the light wave in vacuum and n_1 and n_2 are the refractive index of core and cladding respectively. By dividing the inequality by β_0 leads to:

$$n_2 < \frac{\beta}{\beta_0} < n_1. \tag{2.58}$$

The ratio β/β_0 is called the effective modal index and is indicative of the effective velocity of the wave in the optical fiber. The effective modal index is denoted by n_{eff} and given by,

$$n_2 < n_{eff} < n_1. \tag{2.59}$$

The transverse components of the electric and magnetic fields in the core and cladding can be determined by replacing the obtained longitudinal components in the equations (2.27)–(2.30) Once all the transverse components are determined, they can be substituted in the relation of the boundary conditions at the core-cladding interface, meaning at $r = a$. The applicable boundary conditions are:

- The tangential components of the electric field are continuous across the boundary.
- Since there are no surface currents, the tangential components of the magnetic field are also continuous at the boundary. These can be written as:

$$\overrightarrow{E_{\phi 1}} = \overrightarrow{E_{\phi 2}}, \tag{2.60}$$

$$\overrightarrow{E_{z1}} = \overrightarrow{E_{z2}}, \tag{2.61}$$

$$\overrightarrow{H_{\phi 1}} = \overrightarrow{H_{\phi 2}}, \tag{2.62}$$

$$\overrightarrow{H_{z1}} = \overrightarrow{H_{z2}}. \tag{2.63}$$

Applying the boundary conditions results in four simultaneous equations. However, five unknowns have to be determined namely, the arbitrary constants A, B, C, D as well as the value of the propagation constant β.

For this system of homogeneous equations to accept a non-trivial solution, the determinant of the matrices must be equal to zero. After algebraic development, this condition results in equation (2.64) given by:

$$\left\{ \frac{J'_\nu(ua)}{uJ_\nu(ua)} + \frac{K'_\nu(wa)}{wK_\nu(wa)} \right\} \left\{ \beta_1^2 \frac{J'_\nu(ua)}{uJ_\nu(ua)} + \beta_2^2 \frac{K'_\nu(wa)}{wK_\nu(wa)} \right\} = \frac{\beta\nu}{a} \left\{ \frac{1}{u^2} + \frac{1}{w^2} \right\}^2 . \tag{2.64}$$

where,

$$J'_\nu(x) = \frac{\partial}{\partial x}J_\nu(x) \text{ and } K'_\nu(x) = \frac{\partial}{\partial x}K_\nu(x). \tag{2.65}$$

Equation (2.64) contains all the six components of the electric and magnetic field. Knowing that a mode possessing all the electrical and magnetic field components is called a hybrid-mode, the characteristic equation (2.64) is called the characteristic equation of hybrid-modes.

Using the quantities u and w in (2.50) and (2.51), the value of ν in equation (2.64) is the same quantity that appears in $\varnothing(\phi) \sim e^{j\nu\phi}$. This equation determines the field behaviour in the azimuthal direction. This quantity is an integral constant. If ν takes the value 0, the resulting field is symmetrical circular. Figure 2.12 shows the variation of the field pattern in the azimuthal direction for $\nu = 0$.

The diagram in Figure 2.12 shows that the fields are circularly symmetric in the azimuthal direction with a maximum intensity at the central region of the fiber and gradually decreasing towards the periphery of the core of the fiber. This corresponds to the meridional rays in the ray optic discussion. In short, $\nu = 0$ corresponds to the

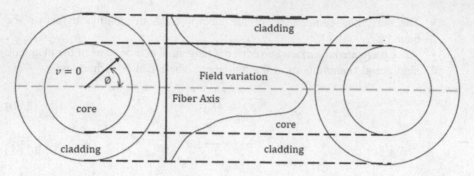

FIGURE 2.12 Light intensity pattern for v =0.

meridional rays and any other value of ν corresponds to skew rays. Therefore, they have no maximum at the axis of the fiber because skew rays spiral around the axis and do not meet at the axis of the fiber. By substituting ν by zero in equation (2.64) the right-hand side of the equation becomes zero and we can write,

$$\left\{ \frac{J_0'(ua)}{uJ_0(ua)} + \frac{K_0'(wa)}{wK_0(wa)} \right\} = 0, \tag{2.66}$$

and,

$$\left\{ \beta_1^2 \frac{J_0'(ua)}{uJ_0(ua)} + \beta_2^2 \frac{K_0'(wa)}{wK_0(wa)} \right\} = 0. \tag{2.67}$$

Equation (2.66) is for TE modes whereas equation (2.67) is for TM modes. The derivative of the Bessel and modified Bessel functions are given by,

$$J_0'(x) = -J_1(x). \tag{2.68}$$

Substituting these derivatives in equation (2.66) yields:

$$\frac{J_1(ua)}{uJ_0(ua)} + \frac{K_1(wa)}{wK_0(wa)} = 0. \tag{2.69}$$

The same reasoning can be followed for TM modes yielding:

$$\beta_1^2 \frac{J_1(ua)}{uJ_0(ua)} + \beta_2^2 \frac{K_1(wa)}{wK_0(wa)} = 0. \tag{2.70}$$

It can be seen TE and TM modes possess field distribution circularly symmetric around the axis of the fiber as they correspond to $\nu = 0$. On the other hand, for values of ν other than zero, we obtain hybrid modes. These modes do not have circularly symmetric field distributions.

Equations (2.66) and (2.67) have multiple solutions because $J_1(x)$ and $J_0(x)$ are oscillatory functions having multiple zeros. Therefore, equations (2.66) and (2.67) will have "m" different solutions depending on the value of "a" in the Bessel function. To fully designate a mode, one needs a set of two parameters (v, m) in which v provides the intensity pattern and "m" gives the number of solutions. For example, the notation TE_{01} means circularly symmetric field of the first solution of the characteristic equation (2.64). In the same way, TM_{02} means circularly symmetric field of the second solution of equation (2.64) etc.

Hybrid modes are also designated in the same way, but the v can be any value except 0. For example, we can have HE_{21}, HE_{54}, HE_{13}, etc. The combination (v,m) is useful for identifying a mode and its corresponding intensity pattern. Reversely, by looking at the intensity pattern of the light one can know v and m, thus defining the mode. The following table shows the light intensity patterns of common modes. (Table 2.1)

2.2.1 NUMBER OF TRANSVERSE MODES

During the wave model analysis of an optical fiber, we introduced two quantities defined by,

$$u^2 = \omega^2 \mu \epsilon_1 - \beta^2, \tag{2.71}$$

$$w^2 = \beta^2 - \omega^2 \mu \epsilon_2. \tag{2.72}$$

Adding the two equations and expressing the core and cladding propagation constants as a function of the propagation constant of the wave in vacuum,

$$u^2 + w^2 = \beta_0^2 (n_1^2 - n_2^2). \tag{2.73}$$

By multiplying both sides of equation (2.73) by a^2, with "a" being the radius of the fiber core, we obtain:

$$a^2(u^2 + w^2) = a^2 \beta_0^2 (n_1^2 - n_2^2). \tag{2.74}$$

Replacing $a^2(u^2 + w^2)$ by V^2, gives:

$$V = \frac{2\pi a}{\lambda}(NA). \tag{2.75}$$

V is called the V-number of the optical fiber. V is a more comprehensive characterization of an optical fiber because it relates all the suitable parameters that describe the fiber namely, core and cladding refractive index, and core radius. The V-number is related to wavelength, hence the frequency of light for a given core

TABLE 2.1

Light Intensity Patterns for Different Values of v and m

Types of Rays	Value of v	Value of m	Transverse intensity pattern
Meridional Rays	0	1	
	0	2	
	0	3	
Skews rays	1	1	
	2	1	
	3	1	
	4	1	
	5	1	

radius. Therefore, the V-number is also known as the Normalized frequency of the fiber.

The quantities 'u' and 'w' were defined by equations (2.50) and (2.51). From equation (2.51) one can see that for $\beta \geq \omega^2 \mu \epsilon_2$, the '$w$' value is no longer purely imaginary. Therefore, the modified Bessel function is no longer valid to describe the behaviour of the field in the cladding. It means that there exists a limit or a cut-off on the value of β beyond which light energy is no more confined inside the core of the optical fiber and starts leaking into the cladding, provoking huge energy losses and preventing light propagation. Physically, β represents the effective propagation constant of the mode inside the fiber. If the value of β approaches β_2, most of the energy propagates into the cladding with refractive index n_2. Similarly, if the value of β is close to β_1, most of the propagation energy is confined to the core with refractive index n_1. It also means that when the value of β lies between β_1 and β_2, a fraction of light propagates inside the core and another fraction inside the cladding. However, because the two mediums have different propagation constants, for light to travel simultaneously in the two mediums, there exists a mutual propagation constant shared by the two mediums. This propagation constant is known as the effective propagation constant and is related to the effective refractive index n_{eff}. We can also introduce a new quantity called the normalized phase constant "b". defined by:

$$b = \frac{n_{eff}^2 - n_2^2}{n_1^2 - n_2^2} \approx \frac{n_{eff} - n_2}{n_1 - n_2}. \tag{2.76}$$

In equation (2.73), the effective index of propagation is given by:

$$n_{eff} = \frac{\beta}{\left(\frac{2\pi}{\lambda}\right)}. \tag{2.77}$$

It can be seen that the value of n_{eff} tends to n_1 when β tends to β_1 and tends to n_2 when β tends to β_2 (cutoff). On the other hand, the value of 'b', the normalized phase constant ranges between 0 and 1. It will be beneficial to study the normalized propagation constant as a function of the normalized frequency instead of propagation constant as a function of angular frequency. Figure 2.13 illustrates the normalized phase constant as a function of normalized frequency.

When the V-number increases, the normalized propagation constant also increases for all the modes. It can also be seen that the HE_{11} mode is the mode that propagates at the lowest frequency. In comparison to the ray optic approach discussed in section 2.2.2, this mode HE_{11} corresponds to the propagating ray in the axis of the fiber. The mode TE_{01} however, which is the first transverse mode, does not propagate until the value of 2.4 is reached. It must be noted that this value of 2.4 corresponds to the first root of the $J_0(x)$ Bessel function. Therefore, the V-number must be greater than 2.4 for the dominant modes TE, TM, and HE to propagate. However, the HE_{11} mode will always propagate because it has the lowest cut-off frequency. The HE_{11} mode is the dominant mode of the fiber and is a hybrid mode, because it contains all the six components of the optical field.

2.2.2 LINEARLY POLARIZED (LP) MODES

Practical optical fibers have very small index differences between core and cladding, as discussed in section 2.2.2. It was shown that for all practical optical fibers, one can define a quantity Δ such that:

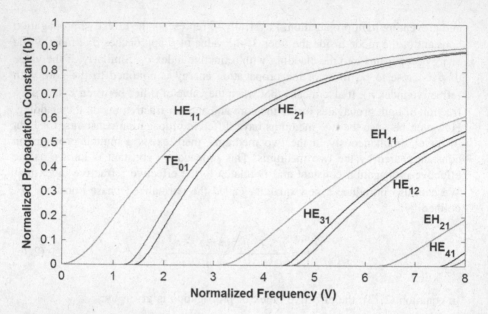

FIGURE 2.13   Normalized propagation constant as a function of V-number.

$$\Delta \equiv \frac{n_1 - n_2}{n_1} \ll 1, \tag{2.78}$$

In the ideal case where $n_1 = n_2$ the core and cladding of the fiber disappear, therefore, the longitudinal components of the electric and magnetic fields vanish and the wave becomes completely transverse electromagnetic. In practical cases, however, n_1 approaches n_2 without reaching it, therefore, the fields are almost transverse and they become linearly polarized, meaning their polarizations remains the same everywhere in the cross-sectional plane of the fiber as shown in Figure 2.14. This situation is known as the weakly guiding approximation.

In the weakly guiding approximation, all the higher mode clusters degenerate into a linearly polarized (LP) mode. This means that in Figure 2.12, the modes TE_{01}, TM_{01}, and HE_{21} degenerate into a single LP mode. For the same reason, the modes HE_{12}, HE_{31}, and HE_{11} will also degenerate into a single LP mode. From Figure 2.12 observations, one can tell that the fundamental mode HE_{11} is already an LP mode.

In the LP mode nomenclature, the fundamental mode HE_{11} is referred to as the LP_{01} mode, where the first index represents the variation of light intensity in the azimuthal direction, and the second index represents the number of zero crossings in the light intensity pattern. The second LP mode to propagate is the LP_{11} mode, which results from the degeneracy of the TE_{01}, TM_{01}, and HE_{21} modes. The simplified b-V diagram for LP modes represented in Figure 2.15 has only three modes, namely only LP_{01}, LP_{11}, and LP_{21} modes.

In conclusion, LP modes are a good approximation to fiber light propagation mode analysis for practical applications. It means that the other electromagnetic

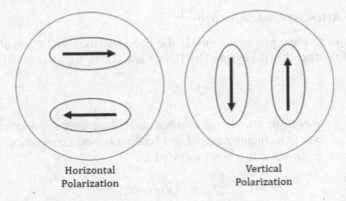

Horizontal
Polarization

Vertical
Polarization

FIGURE 2.14   Polarization of the linearly polarized modes.

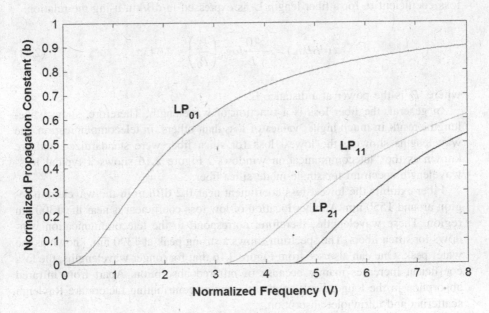

FIGURE 2.15   b-V curve of a fiber in terms of LP modes

modes discussed in the previous section are only important from a theoretical understanding point of view.

2.3   ATTENUATION AND LOSSES IN OPTICAL FIBERS

One of the limiting factors in optical fiber technology is fiber loss. Losses result in attenuation of the optical signal over the transmission distance. A high-quality fiber must have as little loss as possible.

2.3.1 Attenuation Coefficient

As an optical wave propagates inside the fiber, variation of its power along the longitudinal direction is governed by Beer's law (G. P. Agrawal 2012):

$$dP/dz = -\alpha P, \tag{2.79}$$

where α represents the attenuation coefficient. This parameter includes attenuation for various sources including material and bending losses. The solution of equation (2.79) is an exponential function expressed as:

$$P_{(z)} = P_0 exp(-\alpha z), \tag{2.80}$$

where P_0 is the input power, $P_{(z)}$ is the power at a distance z inside the fiber. The loss coefficient α, for a fiber length L, is expressed in dB/km using the relation,

$$\alpha(dB/km) = -\frac{10}{L}log_{10}\left(\frac{P_L}{P_0}\right) \approx 4.343\alpha. \tag{2.81}$$

where, P_L is the power at a distance L.

In general, the fiber loss is a function of wavelength. Therefore, some wavelengths result in much higher values of loss than others. In telecommunication, the wavelengths showing the lowest loss for silica fiber were standardized and are known as the "telecommunication windows". Figure 2.16 shows a typical loss-wavelength spectrum for single-mode silica fiber.

Fiber exhibits the lowest loss coefficient near 0.2 dB/km in the wavelength region around 1550 nm. Another location of low loss coefficient is near the 1300 nm region. These wavelengths, therefore, correspond to the telecommunication windows for silica fibers. The spectrum shows a strong peak at 1390 nm, known as the water peak. One can also see from Figure 2.16 that for longer wavelengths, the loss coefficient increases mainly because of infrared absorption. Apart from infrared absorption in the longer wavelengths, the major contributing factors are Rayleigh scattering and ultraviolet absorption.

2.3.2 Material Absorption

There are two main categories of material absorption namely, intrinsic and extrinsic absorption. The intrinsic absorption is present in any material and is the result of the material-electron and vibrational resonance of the molecules at given wavelengths. For silica fiber, for example, electronic vibration takes place at around 400 nm in the ultraviolet region and vibration resonance takes place at a wavelength longer than 700 nm in the infrared region. Due to the amorphous nature of silica glass, these resonances take the form of the absorption band extending into the visible region. The value of intrinsic absorption remains below 0.1 dB/km in the range between 800 nm and 1600 nm as it can be seen in Figure 2.16 reaching values as

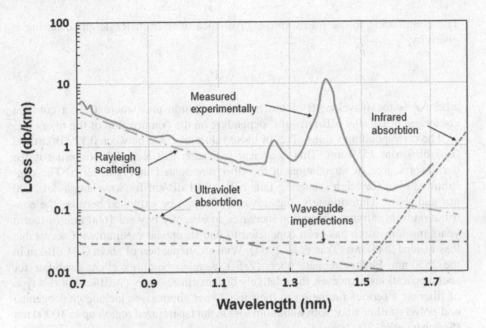

FIGURE 2.16 Loss spectrum of a single-mode silica fiber highlighting the various loss contributions.

low as 0.03 dB/km for wavelength between 1300 nm and 1600 nm. It means that the contribution of intrinsic loss to the total fiber loss is rather marginal. Extrinsic absorption, on another hand, corresponds to losses caused by the presence of impurities within the silica glass matrix. These impurities include Fe, Cu, Mn and exhibit strong absorption in the 600–1600 nm region. In modern fiber where their concentration is reduced to below 1 part per million (ppm), their influence is minor. Much of extrinsic loss in the state of the art fiber has its origin in the presence of water vapour (OH ions). These ions have their vibrational resonance in the 2700 nm region. However, its combination with silica vibration and its harmonic produce peak absorption in the 950, 1240, and 1390 nm wavelengths. These peaks are easily visible in the diagram of Figure 2.16. At the highest peak of 1390 nm, a concentration of 1 part per million may result in attenuation increasing up to 50 dB/km. In the telecommunication industry, the concentration of OH ions can be reduced to less than 10^{-8} ppm making the peak almost disappear. Such fibers are named low water peak fibers (G. Agrawal 2012).

2.3.3 RAYLEIGH SCATTERING

The disordered matrix of amorphous silica glass results in an inhomogeneous density distribution in a scale smaller than the wavelength of the incident radiation, hence inhomogeneous refractive index distribution in the same scale. Rayleigh scattering (Born and Wolf 1999) is due to light scattering within such a medium.

The contribution to the intrinsic loss of a silica fiber from Rayleigh scattering is given by:

$$\alpha_R = C/\lambda^4,$$
(2.82)

where λ is the wavelength of the incident radiation and where C is a constant comprised of 0.7–0.9 (dB/km)-μm^4 depending on the composition of the fiber core.

The corresponding values of α_R of these values of C are between 0.12 dB/km and 0.16 dB/km at 1550 nm. This is a clear indication that Rayleigh scattering is the dominant source of attenuation in the fiber represented in Figure 2.16. The contribution from Rayleigh scattering falls below 0.01 dB/km for wavelengths of 3000 nm and above. Unfortunately, this advantage cannot be exploited because the contribution from infrared absorption increases consistently beyond 1600 nm. In recent years much research has been conducted to find an optical medium with acceptable loss beyond 2000 nm (Tran et al. 1984). With an attenuation of about 0.01 dB/km in the 2550 nm, Fluorozirconate fibers (ZrF_4) seems to be an excellent candidate for such applications. However, the relatively high extrinsic losses (1 dB/km) of this type of fiber set a serious limitation in their use. Other alternatives include chalcogenide and polycrystalline fiber with minimum loss in the far-infrared region up to 10000 nm (Sanghera et al. 2002).

2.3.4 BENDING LOSSES

Bending loss is a type of loss, which is not related to the material but the waveguide imperfection. Macro-bending results from physically bending the fiber at a bending radius lower than a critical bending radius about 10 times the fiber diameter. Macro bending can be explained using the ray optic approach. For total internal reflection to occur, rays of light must hit the core-cladding interface with an angle greater than the critical angle. When the fiber has bent, the incident angle is reduced in such a way that the total internal reflection condition is no longer fulfilled, and light energy leaks out of the fiber core.

A much higher level of fiber loss originates from micro bending. Micro-bending loss appears during cabling when the fiber is pressed against a non-perfectly smooth surface. If no precaution is taken to prevent them, micro bending can reach very high values (Marcuse 1976).

Finally, the last type of loss associated with the imperfection at the core-cladding interface is the Mie scattering losses (Senior and Muhammad 2009). Mie losses are the result of index inhomogeneities on a scale longer than the optical wavelength of the incident optical wave.

2.4 DISPERSION IN SINGLE-MODE FIBERS

Dispersion in optical fiber is the result of the light pulse broadening as it travels through the fiber. Among the most important cause of dispersion is the fiber material itself as well as the waveguide geometry. Dispersion has a net detrimental effect on long-distance communication systems because it reduces signal quality.

2.4.1 MODAL DISPERSION

Modal dispersion is also known as multi-path dispersion, observed only in multi-mode fibers, and it is a result of different modes travelling different path lengths. Because the rays travel at the same velocity, the consequence of modal dispersion will be a delay between different modes. If a pulse is sent through such medium, it is going to broaden after travelling a distance and the longer the distance, the broader the pulse. The extent of pulse broadening can be easily understood by considering the shortest and longest possible paths. The shortest pulse corresponds to an incident angle $\theta_i = 0$ it is, therefore, equal to the fiber length L. The incident angle corresponding to the longest path can be obtained from Snell's equation given by:

$$n_0 \sin\theta_i = n_1 \cos\phi_c = (n_1^2 - n_2^2)^{1/2}. \tag{2.83}$$

The corresponding fiber length will be given by $L/\sin\phi_c$. If the velocity of propagation in the fiber core is given by $v = c/n_1$, the time delay will be given by,

$$\Delta T = \frac{n_1}{c}\left(\frac{L}{\sin\phi_c} - L\right) = \frac{L}{c}\frac{n_1^2}{n_2}\Delta. \tag{2.84}$$

The time delay between the ray travelling the longest and shortest path corresponds to the broadening experienced by a pulse travelling down the fiber. This information is related to the fiber carrying capacity through the bit rate (B). An estimate of the bit rate is often obtained from the condition $B\Delta T < 1$. Replacing this equation in equation 2.84 yields:

$$BL < \frac{n_2}{n_1^2}\frac{c}{\Delta} \tag{2.85}$$

The above condition provides a rough estimate of a fundamental limitation of step-index fibers. The effect of intra-modal dispersion in multi-mode optical fiber is significantly reduced by using graded-index fibers.

Modal dispersion is not possible in single-mode fibers because there is only one propagation mode. However, single-mode fiber is not dispersion free. The type of dispersion affecting single-mode fibers is essentially the intra-modal dispersion or group velocity dispersion (GVD) (Paschotta 2008). Two dispersion mechanisms are combined in intra-modal dispersion namely material dispersion and waveguide dispersion.

2.4.2 GROUP VELOCITY DISPERSION

In an optical fiber, a given spectral component of the frequency ω will travel with a velocity group velocity v_g, given by:

$$v_g = (d\beta/d\omega)^{-1}. \tag{2.86}$$

where the propagation constant β given by,

$$\beta = \bar{n}k_0 = \bar{n}\omega/c. \tag{2.87}$$

It can also be proved by using β in equation (2.86) that $v_g = c/\bar{n}_g$, where $\bar{n}_g$ is the group refractive index. Equation (2.87) shows a clear dependency between the group velocity and the frequency. This will result in pulse broadening as different components travel at different velocities. The extent of pulse broadening is governed by:

$$\Delta T = DL\Delta\lambda. \tag{2.88}$$

In this relation, D is the dispersion parameter expressed in ps/(km-nm) and given by:

$$D = \frac{1}{d\lambda}\left(\frac{1}{v_g}\right) = -\frac{2\pi c}{\lambda^2}\beta_2. \tag{2.89}$$

The distribution parameter has two contributing factors namely, material dispersion and waveguide dispersion. Therefore, it can be written,

$$D = D_M + D_W, \tag{2.90}$$

where the material and waveguide dispersions D_M and D_W, respectively, are given by:

$$D_M = -\frac{2\pi}{\lambda^2}\frac{dn_{2g}}{d\omega} = \frac{1}{c}\frac{dn_{2g}}{d\lambda}, \tag{2.91}$$

and

$$D_W = -\frac{2\pi\Delta}{\lambda^2}\left[\frac{n_{2g}^2}{n_2\omega}\frac{Vd^2(Vb)}{dV^2} + \frac{dn_{2g}}{d\omega}\frac{d(Vb)}{dV}\right], \tag{2.92}$$

where n_{g2} is the group index of the cladding and d is the core diameter of the fiber.

2.4.3 MATERIAL DISPERSION

Material dispersion is a result of the dependency between the fiber material (silica in most cases) and the optical frequency. Figure 2.17 shows the dependence between refractive index and group index as a function of wavelength (or frequency).

The dispersion parameter due to material dispersion is given by the following empirical formula (G. P. Agrawal 2012):

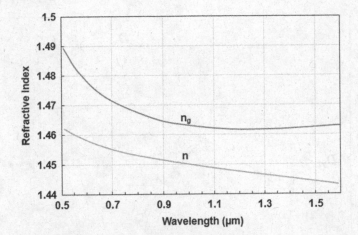

FIGURE 2.17   Refractive index and group index dependency on wavelength for silica fiber.

$$D_M \approx 122(1 - \lambda_{ZD}/\lambda). \qquad (2.93)$$

where λ_{ZD} is called the zero-dispersion wavelength.

It was observed that at the wavelength λ_{ZD}, material dispersion becomes zero. The dispersion parameter D_M has a negative value below this wavelength and becomes positive above it. This wavelength corresponds to 1276 nm for pure silica (G. Agrawal 2012).

2.4.4   WAVEGUIDE DISPERSION

The contribution of waveguide dispersion to the dispersion parameter of the fiber depends on the fiber-normalized frequency V as seen in equation (2.93). Figure 2.18 shows the dependence of material, waveguide, and total dispersion as a function of wavelength for silica fiber. The effect of waveguide dispersion is to shift the position of the zero-dispersion wavelength. For a silica fiber, for example, a shift in wavelength of 30–40 nm was recorded (G. Agrawal 2012), resulting in a zero-wavelength dispersion at around 1310 nm. The dispersion in the region of 1550 nm is rather high with values of 15–18 ps/(km-nm). The waveguide dispersion parameter depends on the core radius and the refractive index difference. It is therefore possible by modifying these parameters, to design fibers with specific characteristics such as dispersion-shifted fibers or dispersion flattened fibers.

2.4.5   POLARIZATION MODE DISPERSION

Polarization Mode Dispersion (PMD) (Gisin et al. 1991) is related to the birefringence of the fiber. The transverse shape of the fiber is never perfectly cylindrical resulting in orthogonally polarized components of the same mode travelling at different group velocities. As for the other type of dispersion, this

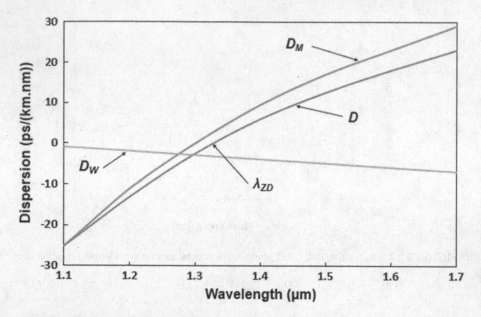

FIGURE 2.18   Total dispersion D and contribution from material and waveguide dispersion of the standard single-mode fiber. The contribution of waveguide dispersion results in the shifting of zero-dispersion wavelength towards 1.31 μm.

situation results in broadened optical pulses. The time delay due to PMD for an ideal fiber with constant birefringence can be written as,

$$\Delta T = \left| \frac{L}{v_{gx}} - \frac{L}{v_{gy}} \right| = L|\beta_{1x} - \beta_{1y}| = L(\Delta\beta_1), \tag{2.94}$$

where $\Delta\beta_1$ represents the difference between group velocities of the two states of polarization of the mode represented by subscripts x and y respectively.

Using ΔT, we can define the quantity $\Delta T/L$ to measure the PMD. The simple case of uniform birefringence along the optical fiber does not comply with practical cases. In fact, in practical cases, birefringence varies along with the fiber, thus the analytical treatment of PMD is quite complex and is largely beyond the scope of this book.

2.5   NON-LINEAR EFFECTS IN OPTICAL FIBER

The response of any dielectric medium to light becomes non-linear for intense electromagnetic fields, and optical fibers are not an exception. Even though silica is intrinsically not a highly non-linear material, the waveguide geometry that confines light to a small cross-section over long fiber lengths makes non-linear effects quite important inside a silica optical fiber (Liang et al. 2015; Saavedra et al. 2017). The most relevant non-linear effects in optical fibers are stimulated Raman, Brillouin scattering, nonlinear phase modulation, and four waves mixing. Stimulated Raman and Brillouin scattering occur due to inelastic scattering phenomena whereas nonlinear phase modulation and four-wave mixing occur due to the change in the refractive index of the medium with optical intensity. The detailed discussion of these effects lies out of the scope of this book and can be found elsewhere in the literature (Carman et al. 1966; R. Stolen 1975; F. S. Ferreira 2011; R. H. Stolen and Lin 1978), therefore, here the authors will only discuss the inelastic scattering phenomena which are more relevant to the subject of fiber lasers.

2.5.1 STIMULATED LIGHT SCATTERING

Rayleigh scattering, discussed in section 2.4.3, is an example of elastic scattering for which the frequency (or the photon energy) of scattered light remains unchanged. By contrast, the frequency of scattered light is shifted downward during inelastic scattering. Raman Scattering and Brillouin scattering (Blow and Wood 1989; Blow and Wood 1989; Pannell et al. 1993) can be understood as the scattering of a photon to a lower energy level such that the energy difference appears in the form of a phonon. The main difference between the two is that optical phonons participate in Raman scattering, whereas acoustic phonons participate in Brillouin scattering. Both scattering processes result in a loss of power at the incident frequency. However, their scattering cross-sections are sufficiently small such that the loss is negligible at low powers levels.

At high power levels, the non-linear phenomena of stimulated Raman scattering (SRS) and stimulated Brillouin scattering (SBS) become important. The intensity of the scattered light in both cases grows exponentially, once the incident power exceeds a threshold value (Smith 1972).

2.5.2 STIMULATED BRILLOUIN SCATTERING

Brillouin scattering is the result of electrostriction (Boyd 2019), which is a property of a dielectric that causes the change in their shape under the influence of an electric field. Experience shows that if an oscillating field at a frequency Ω_p known as the pump frequency is applied to the material, an acoustic wave is generated with a frequency Ω. This acoustic wave has the effect of scattering the original pump wave creating a new wave at the frequency Ω_s. The described scattering process is known as spontaneous Brillouin scattering. Energy conservation imposes that the frequency of the acoustic wave must be exactly equal to the difference between pump frequency and generated Brillouin wave frequency. This frequency difference is known as the "stoke" shift and is written:

$$\Omega = \omega_p - \omega_s \tag{2.95}$$

Similarly, the momentum conservation requires that the wave vectors of the two waves satisfy the condition given by:

$$k_A = k_p + k_s \tag{2.96}$$

where the subscript "A" stands for "acoustic". Knowing the dispersion relation written as (Bennett 1983):

$$|k_A| = \frac{\Omega}{v_A} \tag{2.97}$$

where v_A is the acoustic velocity and assuming that $|k_p| \approx |k_s|$, Ω derived from equation (2.102) can be written as:

$$\Omega = |k_A|v_A = 2v_A|k_p|\sin(\theta/2) \tag{2.98}$$

where θ represent the angle between the pump and scattered waves. It can be easily seen from equation (2.98) that the acoustic wave has a maximum value in the backward propagation direction, whereas its value vanishes in the forward propagation direction. Because light can only propagate in the forward or backward directions, stimulated Brillouin scattering can only take place in that direction with a frequency shift given by $\Omega_B = 2v_A|k_p|$. Expressing the pump wave vector k_p as a function of pump wavelength λ_p, one can write:

$$k_p = 2\pi\bar{n}/\lambda_p \tag{2.99}$$

Using equations (2.104) and (2.103), the Brillouin frequency can be obtained as:

$$v_B = \Omega_B/2\pi = 2\bar{n}v_A/\lambda_p \tag{2.100}$$

where $\bar{n}$ is the modal index of the fiber.

It is worth noting that at sufficiently high intensities, the generated scattered wave will beat with the pump wave creating another wave at a frequency $\omega_p - \omega_s$ which correspond to the frequency of the initial acoustic wave. Therefore, this new beat acts exactly as a source for the acoustic wave increasing its amplitude which in terms increase the amplitude of the scattered wave triggering a positive feedback loop. This positive feedback is the source of stimulated Brillouin scattering whose intensity can be expressed by a system of two coupled differential equations as:

$$\frac{dI_p}{dz} = -g_B I_p I_s - \alpha_p I_p, \tag{2.101}$$

$$-\frac{dI_s}{dz} = +g_B I_p I_s - \alpha_s I_s. \tag{2.102}$$

where I_p and I_s are the intensities of the Stokes waves, g_B is the stimulated Brillouin scattering gain and α_p and α_s are the fiber background loss at the pump and signal wave, respectively.

The SBS gain is a function of the frequency. It was can be shown that it has a Lorentzian spectral profile given by (Pannell et al. 1992; Pratap Singh et al. 2007):

$$g_B(\Omega) = \frac{g_B(\Omega_B)}{1 + (\Omega - \Omega_B)^2 T_B^2} \qquad (2.103)$$

As can be seen from equation (2.103) stimulated Brillouin scattering is a function of frequency. This is due to its dependence on the lifetime of acoustic phonons given by T_B in equation (2.103). It can also be seen that from the same equation the gain of stimulated Brillouin scattering presents a maximum when $\Omega = \Omega_B$. This gain is a function of material parameters such as the elasto-optic coefficient and means the density of the medium (Boyd 2019). The onset of stimulated Brillouin scattering corresponds to a threshold power obtained by solving the coupled equations (2.101) and (2.102) and identifying the values of I_p and I_s where these quantities grow beyond noise levels to significant values.

The threshold power is given by (Pratap Singh et al. 2007),

$$P_{th} = I_p A_{eff}, \qquad (2.104)$$

where A_{eff} is the effective core area given by,

$$A_{eff} \approx \frac{g_B P_{th} L_{eff}}{21}, \qquad (2.105)$$

where L_{eff} is the effective interaction length expressed as,

$$L_{eff} = [1 - exp(-\alpha L)]/\alpha, \qquad (2.106)$$

and α is the fiber losses. For telecommunication systems where the fiber has lengths of tens of kilometres or more, L_{eff} can be approximated as $L_{eff} = 1/\alpha$. Using $A_{eff} = \pi w^2$, where w is the spot size, P_{th} can be as low as 1 mW depending on values of spot size w and background losses α. Therefore, stimulated Brillouin scattering can be a very detrimental effect in optical communication even at low to moderate launched powers.

2.5.3 STIMULATED RAMAN SCATTERING

Spontaneous Raman scattering is the result of pump lightwave scattering by glass molecules. This process can be better understood by a quantum description, represented in Figure 2.19 rather than classical description. In the energy diagram of Figure 2.19 describing, photons at the pump wavelength, ω_p, loose energy to create

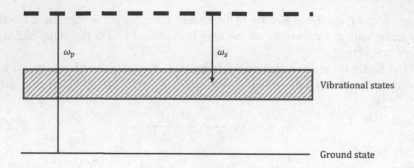

FIGURE 2.19 Energy levels involved in Stimulated Raman scattering process.

photons at a lower frequency ω_s. The lost energy is absorbed by the glass molecules, which move to a vibrational state.

Unlike the case of Brillouin scattering discussed in the previous section, there is no acoustic phonon involved in Raman scattering, therefore, it occurs in all directions.

Similar to stimulated Brillouin scattering, Raman scattering can become a stimulated process if the pump power is increased beyond a threshold value. The interaction between the pump and the scattered light creates a wave at a beat frequency in both forward and backward directions. This wave triggers molecular oscillation, which in turn increases the scattered wave's amplitude. Like in the case of SBS, a positive feedback loop is established. This feedback process is governed by the following coupled differential equations (Pratap Singh et al. 2007),

$$\frac{dI_p}{dz} = -g_R I_p I_s - \alpha_p I_p, \tag{2.107}$$

$$\frac{dI_s}{dz} = g_R I_p I_s - \alpha_s I_s, \tag{2.108}$$

where g_R is the stimulated Raman scattering gain profile shown in Figure 2.19, I_p and I_s the pump and signal intensity respectively, α_p and α_s are the fiber background losses at pump and signal wavelength, respectively as shown in Figure 2.20.

It is interesting to note that, the backward propagating SRS wave is given by an equation identical to that of SBS because of the required minus sign added in front of equation (2.108). Unlike the case of SBS, the spectrum of SRS in an optical fiber is large. Its value can easily exceed 10 THz as can be seen in Figure 2.19 where the spectrum ranges over 42 THz with peak values around 13 THz and 15 THz. This large spectrum is attributed to the amorphous nature of silica glass in which the vibrational energy of molecules merges to form a band.

Similar to the case of stimulated Brillouin scattering, the threshold power of stimulated Raman scattering is given by equation (2.109) (Pratap Singh et al. 2007):

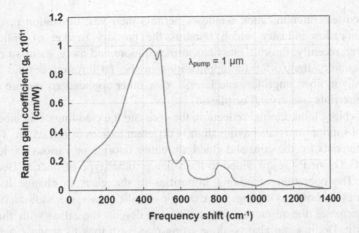

FIGURE 2.20 Raman gain spectrum for silica fiber at around 1 μm.

$$P_{th} = \frac{16A_{eff}}{g_R L_{eff}}, \qquad (2.109)$$

where L_{eff} is the effective length approximated to $1/\alpha$. If also, we approximate A_{eff} by πw^2, where w is the spot size, P_{th} for SRS will be given by:

$$P_{th} \approx 16 \frac{1}{L_{eff}} (\pi w^2)/g_R, \qquad (2.110)$$

Stimulated Raman scattering is more important than stimulated Brillouin scattering in this discussion because of the possibility it offers to construct Raman amplifiers and Raman lasers.

Stimulated Brillouin and stimulated Raman scattering in optical fibers are both inelastic scattering processes. Although the two are quite similar in terms of their origin they also show the following differences:

- Contrary to SRS which takes places in both forward and backward direction, SBS is a strictly backwards process.
- In SBS the frequency of the scattered light undergoes a frequency shift of around 10 GHz whereas the value of the shift is around 13 THz for SRS.
- The Raman-gain spectrum that extends over 20–30 THz is much wider than the Brillouin limited at values below 100 MHz (Kale et al. 2013; Hafizi et al. 2016).

2.6 OPTICAL FIBER MATERIALS

Optical fibers are made of transparent glass consisting of either oxide glasses—the most common of these being Silica—or fluoride glasses—**ZBLAN**—(Saad 2011). To date, the majority of optical fibers are made of silica because their low loss in the so-called telecommunication windows permits their vast utilization in the optical communication industry which remains the primary market of optical fibers. However, recently, fluoride glass has attracted more and more attention because of the possibility they offer for applications in the far-infrared beyond 2000 nm, especially in fiber amplifiers and lasers. Also, other applications require the use of other materials like crystal or plastic.

To achieve total internal reflection, the core and the cladding of the fiber must be made of similar materials having slightly different indices of refraction. To produce such materials for the core and cladding, either fluorine or various oxides such as B_2O_3, GeO_2, or P_2O_5 are added to the silica or other type of matrix (Becker et al. 1999). The operation of adding impurities to the glass to change its physical properties is known as doping. For example, in silica glass, the addition of GeO_2 or P_2O_5 increases the refractive index whereas doping the silica with fluorine decreases it. Doping can also be done using rare-earth ions to produce active fibers used in fiber amplifiers and lasers. In this section we are going to describe the main glasses used to make optical fibers, highlighting their properties and potential applications.

2.6.1 SILICA GLASS

Silica glasses remain the most used host material for optical fibers in applications including telecommunications, fiber amplifiers, fiber lasers, and fiber-optic sensors. Silica glass is formed by a disordered matrix made of strongly covalently bonded atoms. This chemical constitution results in a giant covalent structure with excellent mechanical properties. Silica exhibits a remarkably low absorption in the near-infrared region of the spectrum, with the lowest absorption value of 0.186 dB/km around 1500 nm (Tee et al. 2016). Also, this type of glass has a high melting point at about 1700°C, relatively low refractive index, and low thermal expansion coefficient. Silica glass fibers are mostly obtained by a modified chemical vapour deposition technique; therefore, they show an excellent level of purity. To make active optical fibers used for amplifiers and lasers, silica glasses are also doped with rare-earth ions. The most common doping ion in telecommunication is erbium because its emission range in the infrared matches low loss band of silica glass. The most serious drawback when doping pure silica glass with rare-earth active ions remains the poor solubility of these ions in the glass matrix. Because of this poor solubility, these ions tend to form clusters. Inside the clusters, they exchange energy with one another, thus quenching the overall luminosity inside the gain medium and reducing the efficiency of fiber lasers and amplifiers. To reduce luminosity quenching, silica glasses can be co-doped with compounds like aluminium for example. When network formers such as Al_2O_3 and P_2O_5 are added to silica (Arai et al. 1986), they create the solvation shells necessary for improved rare-earth ions solubility to prevent clustering. However, Aluminosilicate fiber demonstrates a higher Rayleigh scattering loss due to the extra refractive index inhomogeneities introduced by the additional dopants (Saito et al. 2003). Silica glass demonstrates a high threshold for optical damage [50]. Values of approximately 500 MW/cm^2 are not uncommon ensuring robust high-power operation. As in other glasses, second-order non-linearity is very low in silica glasses. However, third-order non-linear processes like Raman scattering can be significantly important because high intensities of light can develop in the core of the fiber at moderate powers due to the fiber length and small core diameter. It is also worth mentioning that owing to the amorphous nature of silica glass, cleaving and fusion splicing of silica fibers is a relatively easy task.

The major limitation of silica remains the relatively narrow transmission band in the infrared spectrum compared to fluoride glass for example. This is the result of the maximum phonon energies of up to $1,100$ cm^{-1} in silica glass (Layne et al. 1977), which sets an upper limit on the emission wavelength. The longest laser wavelength from a silica glass fiber laser ever reported to this day was 2188 nm (S. Sacks et al. 2007). Obtaining lasing at longer wavelength is not possible because of the onset of phonon quenching. By the time of the publication of this book, several commercial applications using silica glass at wavelengths around 2000 nm in Tm^{3+} and Ho^{3+} doped silica glasses are available.

2.6.2 FLUORIDE GLASS AND ZBLAN

Fluoride glass is a type of non-oxide glass composed of fluorides of various metals. The most successful fluoride glasses are the ZBLAN with a composition ZrF_4-BaF_2-LaF_3-AlF_3-NaF. ZBLAN possess interesting optical properties such as a broad optical spectrum extending from 300 nm to approximately 8000 nm, low refractive index, and low dispersion. The most common proportion of ZBLAN elements are 53 mol% ZrF_4, 20 mol% BaF_2, 4 mol% LaF_3, 3 mol% AlF_3 and 20 mol% NaF (Ohsawa et al. 1981). However, to introduce certain desirable characteristics, these proportions can be changed. In this regard viscosity as well as the refractive index can be modified by the addition of PbF_2, ZnF_2 or CaF_2 (Ohsawa et al. 1981).

Compared to silica glass, ZBLAN glass has a remarkably large infrared transmission spectrum. This transmission results from its lower maximum phonon energy (approximately 565 cm^{-1}) (Almeida and Mackenzie 1981) which is a consequence of weak bond strength and a reduced mass of glass atoms. Its internal material loss of less than 0.01 dB/km around 2500 nm is remarkably low, opening the possibility to achieve telecommunication links with extremely low loss. However, this move will mean dropping all together with the use of Erbium-doped fiber amplifier, which constitutes one of the key elements of the modern telecommunication network because their bandwidth is largely below 2500 nm.

The main disadvantage of fluoride glass is the difficulty to process them. In fact, due to their low viscosity, it is not easy to completely avoid crystallization while processing them through the glass transition (or drawing the fiber from the melt). Also, the weaker optical damage threshold of about 25 MW/cm^{-2} for 10 ms pulses around 2.8 μm (Zhu and Jain 2007) results in a much lower achievable power compared to silica glass. In comparison, this type of glass does not experience too much loss to Raman scattering due to its small Raman gain coefficient (Fortin et al. 2011). Power scaling of ZBLAN fiber has been particularly difficult because of the poor physical properties mentioned earlier. To date, the maximum output power reported from a ZBLAN fiber laser does not exceed 20 W at 1.94 μm (Fortin et al. 2011)and 24 W at 2.7 μm (El-Agmy and Al-Hosiny 2010).

2.6.3 OTHER TYPES OF GLASS

Many other types of glasses have been developed and tested mostly for the goal of achieving longer emission wavelengths. Among them are germanate and chalcogenide glasses. Germanate glass shows excellent properties such as mechanical robustness, high phonon energy, and excellent rare-earth ion solubility. These qualities and particularly the rare-earth ion solubility, allow the construction of highly efficient Tm^{3+} doped fiber lasers emitting in the 2000 nm region (Wu et al. 2007). Also, the high solubility allows achieving efficient short cavity fiber lasers with a narrow linewidth output (Geng et al. 2007). Germanate glass is somehow mid-way between silica glass and fluoride glass as they combine the best properties of silica glasses like strength with those of fluoride-based glasses like low nonradiative relaxation rate (Wang et al. 1993). Beyond 2000 nm, however, emission in germanate glasses become hard to realize without annihilating OH^- impurities, which is not an easy task.

Another promising alternative is chalcogenide glasses. Chalcogenide glass contains one or more chalcogens such as Sulphur, Selenium, or Tellurium. These glasses show a high refractive index resulting in large values of absorption and emission cross-section. Their maximum phonon energy is moderately low. Sulfide glass has the highest value (350–425 cm^{-1}) (Barnier et al. 1992) followed by Selenide glasses and Tellurite glasses with values between 250–300 cm^{-1} (Weszka et al. 2000) and 150–200 cm^{-1} (Uemura et al. 1996), respectively. The most important property of chalcogenide glass remains its high upper emission wavelength. Fluorescence values of 15000 nm have been reported for tellurite glasses (Uemura et al. 1996), and for Sulphide and Selenide glasses, the value of 8000 nm is not uncommon (Shaw et al. 2001). However, in practice, it is quite difficult to build fiber lasers emitting at these wavelengths. An Nd^{3+} doped chalcogenide glass fiber laser emitting at 1080 nm was reported (Heo and Chung 2014). This rather low lasing wavelength with respect to what can be achieved with such material was attributed to the addition of La_2O_3 to gallium chalcogen glass to prevent crystallization during fiber manufacturing. The La_2O_3 ions likely increased phonon energy that annihilated longer wavelength emission. No other report of rare-earth-doped chalcogenide glass has been made for the reason that it is very difficult to incorporate rare-earth ions into chalcogenide glass matrix because of its strongly covalently bonded atoms. Studies are being conducted to find suitable co-dopants to improve rare-earth ions solubility and suppress crystallization (J. Sanghera et al. 2001).

2.7 OPTICAL FIBER FABRICATION TECHNIQUES

There is a wide range of fabrication methods for optical fibers. These methods can be divided into two wide categories namely methods involving a preform and direct fiber production. Methods based on a preform are the most used especially in the context of glass fibers, whereas direct methods are common for plastic fibers.

The basic material for optical fiber has been silicon dioxide for years. To guide light by total internal reflection in an optical fiber, several additional dopants such as Germanium or Phosphorus are used to either increase or decrease the core or cladding refractive index of the fiber. In recent years, they have also been significant interest in other glasses such as fluoride glass, mostly for their ability to emit in the far-infrared region.

Although silicon dioxide is found in nature, its high level of impurity makes it impractical for optical fiber manufacturing. Therefore, to achieve required high levels of purity, manufacturing of optical fiber begin with silicon tetrachloride ($SiCl_4$) in the liquid phase. The liquid $SiCl_4$ is then distilled and the obtained vapour is recombined with oxygen at high temperature to produce high purity silicon dioxide.

Most glass fibers are fabricated by pulling a strand from a preform on a fiber-drawing tower which is typically several meters high. The preform is a glass rod with a diameter of about 1 to 10 cm and roughly 1 m in length. This rod has the radial characteristics of the desired fiber (step or graded refractive index profile) so that when the fiber is drawn from its end by melting it using a furnace, it retains the radial characteristics of the preform.

The core and cladding are made of the same type of pure glass (silica in a majority of cases). To increase or decrease the refractive index of the core or the cladding, the glass is doped with impurities. Doping is done by combining volatile organic compounds such as $SiCl_4$, $GeCl_4$, $POCl_3$, $TiCl_4$, and BCl_3 with oxygen as shown in the chemical equations (2.111):

$$SiCl_4 + O_2 \rightarrow SiO_2 + 2Cl_2$$
$$GeCl_4 + O_2 \rightarrow GeO_2 + 2Cl_2$$
$$4POCl_3 + 3O_2 \rightarrow 2P_2O_5 + 6Cl_2 \qquad (2.111)$$
$$TiCl_4 + O_2 \rightarrow TiO_2 + 2Cl_2$$
$$4BCl_3 + 3O_2 \rightarrow 2B_2O_3 + 6Cl_2$$

2.7.1 VAPOUR PHASE DEPOSITION METHODS

The Vapour Phase Deposition techniques for preform fabrication are divided into two main methods, namely flame hydrolysis and chemical vapour deposition techniques. Under chemical vapour deposition are Modified Chemical Vapour Deposition (MCVD), Plasma-activated Chemical Vapour Deposition (PCVD), and Plasma-modified Chemical Vapour Deposition (PCVD). Under the flame hydrolysis techniques are Vapour Axial Deposition (VAD), and Outside Vapour Deposition (OVD). The flow diagram in Figure 2.21 represents these techniques.

2.7.1.1 Outside Vapour Deposition (OVD)

The OVD is the first known mass production process developed by Corning Glassworks. This method involves oxidation of chloride compounds in the vapour phase to form soot particles deposited on cool rotating graphite or ceramic mandrel. The various initial compounds are vaporized through an oxygen flow to remove all

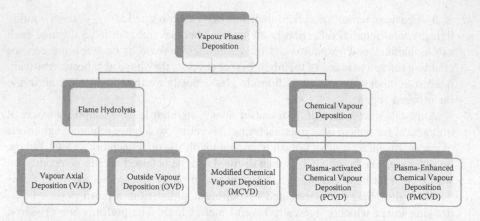

FIGURE 2.21 Classification of fiber pre-form manufacturing processes.

impurities. These vapours are mixed with oxygen and passed between the mandrel and a burner. The burner's flame is reversed back and forth over the length of the mandrel. On each pass of the burner, a small amount of gas reacts with oxygen to create SiO_2 particles called "soot" that are deposited on the mandrel. In this way the diameter of the rod build-up in successive layers. After enough layers are built up, the mandrel is removed; the remaining hollow porous preform is dehydrated and collapsed in a controlled atmosphere, (e.g. helium) into a solid cylindrical preform. The composition of the deposited material is varied by controlling the concentration of the vapour reactants. By adjusting their concentration from layer to layer, precise control of the composition of the entire preform is achieved. Finally, drawing enables the manufacturer to obtain the fiber in the actual size desired and the radial composition equivalent to that of the original preform.

2.7.1.2 Modified Chemical Vapour Deposition (MCVD)

The MCVD is an improved version of OVD developed at Bell Laboratories in 1974 (MacChesney et al. 1974). Unlike the OVD, the reaction takes place on the inner surface of a glass tube. The glass of the original tube will ultimately become the cladding of the drawn fiber. In this process, vapour reactants (chlorides and oxygen) are blown inside a rotating silica substrate tube. A source of heat moves back and forth on the external surface of the tube as illustrated in Figure 2.22. On the action of the heat, chlorine reacts with oxygen to form silica soot, which is deposited inside the tube. The first reaction involves Silicon tetrachloride ($SiCl_4$) with Oxygen to form Silicon Dioxide (SiO_2), and the last Germanium tetrachloride ($GeSl_4$) reacting with Oxygen to form Germanium Dioxide (GeO_2) and increase the refractive index. After the deposition is completed, the temperature is increased to values between 1700°C and 1900°C. The tube is then collapsed to give a solid preform, which may then be drawn into a fiber.

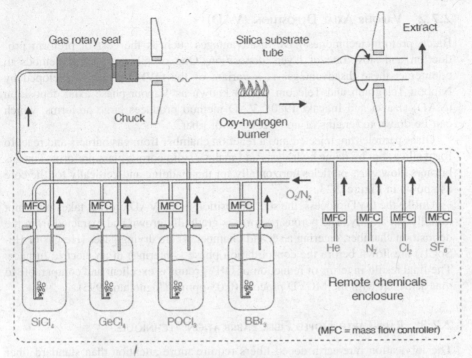

FIGURE 2.22 Schematic representation of the MCVD process showing the glass-working lathe and chemical delivery system.

2.7.1.3   Plasma-Activated Chemical Vapour Deposition (PCVD)

PCVD is a modification of the MCVD method where instead of heating the outer part of the silica tube using a torch, a locally created microwave field provides the energy. This microwave field is produced by a microwave cavity resonator supplied by a 2.45 GHz generator surrounding the silica tube. The microwave field heats the gas plasma directly without heating the tube. Therefore, it can be moved quickly and traverse the tube thousands of times (compared to hundreds for burner) depositing an extremely thin layer of soot each trip, resulting in excellent control of refractive index profile. Also, the tube is maintained around 1200°C to reduce stress in the deposited layers and prevent chlorine implantation. PCVD efficiency is close to 100% and is significantly faster than a traditional MCVD. Finally, there is no need for rotating the tube during deposition. As a result, the preform can be large in volume, therefore very long fibers (50 Km) can be drawn from it.

2.7.1.4   Plasma-Enhanced Chemical Vapour Deposition (PCVD)

The PECVD method is similar to PCVD except that microwave cavity resonator is replaced by coil powered by a 3–5 MHz radiofrequency generator (Cognolato 1995).

2.7.2 Vapour Axial Deposition (VAD)

Batch preform techniques have disadvantages such as the cost of preform production and slow rate of layers deposition. One of the most used methods to counteract these disadvantages is a variant of the OVPO process developed by Nippon Telegraph and Telecom (NTT) known as Vapour-phase axial deposition (VAD) (Izawa and Inagaki 1980). VAD method produces large preforms, which can be drawn to lengths of up to 250 km of fiber.

Glass particles are injected into a reaction chamber from gas burners and react to form silica soot by flame hydrolysis and make a solid porous glass preform. The gas burners blow glass particles horizontally for the cladding and vertically for the core as shown in Figure 2.23.

Unlike the OVD process, the soot deposition in the VAD process takes place not radially but axially. The porous preform is gradually grown and extracted from the deposition chamber, entering a second chamber where drying gases (He-Cl$_2$ or He-SoCl$_2$) are flown before the consolidation phase (sintering) in an electric furnace. The final result, in terms of reduction of OH$^-$ groups is excellent and comparable to that achieved with the MCVD process (0.03 ppm) (Cognolato 1995).

2.7.3 Rare-Earth Doped Fiber Fabrication Techniques

The fabrication rare-earth doped fibers require more attention than standard fiber because of the diversity of material and structures required to achieve better amplification and lasing characteristics. Out of many fabrication techniques described

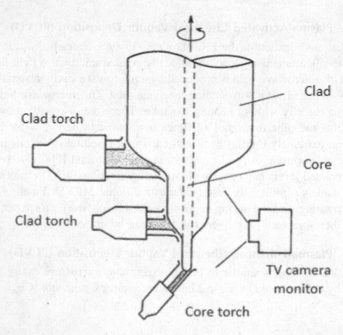

FIGURE 2.23 Vapour Axial Deposition (VAD) method.

in the previous section, MCVD is widely accepted as the best technique for rare-earth-doped fiber fabrication, mainly because of the flexibility it provides in the refractive index profile. Doping silica glass with rare-earth precursors like $ErCl_3$ or $YbCl_3$ cannot be done by deposition methods similar to the ones used for $SiCl_4$, $GeCl_4$, or $POCl_3$ because of the low vapour pressure of these rare-earth precursors at room temperature. Alternative ways to counteract this limitation include Solution doping methods, Sol-gel process, and direct nanoparticle deposition (DND). This list can be completed by the less known methods such as heated frit source delivery, heated source injector delivery, aerosol delivery, and chelate delivery methods.

2.7.3.1   Solution Doping Method

The solution doping method is used along the traditional MCVD process. It begins with the deposition of soot in the inner surface of the substrate silica tube. The soot deposition is done at moderate temperatures to transform silicon tetrachloride into silicon dioxide without consolidation of the resultant soot. The substrate tube is then soaked in a solvent solution containing salts of rare-earth or rare-earth with additional dopants like Aluminium. After being removed from the solvent, the substrate is heated in the presence of oxygen to trigger the formation of rare-earth oxides. The heating step is important to prevent the evaporation of rare-earth salts during subsequent processing stages at high temperatures that can result in rare-earth density reduction in an uncontrolled manner. The obtained porous layer then undergoes dehydration by flowing chlorine gas through the substrate tube at about 600°C. This process is necessary to eliminate OH⁻ compounds absorbed in the soot during the solution doping stage. In this reaction, chlorine reacts with water molecules resulting in the liberation of hydrogen chloride in the vapour phase and oxygen as in equation (2.17) below:

$$2H_2O + 2Cl_2 \rightarrow 4HCl + O_2 \qquad (2.112)$$

After dehydration, sintering the soot layer takes place at temperatures around 2000°C. Finally, the substrate layer is collapsed into the rod pre-form in the usual manner. The sintering is done with successive passes with gradual temperature increment of the burner to ensure smooth sintering process and prevent the undesirable formation of imperfections inside the layer.

Other methods of rare-earth doping include vapour-phase deposition which has hugely contributed to the advances in large mode area (LMA) fibers used in high power fiber lasers applications (Barnini et al. 2018), heated frit (Digonnet 2001), heated source (Poole et al. 1985), heated source injector(Digonnet 2001), aerosol delivery (Morse et al. 1991; Morse et al. 1991), and organometallic chelate delivery (Digonnet 2001).

2.7.4   Fiber Drawing from a Preform

After the solid glass pre-form has been fabricated and tested for quality, the next step consists of converting it into a fiber using a drawing tower. On the drawing

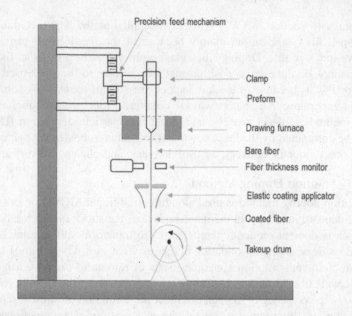

Precision feed mechanism

Clamp

Preform

Drawing furnace

Bare fiber
Fiber thickness monitor

Elastic coating applicator

Coated fiber

Takeup drum

FIGURE 2.24   Diagram of a fiber-drawing tower used to draw optical glass fibers from a pre-formed blank.

tower, illustrated in Figure 2.24, the pre-form is held in a feed at the bottom of which a graphite furnace heats it at approximately 1900 to 2200°C. The tip of the pre-form is melted until a molten relatively large drop, falls due to gravity. While dropping, it cools down and forms a thread. The tip is then pulled using a tractor-controlled spool after passing through a series of coating cups to put the buffer coating on top of the cladding and ultraviolet light curing ovens.

During the drawing stage, fiber geometric properties like fiber diameter are also controlled using a laser micrometre. The speed of the fiber draw depends on the preform, fiber type, and available equipment as well as specific applications (Roy Choudhury and Jaluria 1998; Jochem and van der Ligt 1985). The drawing speed can range from a few meters per minute up to 2500 meters per minute. For example, the drawing speed will depend on whether the drawing tower is designed for manufacturing or research. For manufacturing, where hundreds of kilometres of fibers must be produced daily, the drawing speed can be beyond 600 meters per minute to lower operational cost. The drawing speed also dictates the tower height. At high drawing speed, the fiber does not have enough time to cool down and be ready for coating. Therefore, the tower height must be increased to allow the fiber to cool down. The towers as high as 30 meters are not uncommon. However, even several meters away from the furnace exit, the temperature can still be above 100°C at a draw speed of 600 meters per minute or more (Jochem and van der Ligt 1985). Thus, forced cooling of the fiber may be required (Jochem and van der Ligt 1985).

After the drawing process is finished, fibers are tested to verify properties. These properties include attenuation, bending losses, cut-off wavelength, mode field diameter, chromatic dispersion, polarization mode dispersion, tensile proof test, glass geometry.

2.8 SUMMARY

In this chapter, an introduction to optical fiber was provided. Optical fiber in itself is a vast subject that can make an object of an entire book. The goal of this chapter was to provide the reader with an insight into an optical fiber to help him understand subsequent chapters. The development of the subject was done gradually increasing the level of abstraction. In this regard, a ray optic analysis was first provided to help the reader to understand important concepts such as numerical aperture and propagating modes. The ray optic model is intuitive in understanding phenomena of light propagation but is not a complete model because it cannot explain notions like evanescent fields or transverse distribution of light intensity in the fiber. To overcome the limitation of the ray optic model, the wave optic model introduced. This model starts with derivation of the wave equation from Maxwell equations. The wave equation was then written in cylindrical coordinates and solved with appropriate boundary conditions to obtain important fiber characteristics like propagation constant of the mode and normalized frequency. The origin and influence of loss were explained as well as other limiting factors like dispersion and nonlinear effects. The detailed composition of optical fiber in terms of the material was presented, in particular, properties of silica glass, fluoride glass and other types of glass were reviewed and their advantages and disadvantages for given applications highlighted. Finally, fabrication techniques for undoped and rare-earth-doped fibers were presented. The next chapter will focus on the physical properties of rare-earth ions and basic operation of rare-earth-doped fiber lasers.

REFERENCES

A.W. Snyder and J.D. Love. *Optical Waveguide Theory*. 1st Ed. Springer, US, 1983

Aerdams, M.J. *An Introduction to Optical Waveguides, John Wiley Sons Ltd, New York, 1981*.

Agrawal, G. *Fiber-Optic Communication Systems: Fourth Edition. Wiley, 2012*.

Agrawal, Govind P. 2012. *Fiber-Optic Communication Systems*. Vol. 222. John Wiley & Sons.

Almeida, Rui M. and John D. Mackenzie. 1981. "Vibrational Spectra and Structure of Fluorozirconate Glasses." *The Journal of Chemical Physics* 74 (11): 5954–5961. 10.1 063/1.441033.

Arai, Kazuo, Hiroshi Namikawa, Ken Kumata, Tatsutoku Honda, Yoshiro Ishii, and Takashi Handa. 1986. "Aluminum or Phosphorus Co-doping Effects on the Fluorescence and Structural Properties of Neodymium-doped Silica Glass." *Journal of Applied Physics* 59 (10): 3430–3436. 10.1063/1.336810.

Baird, J.L. 1927. "British Patent 285."

Barnier, S., M. Guittard, M. Palazzi, M. Massot, and Christian Julien. 1992. "Raman and Infrared Studies of the Structure of Gallium Sulphide Based Glasses." *Materials Science and Engineering B-Advanced Functional Solid-State Materials* 14 (September): 413–417. 10.1016/0921-5107(92)90084-M.

Barnini, Alexandre, Daniel Caurant, Thierry Gotter, Arnaud Laurent, Pascal Guitton, Cédric Guyon, Ronan Montron, Thierry Robin, Gerard Aka, and Carine Ranger. 2018.

Towards Large-Mode-Area Fibers Fabricated by the Full Vapor-Phase SPCVD Process. 10.1117/12.2290219.

Becker, Philippe M., Anders A. Olsson, and Jay R. Simpson. 1999. *Erbium-Doped Fiber Amplifiers: Fundamentals and Technology.* Elsevier.

Bennett, M.J. 1983. "Dispersion Characteristics of Monomode Optical-Fiber Systems." In *IEE Proceedings H (Microwaves, Optics and Antennas)*, 130:309–314. IET.

Blow, K.J. and D. Wood. 1989. "Theoretical Description of Transient Stimulated Raman Scattering in Optical Fibers." *IEEE Journal of Quantum Electronics* 25 (12): 2665–2673. 10.1109/3.40655.

Born, M. and E. Wolf. 1999. *Principles of Optics*, 7th Ed. New York: Cambridge University Press.

Boyd, Robert W. 2019. *Nonlinear Optics.* Academic Press.

Buck, J.A. *Fundamentals of Optical Fibers*, 2nd Ed. 2004

C.N. Pannell, P.St. J. Russell, and T.P. Newson. 1992. "Stimulated_Brillouin_scattering_in_optical_fibers_.Pdf."

Cancellieri, G. 1991. *Single-Mode Optical Fibers, Ancona: Elsevier Ltd.*

Carman, R.L., R.Y. Chiao, and P.L. Kelley. 1966. "Observation of Degenerate Stimulated Four-Photon Interaction and Four-Wave Parametric Amplification." *Physical Review Letters* 17 (26): 1281–1283. 10.1103/PhysRevLett.17.1281.

Cognolato, L. 1995. "Chemical Vapour Deposition for Optical Fiber Technology." *Le Journal de Physique IV* 5 (C5): C5–975.

Digonnet, Michel. 2001. "Rare-Earth-Doped Fiber Lasers and Amplifiers." *Inc. New York EUA*, January, 172–184. 10.1201/9780203904657.

Ed, T. Li. 1985. *Optical Fiber Communications,1st Ed. Academic press.*

El-Agmy, R.M. and N.M. Al-Hosiny. 2010. "2.31 Mm Laser under up-Conversion Pumping at 1.064 Mm in Tm3+:ZBLAN Fiber Lasers." *Electronics Letters* 46 (13): 936–937. 10.1049/el.2010.1248.

F. S. Ferreira, Mário. 2011. "Nonlinear Phase Modulation." 85–109. 10.1002/9781118003398.ch5.

Fortin, Vincent, Martin Bernier, Julien Carrier, and Réal Vallée. 2011. "Fluoride Glass Raman Fiber Laser at 2185 Nm." *Optics Letters* 36 (21): 4152–4154. 10.1364/OL.36.004152.

Geng, Jihong, Jianfeng Wu, Shibin Jiang, and Jirong Yu. 2007. "Efficient Operation of Diode-Pumped Single-Frequency Thulium-Doped Fiber Lasers near 2 Mm." *Optics Letters* 32 (4): 355–357. 10.1364/OL.32.000355.

Gisin, N., J.- Von der Weid, and J.- Pellaux. 1991. "Polarization Mode Dispersion of Short and Long Single-Mode Fibers." *Journal of Lightwave Technology* 9 (7): 821–827. 10.1109/50.85780.

Hafizi, B., J.P. Palastro, J.R. Peñano, T.G. Jones, L.A. Johnson, M.H. Helle, D. Kaganovich, Y.H. Chen, and A.B. Stamm. 2016. "Stimulated Raman and Brillouin Scattering, Nonlinear Focusing, Thermal Blooming, and Optical Breakdown of a Laser Beam Propagating in Water." *Journal of the Optical Society of America B* 33 (10): 2062–2072. 10.1364/JOSAB.33.002062.

Hansell, C.W. 1930. "U.S. Patent 1, 751."

Heel, A.C.S. VAN. 1954. "A New Method of Transporting Optical Images without Aberrations." *Nature* 173 (4392): 39. 10.1038/173039a0.

Heo, J. and W.J. Chung. 2014. "11 – Rare-Earth-Doped Chalcogenide Glass for Lasers and Amplifiers." In *Chalcogenide Glasses Zhang*, edited by Jean-Luc Adam and Xianghua B.T., 347–380. Woodhead Publishing. 10.1533/9780857093561.2.347.

Izawa, T. and N. Inagaki. 1980. "Materials and Processes for Fiber Preform Fabrication—Vapor-Phase Axial Deposition." *Proceedings of the IEEE* 68 (10): 1184–1187. 10.1109/PROC.1980.11827.

Izawa, T. and S. Sudo. 1987. *Optical Fibers: Materials and Fabrication.* 1st Ed. Springer Netherland

Jeunhomme, Luc B. 1990. *Single-Mode Fiber Optics: Principles and Applications.* New York: Marcel Dekker.

Jochem, C.M.G. and J.W.C. van der Ligt. 1985. "Method for Cooling and Bubble-Free Coating of Optical Fibers at High Drawing Rates." *Electronics Letters* 21 (18): 786–787. 10.1049/el:19850554.

Kale, R.U., P.M. Ingale, and R.T. Murade. 2013. "Comparison of SRS & SBS (Non Linear Scattering) In Optical Fiber." *International Journal of Recent Technology & Engineering* 2 (1): 118–122.

Kao, K.C. and G.A. Hockham. 1966. "Dielectric-Fiber Surface Waveguides for Optical Frequencies." *Proceedings of the Institution of Electrical Engineers* 113 (7): 1151–1158. 10.1049/piee.1966.0189.

Kapany, N.S. 1967. *Fiber Optics: Principles and Applications, Academic press.*

Kapany, N.S. 1959. "Fiber Optics. VI. Image Quality and Optical Insulation*†." *Journal of the Optical Society of America* 49 (8): 779–787. 10.1364/JOSA.49.000779.

Kapron, F.P., D.B. Keck, and R.D. Maurer. 1970. "Radiation Losses in Glass Optical Waveguides." *Applied Physics Letters* 17 (10): 423–425. 10.1063/1.1653255.

Lamm, H. 1930. "Instrumentek 50."

Layne, C.B., W.H. Lowdermilk, and M.J. Weber. 1977. "Multiphonon Relaxation of Rare-Earth Ions in Oxide Glasses." *Physical Review B* 16 (1): 10–20. 10.1103/PhysRevB.16.10.

Liang, X., S. Kumar, and J. Shao. 2015. "Mitigation of Fiber Linear and Nonlinear Effects in Coherent Optical Communication Systems." In *2015 49th Asilomar Conference on Signals, Systems and Computers*, 1003–1006. 10.1109/ACSSC.2015.7421289.

MacChesney, J.B., P.B. O'Connor, and H.M. Presby. 1974. "A New Technique for the Preparation of Low-Loss and Graded-Index Optical Fibers." *Proceedings of the IEEE* 62 (9): 1280–1281. 10.1109/PROC.1974.9608.

Marcuse, D. 1991. *Theory of Dielectric Optical Waveguides*, 2nd Ed. Academic Press.

Marcuse, D. 1976. "Microbending Losses of Single-Mode, Step-Index and Multimode, Parabolic-Index Fibers." *The Bell System Technical Journal* 55 (7): 937–955. 10.1002/j.1538-7305.1976.tb02921.x.

Miya, T., Y. Terunuma, T. Hosaka, and T. Miyashita. 1979. "Ultimate Low-Loss Single-Mode Fiber at 1.55 Mm." *Electronics Letters* 15 (4): 106–108. 10.1049/el:19790077.

Morse, T.F., A. Kilian, L. Reinhart, and J.W. Cipolla. 1991. "Aerosol Transport for Optical Fiber Core Doping: A New Technique for Glass Formation." *Journal of Aerosol Science* 22 (5): 657–666. 10.1016/0021-8502(91)90018-D.

Morse, T.F., A. Kilian, L. Reinhart, W. Risen, and J.W. Cipolla. 1991. "Aerosol Techniques for Glass Formation." *Journal of Non-Crystalline Solids* 129 (1): 93–100. 10.1016/0022-3093(91)90083-I.

Neumann, E.G. 1988. *Single-Mode Fibers.* 1st Ed. Springer-Verlag Berlin Heidelberg

Ohsawa, K., T. Shibata, K. Nakamura, and S. Yoshida. 1981. "Fluorozirconate Glasses for Infrared Transmitting Optical Fibers." In *Proceedings of the7th European Conference on Optical Communication (ECOC)*, 1.

Okoshi, T. 1982. *Optical Fibers. 1st Ed. Academic Press.*

Pannell, Chris, P St. J. Russell, and T.P. Newson. 1993. "Stimulated Brillouin Scattering in Optical Fibers: The Effects of Optical Amplification." *Journal of The Optical Society of America B-Optical Physics - J OPT SOC AM B-OPT PHYSICS* 10 (April). 10.1364/JOSAB.10.000684.

Paschotta, R. 2008. "Encyclopedia of Laser Physics and Technology." Wiley.

Poole, S.B., D.N. Payne, and M.E. Fermann. 1985. "Fabrication of Low-Loss Optical Fibers Containing Rare-Earth Ions." *Electronics Letters* 21 (17): 737–738. 10.1049/el:19850520.

Pratap Singh, Sunil, Ramgopal Gangwar, and Nar Singh. 2007. "Nonlinear Scattering Effects in Optical Fibers." *Progress in Electromagnetics Research-Pier – PROG ELECTR-OMAGN RES* 74 (January): 379–405. 10.2528/PIER07051102.

Roy Choudhury, S. and Yogesh Jaluria. 1998. "Practical Aspects in the Drawing of an Optical Fiber." *Journal of Materials Research* 13 (February): 483–493. 10.1557/JMR.1998.0063.

S. Sacks, Zachary, Zeev Schiffer, and Doron David. 2007. "Long Wavelength Operation of Double-Clad Tm:Silica Fiber Lasers." *Proceedings of SPIE – The International Society for Optical Engineering* 6453 (February). 10.1117/12.700525.

Saad, M. 2011. "Fluoride Glass Fibers." In *2011 IEEE Photonics Society Summer Topical Meeting Series*, 81–82. 10.1109/PHOSST.2011.6000055.

Saavedra, G., M. Tan, D.J. Elson, L. Galdino, D. Semrau, M.A. Iqbal, I.D. Phillips, et al. 2017. "Experimental Analysis of Nonlinear Impairments in Fiber Optic Transmission Systems up to 7.3 THz." *Journal of Lightwave Technology* 35 (21): 4809–4816. 10.11 09/JLT.2017.2760138.

Saito, K., M. Yamaguchi, H. Kakiuchida, A.J. Ikushima, K. Ohsono, and Y. Kurosawa. 2003. "Limit of the Rayleigh Scattering Loss in Silica Fiber." *Applied Physics Letters* 83 (25): 5175–5177. 10.1063/1.1635072.

Sanghera, Jas S., L. Brandon Shaw, and Ishwar D. Aggarwal. 2002. "Applications of Chalcogenide Glass Optical Fibers." *Comptes Rendus Chimie* 5 (12): 873–883. 10.101 6/S1631-0748(02)01450-9.

Sanghera, Jasbinder, Ishwar Aggarwal, L.B. Shaw, L.E. Busse, P. Thielen, V.Q. Nguyen, P. Pureza, Shyam Bayya, and Frederic Kung. 2001. "Applications of Chalcogenide Glass Optical Fibers at NRL." *Journal of Optoelectronics and Advanced Materials* 3 (September).

Senior, M.J. and J. Muhammad. 2009. *Optical Fiber Communications: Principles and Practice*. 3rd Ed. Prentice Hall

Shaw, L.B., B. Cole, P.A. Thielen, J.S. Sanghera, and I.D. Aggarwal. 2001. "Mid-Wave IR and Long-Wave IR Laser Potential of Rare-Earth Doped Chalcogenide Glass Fiber." *IEEE Journal of Quantum Electronics* 37 (9): 1127–1137. 10.1109/3.945317.

Smith, R.G. 1972. "Optical Power Handling Capacity of Low Loss Optical Fibers as Determined by Stimulated Raman and Brillouin Scattering." *Applied Optics* 11 (11): 2489–2494. 10.1364/AO.11.002489.

Stolen, R. 1975. "Phase-Matched-Stimulated Four-Photon Mixing in Silica-Fiber Waveguides." *IEEE Journal of Quantum Electronics* 11 (3): 100–103. 10.1109/JQE.1975.1068571.

Stolen, R.H. and Chinlon Lin. 1978. "Self-Phase-Modulation in Silica Optical Fibers." *Physical Review A* 17 (4): 1448–1453. 10.1103/PhysRevA.17.1448.

Tee, Din Chai, Nizam Tamchek, and Raymond Ooi. 2016. "Numerical Modelling of Fundamental Characteristics of ZBLAN Photonic Crystal Fiber for Communication in 2-3μm Mid-Infrared Region." *IEEE Photonics Journal* 8 (March): 1. 10.1109/JPHOT.2016.2536940.

Tran, D., G. Sigel, and B. Bendow. 1984. "Heavy Metal Fluoride Glasses and Fibers: A Review." *Journal of Lightwave Technology* 2 (5): 566–586. 10.1109/JLT.1984.1073661.

Tyndall, J. 1854. "On Some Phenomena Connected with the Motion of Liquids." *Proceedings of the Royal Institution of Great Britain* 1: 446.

Uemura, O., N. Hayasaka, S. Tokairin, and T. Usuki. 1996. "Local Atomic Arrangement in Ge-Te and Ge-S-Te Glasses." *Journal of Non-Crystalline Solids* 205–207: 189–193. 10.1016/S0022-3093(96)00376-6.

Wang, Ji, J.R. Lincoln, William Brocklesby, R.S. Deol, C.J. Mackechnie, A. Pearson, A.C. Tropper, David Hanna, and D.N. Payne. 1993. "Fabrication and Optical Properties of

Lead-Germanate Glasses and a New Class of Optical Fibers Doped with Tm3." *Journal of Applied Physics* 73 (July): 8066–8075. 10.1063/1.353922.

Weszka, J., Ph. Daniel, A. Burian, A.M. Burian, and A.T. Nguyen. 2000. "Raman Scattering in In2Se3 and InSe2 Amorphous Films." *Journal of Non-Crystalline Solids* 265 (1): 98–104. 10.1016/S0022-3093(99)00710-3.

Wu, Jianfeng, Zhidong Yao, Jie Zong, and Shibin Jiang. 2007. "Highly Efficient High-Power Thulium-Doped Germanate Glass Fiber Laser." *Optics Letters* 32 (6): 638–640. 10.13 64/OL.32.000638.

Zhu, Xiushan, and R.K. Jain. 2007. "10-W-Level Diode-Pumped Compact 2.78 Microm ZBLAN Fiber Laser." *Optics Letters* 32 (February): 26–28. 10.1364/OL.32.000026.

APPENDIX 2.1   MATLAB® PROGRAM FOR PROPAGATION CONSTANT B AS A FUNCTION OF NORMALIZED FREQUENCY V

```
% function which plots b-V diagram of a planar slab
waveguide
clear all
close all
clc
N = 400;          % Plot's x resolution
b = linspace(0,1.0,N);
hold on
for nu = [0, 1, 2, 3]
  for a = [0.0, 8.0,50.0]; % asymmetry coefficient
    % determine V
    V1 = atan(sqrt(b./(1-b)) );
    V2 = atan(sqrt((b+a)./(1.0-b)));
    V3 = 1./sqrt(1.0-b);
    v = (nu*pi + V1 + V2).*V3;
    %
    plot(v,b,'LineWidth',2)
    axis([0.0 8.0 0.0 1.0])
  end
end
box on
xlabel ('Normalized Frquency (V)','FontSize',12);
ylabel('Normalized Propagation Constant (b)','Font-
Size',12);
```

APPENDIX 2.2   MATLAB PROGRAM FOR BESSEL FUNCTION OF THE FIRST KIND

```
% Plot of Bessel functions of the first kind J_m(z)
clear all
close all
clc
N = 101;          % Plot x resolution
x = linspace(0,10,N); % creation of x arguments
%
hold on
for m = [0 1 2] % m - order of Bessel function
  J = besselj(m,x);
  h = plot(x,J);
  box on
  syms F_a F_b
  xlabel('qr','FontSize',22);
  ylabel('J_v(qr)','FontSize',22);
  %
   set(h,'LineWidth',2.5); % new thickness of plotting
lines
   set(gca,'FontSize',22); % new size of tick marks on
both axes
end
```

APPENDIX 2.3   MATLAB PROGRAM FOR BESSEL FUNCTION OF THE SECOND KIND (NEUMANN FUNCTIONS)

```
% Plot of modified Bessel functions of the second kind
K_m(z)
clear all
close all
clc
N = 101;          % Plot's x resolution
x = linspace(0.01,3,N); % Creation of x arguments
%
hold on
for m = [0 1 2] % m - order of Bessel function
  K = besselk(m,x);
  h = plot(x,K);
```

```
   xlabel('qr','FontSize',22);
   ylabel('K_v(qr)','FontSize',22);
   axis([0 3 0 5]);
   box on
   %
   set(h,'LineWidth',2.5);
   set(gca,'FontSize',22);
end
```

APPENDIX 2.4   MATLAB PROGRAM FOR MODIFIED BESSEL FUNCTION OF THE FIRST KIND

```
% Plot of modified Bessel functions of the first kind I_m(z)
clear all
close all
clc
N = 101;          % Plot x resolution
x = linspace(0,3,N); % creation of x arguments
%
hold on
for m = [0 1 2]       % m - order of Bessel function
  Y = besseli(m,x);
  h = plot(x,Y);
  box on
  xlabel('qr/j','FontSize',22);
  ylabel('I_m(qr/j)','FontSize',22);
  %
  set(h,'LineWidth',2);
  set(gca,'FontSize',22);
end
```

APPENDIX 2.4   MATLAB PROGRAM FOR MODIFIED BESSEL FUNCTION OF THE SECOND KIND

```
% Plot of Bessel functions of the second kind Y_m(z)
clear all
close all
```

```
clc
N = 101;            % Plot's x resolution
x = linspace(0.01,10,N); % Creation of x arguments
%
hold on
for m = [0 1 2] % m - order of Bessel function
  Y = bessely(m,x);
  h = plot(x,Y);
  xlabel('qr','FontSize',22);
  ylabel('N_v(qr)','FontSize',22);
  axis([0 10 -5 1.5]);
  box on
  %
  set(h,'LineWidth',2.5);
  set(gca,'FontSize',22);
end
```

3 Rare-Earth Ions and Fiber Laser Fundamentals

3.1 INTRODUCTION

This chapter is concerned with the physical properties of rare-earth ions for application to fiber lasers. The working principle of rare-earth doped fiber laser is based on the interesting properties of rare-earth ions. The most important of these properties is their relatively long metastable lifetime, which allows achieving stimulated emission very easily. The properties of rare-earth ions are a consequence of their atomic structure. For this reason, the atomic structure of rare-earth atoms will be presented first. Energy levels, intensities of transitions, influence of the solid host on these transitions will be derived from well-known quantum mechanics principles. Deep details of development are not always presented because it is assumed that the reader possesses some quantum mechanics prerequisite. Nevertheless, the notions will be presented in an understandable way without falling into oversimplification. Understanding this chapter is indispensable for the rest of the book as most of the notions used in rare-earth doped fiber amplifiers and lasers are defined here.

3.2 GENERAL PROPERTIES AND ELECTRONIC STRUCTURE OF RARE-EARTHS

Rare-earth elements are classified into two groups, namely lanthanides with atomic numbers ranging from 57 to 71 and actinides with atomic numbers ranging from 83 to 103. Because of their interesting optical properties, lanthanides are more important in solid-state laser technology than actinides; therefore, this chapter focuses on lanthanides rather than actinides. The special optical behaviour of lanthanide is a consequence of their atomic structure represented in Figure 3.1. The study of this atomic structure started at the beginning of the 20th century when Becquerel observed sharp absorption lines in the spectrum of rare-earth salts at temperatures lower than 100°K (Jean Becquerel 1907; J Becquerel 1906). The origin of this behaviour remained unclear until almost 40 years later when A. Mayer provided a theoretical explanation by calculating the atomic structure of the lanthanides from first principles (Mayer 1941).

In a classical representation, an atom is made up of a nucleus surrounded by electrons placed on shells of gradually increasing radii as the atomic number increases. However, an abrupt contraction is observed at atomic number $Z = 57$, where the 5s and 5p shells are first fully populated before 4f shell. The 4f shell is then filled progressively with 1 to 14 electrons as the atomic number grows from 58

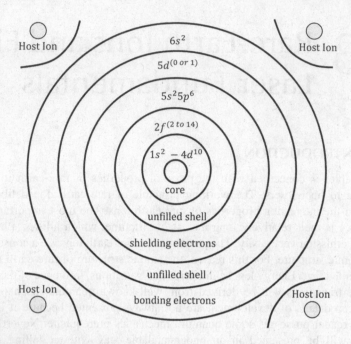

FIGURE 3.1   Graphical representation of the rare-earth atomic structure.

to 71. Therefore, the 4f shell contracts and ends up being bonded by shells 5s and 5p of larger radius.

The shielding of the 4f electron shell by the 5s and 5p shells is at the origin of the unusual spectrum that has been observed in the rare-earth. Lanthanides elements in their neutral state have the atomic configuration $[Xe]4f^N6s^2$ and $[Xe]4f^{N-1}5d^16s^2$, where $[Xe]$ the electronic configuration of Xenon given by $1s^22s^22p^63s^23p^63d^{10}4s^24p^64d^{10}5s^25p^6$ and N the number of electrons inside the 4f shell varying from 1 to 14 (Becker et al. 1999). However, these elements are commonly found in the ionic state as trivalent ions in the form of Ln^{3+}. Neutral rare-earth atoms achieve ionisation by losing the two bonding electrons of 6s shell and one electron of whether 5d or 4f shell; therefore, the rare-earth trivalent ions have the atomic configuration $[Xe]4f^N$ as can be seen in Table 3.1.

The $5s^2$ and $5p^2$ shells behave like a metallic sphere that shield the 4f electrons from the external environment. This shielding is responsible for the atomic-like properties of the lanthanides when incorporated in a crystal or glass host. In Figure 3.2 representing the radial distribution function of the 4f, 5s, 5p, 5d and 5g orbitals of the Pr^{3+} free ion, the shielding is observed because of the 4f orbital closer to the nucleus than the 5s and 5p orbitals (Becker et al. 1999; Marks and Fischer 1979; Malta and Carlos 2003; Walsh 2006). It is important mentioning that the radial distribution function is the square of the wave function of the orbital and represents the probability to find an electron at a specific coordinate in 3D space.

In Figure 3.2, the y-axis represents the probability distribution of electrons and the x-axis the distance from the nucleus in units of Bohr radius.

TABLE 3.1
Electronic Configuration of Lanthanide Atoms and Ions

Z	Element	Inner orbitals	Electronic configurations of neutral atoms					Electronic configurations of trivalent ions
			4f	5s	5p	5d	6s	
57	La	Fully filled inner	0	2	6	1	2	[Xe]4f^0
58	Ce	orbital with the	1	2	6	1	2	[Xe]4f^1
59	Pr	possible 46	3	2	6		2	[Xe]4f^2
60	Nd	electrons	4	2	6		2	[Xe]4f^3
61	Pm		5	2	6		2	[Xe]4f^4
62	Sm		6	2	6		2	[Xe]4f^5
63	Eu		7	2	6		2	[Xe]4f^6
64	Gd		7	2	6	1	2	[Xe]4f^7
65	Tb		9	2	6		2	[Xe]4f^8
66	Dy		10	2	6		2	[Xe]4f^9
67	Ho		11	2	6		2	[Xe]4f^{10}
68	Er		12	2	6		2	[Xe]4f^{11}
69	Tm		13	2	6		2	[Xe]4f^{12}
70	Yb		14	2	6		2	[Xe]4f^{13}
71	Lu		14	2	6	1	2	[Xe]4f^{14}

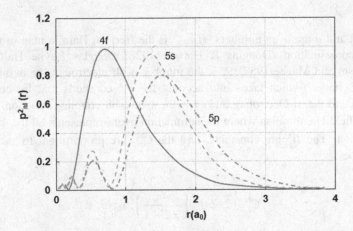

FIGURE 3.2 Radial distribution functions of the 4f, 5s, 5p, 5d, and 5g orbitals for the Pr^{3+} free ion, from Hartree-Fock calculations (Douglas R Hartree 1928).

3.3 ENERGY LEVELS OF RARE-EARTH IONS

3.3.1 Atomic Interactions of the Free Ions and Crystal Field Influence

The possible energy levels of a free rare-earth ion are the results of the interaction between the N electrons of the 4f, their interaction with the nucleus as well as their interactions with the electrons of other shells. These energy levels are solutions of the time-independent Schrödinger equation of the isolated ion given by:

$$H\psi = E\psi \tag{3.1}$$

where ψ is the wavefunction of the 4f electrons, E the energies of the electronic levels and H the Hamiltonian operator. This Hamiltonian operator contains terms that describe the interaction of the free ion and terms that describe the interaction of the electrons with the crystal field as shown by equation (3.2).

$$H = H_{atomic} + H_{cf} \tag{3.2}$$

where H_{atomic} is the Hamiltonian of the free ion written as:

$$H_{atomic} = -\frac{h^2}{2m}\sum_{i=1}^{N}\nabla_i^2 - \sum_{i=1}^{N}\frac{Z^*e^2}{r_i} + \sum_{i<j}^{N}\frac{e^2}{r_{ij}} + \sum_{i=1}^{N}\xi(r_i)\vec{S_i}.\vec{l_i} \tag{3.3}$$

and H_{cf} account for the crystal field Hamiltonian given by:

$$H_{cf} = \sum_{k,q,i} B_q^k C_q^{(k)}(i). \tag{3.4}$$

where k and q are even numbers. H_{atomic} is the free-ion Hamiltonian in the central field approximation (Douglas R Hartree 1928; Douglas Rayne Hartree 1928; Funabashi and Magee 1957). N is the number of 4f electrons, Z^*e is the effective nuclear charge (which takes into account the closed shells that lie between the nucleus and the 4f electrons), and $\xi(r_i)$ the spin-orbit coupling function. H_{cf} is the crystal-field Hamiltonian where the summation over i represents all the 4f electrons of the ion. The B_q^k are constants, and the $C_q^{(k)}$ are proportional to the spherical harmonics:

$$C_q^{(k)}(\theta, \phi) = \left(\frac{4\pi}{2k+1}\right)^{\frac{1}{2}} Y_{kq}(\theta, \phi) \tag{3.5}$$

where the B_q^k 's (k = 2, 4 and 6) are the so-called ligand field parameters of even rank and $\mathbf{C}^{(k)}$ is a Racah tensor operator of rank k (Kenyon 2002).

An exact analytical solution of the Schrödinger equation (3.1) is impossible to find; therefore, approximations and empirical methods are inevitable. One commonly used approximation is the "central field" approximation in which each electron is considered to be moving independently in the field of the nucleus and sees the identical potential that is a function only of the distance from the nucleus. The first term of H_{atomic} is the sum of the kinetic energies of all the electrons of the 4f shell; the second term is the potential energy of all the electrons in the field of the nucleus or in other words, the Coulomb interaction between the nucleus and the 4f atoms. In the central field approximation, the two first terms are spherically symmetric and do not lift the degeneracy of the $4f^N$ configuration. The third term represents the mutual repulsion between the electrons within the 4f shell, and the last term is the spin-orbit interaction (Winkler et al. 2003), which accounts for the coupling between the spin angular momentum and the orbital angular momentum of the electron. These two last terms of the Hamiltonian are responsible for the spread of the $4f^N$ free-ion levels over tens of thousands of wavenumbers.

For systems with high atomic numbers such as rare-earth materials, calculation of H and the resulting 4f energy from first principles become unpractical. Therefore, H is written in a parameterized form and the parameters are obtained by fitting them with experimentally measured energy levels.

A free rare-earth ion incorporated in a solid host undergoes repulsion from the host electrons and attraction from its nuclei, creating a net electric field in the close environment of the ion known as the crystalline field. This crystalline field is represented by additional terms that often complete the Hamiltonian of the free ion. In general, for the lanthanides this crystal field is several orders of magnitude weaker than the spin-orbit coupling; though, it plays a central role in making many 4f transitions possible as is going to be seen in a subsequent section of this chapter.

3.3.2 TERMS SYMBOLS—SPIN-ORBIT COUPLING

There are several schemes for describing the states of a many-electron system. Because the spin-orbital coupling of lanthanides is stronger than the Coulomb force between electrons, it is customary to use Russell-Saunders (Hyde 1975), or LS, coupled states to describe their energy levels. In this scheme atomic states are described by term symbols of form $^{2S+1}L_J$, where L is the total orbital quantum number, S the total spin quantum number, J the total angular momentum and the quantity $^{2s+1}L_J$ is called spin multiplicity. The values of L, S and J are related to the four quantum numbers (Griffiths and Schroeter 2018; Jammer 1966) required to describe the state of a unique electron namely:

- Principal quantum number n which can take values 1, 2, 3, 4…
- Orbital quantum number l which can take values 0, 1, 2, 3,…,n-1
- Magnetic quantum number m_l which can take values 0, ±1, ±2, ±3… ±l
- Spin quantum number s which can take values ±1/2
- Spin magnetic quantum number m_s

The distribution of the 4f electrons in the orbitals when the lanthanide elements are in their ground state is shown in Table 3.2, where Δ represents the energy difference between the ground state and the J multiple states that lies right above the ground

TABLE 3.2
Electronic Configurations and Spectral Terms of Trivalent Lanthanide Ions in the Ground State

ion	$4f^a$	Magnetic quantum number of 4f orbital							L	S	J	Ground state spectral term	Δ (cm^{-1})	ξ_{4f} (cm^{-1})
		3	2	1	0	-1	-2	-3						
											$J = L - S$			
La^{3+}	0								0	0	0	$^1S_0$		
Ce^{3+}	1	↑							3	1/2	5/2	$^2F_{5/2}$	2200	640
Pr^{3+}	2	↑	↑						5	1	4	$^3H_4$	2150	750
Nd^{3+}	3	↑	↑	↑					6	3/2	9/2	$^4I_{9/2}$	1900	900
Pm^{3+}	4	↑	↑	↑	↑				6	2	4	$^5I_4$	1600	1070
Sm^{3+}	5	↑	↑	↑	↑	↑			5	5/2	5/2	$^6H_{5/2}$	1000	1200
Eu^{3+}	6	↑	↑	↑	↑	↑	↑		3	3	0	$^7F_0$	350	1320
											$J = L + S$			
Gd^{3+}	7	↑	↑	↑	↑	↑	↑	↑	0	7/2	7/2	$^8S_{7/2}$		1620
Tb^{3+}	8	↑↓	↑	↑	↑	↑	↑	↑	3	3	6	$^7F_6$	2000	1700
Dy^{3+}	9	↑↓	↑↓	↑	↑	↑	↑	↑	5	5/2	15/2	$^6H_{15/2}$	3300	1900
Ho^{3+}	10	↑↓	↑↓	↑↓	↑	↑	↑	↑	6	2	6	$^5I_8$	5200	2160
Er^{3+}	11	↑↓	↑↓	↑↓	↑↓	↑	↑	↑	6	3/2	7/2	$^4I^{15}$	6500	2440
Tm^{3+}	12	↑↓	↑↓	↑↓	↑↓	↑↓	↑	↑	5	1	0	$^2F_{7/2}$	8300	2640
Yb^{3+}	13	↑↓	↑↓	↑↓	↑↓	↑↓	↑↓	↑	3	1/2	7/2	$^2F_{7/2}$	10300	2880
Lu^{3+}	14	↑↓	↑↓	↑↓	↑↓	↑↓	↑↓	↑↓	0	0	0	$^1S_0$		

state and ξ_{4f} is the spin-orbital coupling coefficient. The orbital quantum number of the 4f shell is l=3, therefore it contains 7 orbitals that can be filled with a maximum of 14 electrons because each orbital can accommodate 2 electrons. The magnetic quantum numbers are −3, −2, −1, 0, 1, 2 and 3 respectively. The electrons in orbitals are represented by vertical arrows. An arrow pointing upwards represents an electron with a positive spin quantum number $m_s = +1/2$ whereas an arrow pointing downward represents as negative spin quantum number $m_s = −1/2$.

3.3.2.1 Total Orbital Quantum Number L

L is the total orbital quantum number that can take values 0, 1, 2, 3, 4, 5 and 6 represented by capital letters S, P, D, F, G, H, and I respectively. L is obtained by adding the magnetic quantum numbers m_l for each electron, therefore if there are two electrons in the same orbital the magnetic quantum number for that orbital will be twice the value of m_l. In Table 3.2 for example, the 4 first orbitals of Erbium with magnetic quantum numbers 3, 2, 1 and 0 are occupied by 2 electrons each, so the overall magnetic quantum number for these 4 orbitals will be (6 + 4 + 2 + 0 = 12).

The three remaining orbitals with magnetic quantum numbers $-1, -2, -3$ are populated by a single electron, so the overall magnetic quantum number for these orbitals will be $(-1-2-3 = -6)$. Summing up the two values yields a total orbital quantum number of 6 which correspond to the capital letter I.

3.3.2.2 Total Spin Quantum Number S

The total spin quantum number S is calculated by adding the spin quantum number for each electron. According to Hund's first rule, the ground state has all unpaired electron spins parallel with the same value of m_s, conventionally chosen as $+1/2$. The overall S is then $1/2$ times the number of unpaired electrons. In Table 3.2, Erbium has three unpaired electrons, so the total spin quantum number will be $3 \times \frac{1}{2} = \frac{3}{2}$.

3.3.2.3 Total Angular Momentum Quantum Number J

The total angular momentum quantum number takes the minimum value $J = |L - S|$ if less than half of the subshell is occupied and the minimum value of $J = |L + S|$ if the subshell is more than half-filled. For excited states, J takes values $L + S, L + S - 1,\dots$ $L - S$ when $L \geq S$, and S takes values $S + L, S + L - 1,\dots S-L$ when $L \leq S$. From Table 3.2, one can calculate the value of J for Erbium as $(6 + 3/2 = 15/2)$, were $J = S + L$ because, with 11 electrons, the subshell is more than half-filled.

3.3.2.4 Total Spin Multiplicity

The total spin multiplicity is given by $2S+1$, which in Table 3.2 gives $(2 \times 3/2 + 1 = 4)$. From the above considerations, the RS notation for Erbium ions in the ground state will be $^4I_{15/2}$.

3.3.3 ATOMIC INTERACTIONS AND ENERGY LEVELS SPLITTING

The spread in the $4f^N$ energy levels is the result of diverse interactions taking place between the electrons as demonstrated by equation (3.2). Energy levels of free ions are affected by four main fields. First is the central field (H_0) which describes the interaction between the electrons and the nucleus and separates the various configurations (5p, 4f, 5d, etc.) in about 10^5 cm^{-1} gaps in energy. The central field does not remove the degeneracy of 4f configurations because of its spherical symmetry. Secondly, the Coulomb repulsion (H_{e-e}) between electrons which is inherent to multi-electronic systems removes the degeneracy of 4f configurations. As a result, 4f energy states are split into $^{(2S+1)}L$ levels, separated by energies gaps of about 10^4 cm^{-1}. It is worth mentioning that low multiplicity levels have lower energies than the higher energy counterparts. Each of the $^{(2S+1)}L$ levels is further affected by the spin-orbit coupling term (H_{SO}), which lift the degeneracy of $^{(2S+1)}L$ to some degree resulting in multiple $^{(2S+1)}L_J$ levels with an energy separation of about 10^3 cm^{-1} among them. In short, the electrostatic interaction splits the configuration into terms ^{2S+1}L, and the spin-orbit interaction lifts the degeneracy with respect to J and splits the terms into levels $^{2S+1}L_J$.

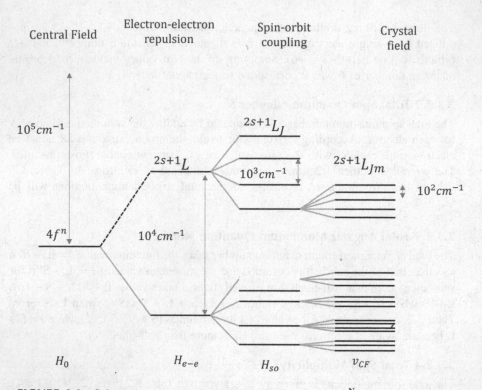

FIGURE 3.3 Schematic representation of the splitting of the $4f^N$ ground configuration under the effect of progressively weaker perturbations, namely the central field first, followed by the Coulomb repulsion between electrons, the spin-orbit coupling of the electrons and finally the crystal field.

When the free ion is placed in a crystalline host, a further splitting of the energy levels occurs because the spherically symmetric environment of the free ion is broken by the crystal field of the host. As a consequence, the degeneracy of the 4f state is removed and the $^{2S+1}L_J$ components are split into several energy levels separated by energies gaps of about 10^2 cm^{-1}. This last splitting of the energy levels is referred to as Stark splitting and the resulting states are called "Stark components" of the parent manifold. Figure 3.3 shows a schematic representation of the successive splitting of the $4f^N$ configuration of lanthanides because of the influence of various fields.

It is clear from Figure 3.3 and explanations provided in the previous paragraph that for 4f electrons, the crystal-field Hamiltonian is a small perturbation compared to the Coulomb and spin-orbit interactions since it is roughly 100 times smaller in magnitude than the first and about 10 times smaller than the second. In contrast, the crystal-field splitting of the excited configurations are quite large,—on the order of 20,000 cm^{-1}—as the excited orbitals (e.g. 5d or 5g) are not shielded as the 4f orbitals are. Since the crystal field of the host lattice is usually such a small perturbation, basic properties regarding 4f configurations can be represented for each lanthanide in a practically universal scheme independently of the ionic crystal or glass host.

3.4 LIGHT EMISSION AND ABSORPTION BY LANTHANIDES—BASIS ASPECTS

3.4.1 SELECTION RULES

It is well known that because of the conservation of energy, an incident photon can trigger a radiative transition between 2 energy levels only if the equation $h\nu = E_2 - E_1$ (where h is the planck constant, E_1 and E_2 the energies of the lower and upper energy levels involved in the transition respectively) is verified. However, the energy requirement is not enough for radiative transitions and additional conditions known as selection rules must be fulfilled. When an atom is under the influence of electromagnetic radiation, it can behave like an oscillating electric dipole or an oscillating magnetic dipole or even as an electric or magnetic quadrupole. The selections rules apply differently to the electric dipole, magnetic dipoles, and electric quadrupoles. The most popular of these selection rules are known as the Laporte rule (Laporte 1924) and says that transition between the quantum state of same parity are forbidden for electric dipole and transition between quantum states of odd parity are forbidden for the magnetic dipole. Another way of saying this is that the algebraic sum of the angular momenta of the electrons in the initial and final state must change by an odd integer. In quantum mechanics terms, this means that for a transition to be allowed, the change in total spin quantum number S must be equal to zero, the change in the total orbital quantum number L must be 0, +1 or −1 and the change in total angular momentum J must be either 0, +1 or −1 as summarized in Table 3.3.

where ΔS, ΔL and ΔJ represent the variation in total spin quantum number, total orbital quantum number and total angular momentum, respectively. One has to understand that Laporte rules apply only for centrosymmetric interactions. According to Laporte rules, inside the $4f$ shell, electric dipole transitions are forbidden, whereas magnetic dipole and electric quadrupole transitions are allowed. It is worth mentioning that the terms "forbidden" and "allowed" are not strictly accurate as they only mean the probability of the transition is low.

TABLE 3.3
Selection rules with the Russell-Saunders notations

	S	L	J(no 0 ↔0)	Parity
Electronic dipole	$\Delta S = 0$	$\Delta L = 0, \pm 1$	$\Delta J = 0, \pm 1$	opposite
Magnetic dipole	$\Delta S = 0$	$\Delta S = 0$	$\Delta J = 0, \pm 1$	same
Electric quadrupole	$\Delta S = 0$	$\Delta L = 0, \pm 1, \pm 2$	$\Delta L = 0, \pm 1, \pm 2$	same

3.4.2 The "Puzzle" of 4f Electron Optical Spectra—Selection Rules

Long before the demonstration of the first working laser, rare-earth materials were a subject of intense research interest because of their optical properties. In fact, observation showed that rare-earths materials exhibited sharp intense spectral lines suggesting that radiative transitions were taking place within the 4f electronic shell, contradicting Laporte selection rules. In 1937 J.H van Vleck published an article (Vleck 1937) in which he reported this "bizarre" phenomenon which he referred to as "The puzzle of the Rare-earth spectra in solids". In these conditions, to explain the relatively strong intensity and spectral features of spectra lines observed, it was suggested that this "puzzle" originated from four possibilities, namely, 4f to 5d transitions, electric dipole transition, magnetic dipole transition, or electric quadrupole transition. A transition from 4f to 5d would result in broad spectral line contrary to the sharp lines observed, therefore this possibility was eliminated. Magnetic transition on the other hand account only for a few numbers of transition and cannot be retained as responsible for the observed spectra. Quadrupole radiation account for all transitions but their rather weak strengths do not justify the relatively high intensities observed. The only possibility left is electric dipole transition which is in principle forbidden by Laporte selection rules as it was reported earlier. Van Vleck in 1937 predicted that the crystal host was responsible for the "puzzle"(Vleck 1937). The prediction was later confirmed by Broer et al. in 1945 (Broer et al. 1945).

In fact, if the crystal field surrounding the rare-earth ion is non-centrosymmetric (meaning it lacks a centre of symmetry at the equilibrium position.), the ligand-field interactions mix electronic states of opposite parity into 4f wavefunctions, which relaxes the selection rule and the transitions become partially allowed. The transitions allowed by the crystal field action are known as induced or forced electric dipole transitions. Other sources were also reported to contribute to their intensity such as ligand-to-metal charge transfer (LMCT) states, and/or vibrational levels (Nazeeruddin et al. 1993; Wang et al. 2014).

3.5 INTENSITIES OF ONE-PHOTON TRANSITIONS—JUDD-OFELT THEORY

The theory of intensities of 4f–4f transitions in rare-earths was introduced independently by Judd and Ofelt in 1962 (Ofelt 1962; Judd 1962) and has been known ever since as Judd-Ofelt theory. Judd-Ofelt theory describes mostly the intensity of electric dipole transitions as it has been shown that the other transition does not contribute significantly to the observed spectral rays of rare-earths. The intensity of transitions between two states is characterized by its oscillator-strength, which is a dimensionless quantity that expresses the probability of absorption or emission of electromagnetic radiation between the two states. The oscillator transition is a key property because important parameters such as the radiative transition rate and a lifetime can be derived from it. According to Judd-Ofelt theory, the oscillator strength of an electronic transition between two energy states within the 4f shell of rare-earth ions incorporated in a solid or solution can be described as:

$$f_{ij}^{ED} = \frac{8\pi^2 m \nu}{3h(2J+1)} \frac{(n^2+2)^2}{9n} \sum_{\lambda=2,4,6} \Omega_\lambda |f^N J| U_\lambda |f^N J'|^2, \tag{3.6}$$

Where i and j are the initial and final states, respectively, ν is the frequency of the transition between the two states, n is the refractive index of the host material, m the electron mass, and $f^N J| U_\lambda |f^N J'$ is the reduced matrix elements for intermediate coupling. Intermediate coupling represents a situation whereby the electron-electron repulsion magnitude becomes comparable to that of spin-orbit coupling, situation possible in the case of heavy metals like rare-earth. The elements of the matrix are constant, independent of the crystal host and have been tabulated by Nielsen and Kolster (Nielson and Koster 1963).

The Judd-Ofelt parameters, usually labelled, Ω_1, Ω_4, and Ω_6 are obtained by fitting the experimental absorption or emission measurements in a least-squares difference sum with Judd-Ofelt equation (3.6). The parameters are then subsequently used to calculate the transition probabilities of all the excited states A_{ij} from which the radiative lifetime is derived. Judd-Ofelt theory has been widely used despite the rough approximations involved in the calculations which often result in errors of up to 50% when using the parameters to describe the transition rates (Peacock 1975; Jørgensen and Judd 1964; Mason et al. 1975). For example, in fluoride glass the reported values of Judd-Ofelt parameters and error margin were $\Omega_2 = 1.54 \pm 0.25$, $\Omega_4 = 1.13 \pm 0.40$, and $\Omega_6 = 1.19 \pm 0.20$ (Reisfeld et al. 1983). Table 3.4 list the Judd-Ofelt parameters for Er^{3+} in different hosts. In an unordered structure like glass, the calculated Judd-Ofelt parameters are the average values of the parameters obtained in each location of the rare-earth ion since the variations from one location to another is large. Among all parameter, Ω_2 has been reported to be the most location-dependent (Di Bartolo and Armagan 2012).

TABLE 3.4
Judd-Ofelt Parameters Ω_2, Ω_4, and Ω_6 for Er^{3+} in Various Glass and Crystalline Hosts, in units of 10^{-20} cm^2

Host matrix	Ω_2	Ω_4	Ω_6
Phosphate	9.92	3.74	7.36
Borate	11.36	3.66	2.24
Germanate	6.40	0.75	0.34
Tellurite	7.84	1.37	1.14
ZBLA	3.26	1.85	1.14
ZBLA	2.54	1.39	0.97
Fluoride glass	1.54	1.13	1.19
ZBLAN	2.3	0.9	1.7
LaF$_3$	1.1	0.3	0.6
Y$_2$SiO$_5$	2.84	1.42	0.82

The magnetic dipole transition within the 4f shell is always allowed as dictated by the Laporte rule. Its oscillator strength is given by the expression (3.7).

$$f_{ij}^{MD} = \frac{8\pi^2 m\nu}{3he^2(2J+1)} n^3 |\mu_B|^2 |f^N J \| \vec{L} + 2\vec{S} \| f^N J'|^2. \tag{3.7}$$

where μ_B is the Bohr magnetron, given by $e\hbar/2m$, and $\vec{L} + 2\vec{S}$ the magnetic dipole operator. In the normal case, its contribution is weak and it can be neglected in the calculations. However, when the electric dipole matrix becomes small, its contribution becomes significant and it cannot be neglected without severely impacting the accuracy of the Judd-Ofelt model. A good example of such case is the Er^{3+} transition between $^4I_{13/2}$ and $^4I_{15/2}$, where the computed spontaneous emission probability for the magnetic dipole of 34.7 s^{-1}, which is almost half of the 73.5 s^{-1} obtained for the electric dipole. The overall spontaneous emission lifetime found from these values was 9.2 ms in ZBLAN glass.

Finally, it is important to mention that the Judd-Ofelt theory yields the total oscillator strength for a transition between two $^{2S+1}L_J$ multiplets and not between their respective stark components. Therefore, before estimating the oscillator strength between two multiplets, the summation over all stark components must be performed first.

3.6   LIGHT–MATTER INTERACTION

3.6.1   BLACKBODY RADIATION

When electromagnetic radiation in an isothermal enclosure, or cavity, is in thermal equilibrium at temperature T, the distribution of radiation density $\varrho(\nu)d\nu$, contained in a bandwidth $d\nu$ is given by Planck's law,

$$\varrho(\nu)d\nu = \frac{8\pi\nu^2 d\nu}{c^3} \frac{h\nu}{e^{h\nu/kT} - 1}. \tag{3.8}$$

where $\varrho(\nu)$ is the radiation density per unit frequency [Js/cm^3], k is the Boltzmann's constant, and c is the velocity of light. The spectral distribution of thermal radiation vanishes at $\nu = 0$ and $\nu \to \infty$ and has a peak that depends on the temperature.

In equation (3.8), the density of radiation per unit volume and unit frequency interval is defined by:

$$\frac{8\pi\nu^2}{c^3} = p_n \tag{3.9}$$

The factor p_n can also be interpreted as the number of degrees of freedom associated with a radiation field, per unit volume, per unit frequency interval. The expression for the mode density p_n [modes s/cm^3] plays an important role in connecting the spontaneous and the induced transition probabilities.

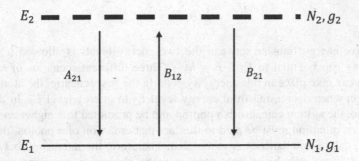

FIGURE 3.4 Two energy levels with a population N_1, N_2, and degeneracies g_1, g_2, respectively.

3.6.2 BOLTZMANN'S STATISTICS

Let us consider an atomic system with many discrete energy levels. If level 1 and 2 are two of these levels with E_1 and E_2 their respective energies such that $E_2 > E_1$ as shown in Figure 3.4.

It comes from statistical mechanics (Landau and Lifshitz 1980) that when many identical particles are in thermal equilibrium at temperature T, occupying these two energy levels, their relative populations, are distributed following the Boltzmann law according to:

$$\frac{N_2}{N_1} = exp\left(\frac{-(E_2 - E_1)}{kT}\right), \tag{3.10}$$

In this relation, N_1 and N_2 are the numbers of particles (atoms or ions) in the energy levels E_1 and E_2, respectively. If the energy gap is large enough ($E_2 - E_1 = h\nu_{21} \gg kT$), the limit of equation (3.10) tends to zero. The physical meaning of this is that most particles are in the lower energy level at thermal equilibrium.

3.6.3 RADIATION-MATTER INTERACTION—EINSTEIN COEFFICIENTS

According to the 1916 Einstein Hypothesis (Weinberg and Weinberg 2015) there are three processes involved in the formation of spectral lines which are absorption, spontaneous emission, and stimulated emission. To quantify these three processes, Einstein introduced three coefficients, which have been ever since known as the Einstein coefficients.

These Einstein coefficients can be derived by loosely following Einstein's original derivation achieved by simply considering an ideal material with only two non-degenerate energy levels E_1 and E_2 with atomic populations N_1 and N_2 respectively. Since there are only two energy levels, the total population of this system will remain constant and will be the sum of N_1 and N_2 as in equation (3.11),

$$N_1 + N_2 = N_{tot}. \tag{3.11}$$

Radiative energy transfer between the two energy levels is allowed if the corresponding gap is equal to $E_2 - E_1 = h\nu_{21}$. Three different scenarios of energy exchange can take place in this energy system. In the first scenario, the atom can emit a photon when decreasing from energy level E_2 to energy level E_1. In the second scenario, the system can absorb a photon and be promoted to a higher energy level. However, attention must be paid to the fact that emission of a photon, thus loss of energy, can be spontaneous or stimulated. Therefore, the interaction between light and matter can only be of three types namely, absorption, spontaneous emission, and stimulated emission. These are described in more detail in the following section.

3.6.3.1 Absorption

Absorption occurs when an incident photon of energy $h\nu$ (where ν is the photon frequency and h is Planck's constant) is absorbed by an atom or ion. Upon absorbing the photon, the ion or atom gains energy and an electron is promoted to a higher energy level. If a quasi-monochromatic electromagnetic wave made of many photons of frequency ν_{21} travels into a material made of multiple atoms with N_1 and N_2 as populations of lower and higher energy levels respectively, then the lower energy level will be depopulated at a rate proportional to the radiation density $\varrho(\nu)$ as well as its population N_1. This is given by the following equation:

$$\frac{\partial N_1}{\partial t} = -B_{12}\varrho(\nu)N_1, \tag{3.12}$$

where B_{12} is a constant of proportionality with dimensions cm^3/s^2 J.

Physically the product $B_{12}\varrho(\nu)$ can be thought of as the probability that the transition will take place.

3.6.3.2 Spontaneous Emission

Let us consider the previous system, which has absorbed energy and has some of its atoms promoted to the energy level E_2. These atoms, however, are not remaining indefinitely in the excited state. After a certain amount of time corresponding to the lifetime of the E_2 energy level, the atoms are going to relax to the E_1 energy level releasing the accumulated energy. This process is known as the spontaneous emission and its rate is proportional to the population of the E_2 level as well as a rate of proportionality as given by equation (3.13):

$$\frac{\partial N_2}{\partial t} = -A_{21}N_2, \tag{3.13}$$

where A_{21} is a constant of proportionality with dimension s^{-1}. Physically, the quantity A_{21} expresses the probability of an atom to relax to the lower energy state within a certain amount of time. This quantity depends strongly on the two energy levels involved in the transition.

Spontaneous emission is a random process and is therefore characterized by a total absence of phase between emitted photons. In other words, the electromagnetic radiation emitted is incoherent. For this reason, we will show later that spontaneous emission does not contribute to the gain of optical amplifiers or lasers and constitute undesirable noise.

Solving equation (3.13) yields,

$$N_2(t) = N_2(0)exp\left(\frac{-t}{\tau_{21}}\right),\qquad(3.14)$$

where τ_{21} is the lifetime for the spontaneous radiation of the transition from level E_2 to level E_1. As we have already mentioned it in the previous section, this radiation lifetime is equal to the reciprocal of the Einstein coefficient,

$$\tau_{21} = A_{21}^{-1}\qquad(3.15)$$

3.6.3.3 Stimulated Emission

If an atomic system with several atoms excited to a higher energy level is acted upon by electromagnetic radiation of the appropriate frequency (which corresponds to the energy gap between the two energy levels involved in the transition), its excited atoms will be demoted to the lower-lying energy level E_1 releasing photons with the same frequency and phase as the incident photons. This process is known as stimulated emission defined by:

$$\frac{\partial N_2}{\partial t} = -B_{21}\varrho(\nu_{21})N_2,\qquad(3.16)$$

Similar to equation (3.13), B_{21} is a constant of proportionality. Unlike the case of spontaneous emission, this proportionality constant depends on N_2, the population of the upper energy level involved in the transition as well as the incident radiation. It is easy to understand that the total emitted radiation is made of both spontaneous and stimulated emission contributions. The beneficial parameter for amplifier and laser operation is the B_{21} coefficient whereas the A_{21} coefficient is detrimental because it represents a loss term. As we have previously mentioned, the spontaneous emission parameter represents a source of noise for both amplifiers and lasers.

3.6.3.4 Relations between Einstein Coefficients

Combining the previous expressions for absorption, spontaneous emission, and stimulated emission makes it possible to write the rate equations for a two-level system which are ordinary differential equations describing the change of population density of a given energy level as a function of time. These equations will be very useful in modelling the behaviour of a given energy system. The rate equations of a two-level system is written as follows,

$$\frac{\partial N_1}{\partial t} = -\frac{\partial N_2}{\partial t} = B_{21}\varrho\,(\nu_{21})N_2 - B_{12}\varrho\,(\nu_{21})N_1 + A_{21}N_2. \tag{3.17}$$

The reader will notice that equations for the two levels E_1 and E_2 are identical except for the sign. This only means depopulating lower level always corresponds to populating the higher level. If one considers the case of thermal equilibrium, one can easily say that the number of transitions per unit time from E_1 to E_2 is equal to the number of transitions from E_2 to E_1. Physically, it means that there is no change in population density of the two-level at thermal equilibrium. Mathematically, the derivatives of both N_1 and N_2 populations are going to vanish as obtained in equation (3.18) given by:

$$\frac{\partial N_1}{\partial t} = \frac{\partial N_2}{\partial t} = 0. \tag{3.18}$$

Thus, equation (3.17) becomes,

$$N_2 A_{21} + N_2\varrho\,(\nu_{21})B_{21} = N_1\varrho\,(\nu_{21})B_{12}. \tag{3.19}$$

Using Boltzmann's equation (3.10) for the ratio N_2/N_1, one can write equation (3.19) as:

$$\varrho\,(\nu_{21}) = \frac{(A_{21}/B_{21})}{(B_{12}/B_{21})\,exp\,(h\nu_{21}/kT) - 1} \tag{3.20}$$

Comparing this expression with the blackbody radiation law (3.8) yields:

$$\frac{A_{21}}{B_{21}} = \frac{8\pi h\nu^3}{c^3}, \tag{3.21}$$

and

$$B_{21} = B_{12}. \tag{3.22}$$

The parameters A and B are known as Einstein coefficients and the relations between them are Einstein's relations. The factors $8\pi\nu^2/c^3$ linking A_{21} and B_{21} is the mode density p_n obtained in expression (3.9).

In the ideal case of a simple system without degeneracy, the terms g_1 and g_2 will be equals. This implies that B_{21} will also be equal to B_{12}. Physically, this means that for such a system, the Einstein coefficients for absorption and stimulated emission are equal. However, this will never be the case in practice because of the atomic interactions and external field which lifted the degeneracy to some degree.

Case of degenerate energy levels.

In the case of degenerate energy levels (rare-earth ions in a solid), the Boltzmann distribution becomes,

$$\frac{N_2}{N_1} = \frac{g_2}{g_1} exp\left(\frac{-(E_2 - E_1)}{kT}\right). \tag{3.23}$$

One can show that if the sub-levels are equally populated (such that for each sub-level we have $n_1 = N_1/g_1$ and $n_2 = N_2/g_2$, where N_1 and N_2 are the total populations of level 1 and 2 and g_1 and g_2 their respective degeneracies) following the same derivation as in the case of the non-degenerate energy, the Einstein relation between the B coefficients becomes,

$$g_2 B_{21} = g_1 B_{12}. \tag{3.24}$$

3.6.4 TRANSITION CROSS SECTION

The cross-section of an ion is an important parameter in that it quantifies the ability of the ion to absorb or emit photons of light. Mathematically, the cross-section of an absorption or emission transition can be understood as the probability for that transition to occur. More precisely, if a transition involving two energy levels E_2 and E_1 (with E_2 of greater energy than E_1) is taking place, the probability of that transition will be directly proportional to the absorption or emission cross-section depending on the direction of the transition. The cross-section has a dimension area. For a given two-level system such as the one represented in Figure 3.4, where the energy levels are E_2 and E_1 for levels 2 and 1 respectively, one can show that the transition probability for the emission or absorption of a photon with energy $\hbar = E_2 - E_1$ is directly proportional to the absorption or emission cross-section σ_{12} and σ_{21}, as well as to the intensity of the incident light. The number of photons absorbed per unit time N_{abs} can be written as:

$$N_{abs} = \sigma_{12} \phi(\omega). \tag{3.25}$$

In the above relation $\phi(\omega)$ is the photon flux which is the number of photons per unit area per unit time.

The power absorbed is given by,

$$P_{abs} = \sigma_{12} I, \tag{3.26}$$

where I, is the intensity of the light incident upon the ion. The number of absorbed photons is linked to the absorbed power as shown in equation (3.26),

$$N_{abs} = \sigma_{12} \frac{I}{\hbar \nu} = \sigma_{12} \phi(\nu). \tag{3.27}$$

Similarly, the power of light emitted by stimulated emission from an ion interaction by an incident light of intensity I is given by,

$$P_{em} = \sigma_{21} I. \tag{3.28}$$

For better understanding, the absorption cross-section can be pictured as an area capable of capturing photons. Therefore, the bigger the area, the higher the number of photons that can be captured by it. The emission cross-section can be interpreted in the same way as an area capable of emitting photons of light. In this case also, the bigger the area, the higher the number of photons that can be emitted. In the case of a two-level system with ions populations of N_1 and N_2 (N_1 being the population of lower energy state and N_2 the population of higher energy state), the variation of power of light travelling inside this system can be written as,

$$\Delta P = P_{em} - P_{abs} = (N_2 \sigma_{21} - N_1 \sigma_{12}) I. \tag{3.29}$$

This quantity represents the amount of light emitted by the medium. In the case of absorption, the same quantity will be exchanged, but with a negative sign. One of the most important things to know is that both the absorption and emission probabilities depend on the light intensity and not its power. It means that for the same amount of power, the higher intensity is reached for smaller areas resulting in a higher probability of the transition. This is the main reason behind the fiber laser's low threshold values and high efficiency.

The absorption and emission cross-sections are equals for non-degenerate states. This picture is often used to derive the expressions linking cross-sections, Einstein's A and B coefficients as well as radiative lifetime. In practice, however, the emission and absorption cross-sections are different. The difference comes from the fact that when rare-earth ions are hosted in solids; the upper and lower states are no longer made up of a unique energy level but rather made of a comb of sub-levels very close to one another. These sub-levels are populated differently following the Boltzmann distribution. As a result, the cross-section must be specified for each frequency and the cross-section becomes a function of frequency (or wavelength) within the spectral bandwidth of the transition.

3.6.5 LADENBURG-FUCHTBAUER RELATION

In fiber amplifier and laser design, accurate knowledge of the cross-sections is very important. However, measuring directly these values is not possible. Therefore, one always uses an indirect method to obtain these values. The Fuchtbauer-Ladenburg is a relation used in the procedure for determining emission cross-sections of a given amplifying medium. The entire procedure consists of measuring the fluorescence of an electronic transition between two energy levels. The intensity of the fluorescence is proportional to the emission cross-section. The fluorescence is relatively easy to measure. However, deriving the cross-section from this measurement is a bit more challenging. The Fuchtbauer-Ladenburg equation does it based on the fact that the quantum efficiency of a laser transition is close to unity. Physically, this means that the upper-state lifetime is almost equal to the radiative lifetime, which is, in turn, is determined by the emission cross-section to lower energy levels. In this regard, it can be seen that the Fuchtbauer-Ladenburg (Fowler and Dexter 1962) equation relates cross-section to the lifetime of the upper energy level.

The emission and absorption cross-sections are equal within a simple system of two non-degenerate energy systems. However, this case is only ideal, and complications often arise in the case of rare-earth ions in a solid host. In this case, one has to consider the Stark sub-levels resulting from the crystal field of the host. The two states involved in the transitions become a comb of sub-levels and their respective populations vary greatly in proportion with their thermal distribution. The cross-section will then have the meaning of cross-section at a given frequency within the spectral range of the transition. The absorption and emission cross-section will then be equal only under a condition that the sub-levels are equally populated or the oscillator strength of the transition between any of the individual levels are equal. We first discuss this case here and later discuss the general case, where the populations of the sub-levels, as well as the strengths between transitions, are not equal.

We consider the simple two-level system shown in Figure 3.4 the probabilities for spontaneous emission stimulated emission and stimulated absorption is A_{21}, B_{21}, and B_{12}. In equilibrium,

$$n_1 \rho(\nu) B_{12} = n_2 \rho(\nu) B_{21} + n_2 A_{21}, \qquad (3.30)$$

where $\rho(\nu)$ is the photon energy density, and n_1 and n_2 are the ground and metastable population densities.

Solving for $\rho(\nu)$,

$$\rho(\nu) = \frac{A_{21}/B_{21}}{B_{12} n_1 / B_{21} n_2 - 1}. \qquad (3.31)$$

By making use of Boltzmann statistics, we can write,

$$\frac{n_2}{n_1} = \frac{g_2}{g_1} exp(h\nu/kT). \qquad (3.32)$$

Substitution into (3.31) yields,

$$\rho(\nu) = \frac{A_{21}/B_{21}}{g_2 B_{12}/g_2 B_{21} exp(h\nu/kT) - 1}. \qquad (3.33)$$

Now, Planck's law can be stated as,

$$\rho(v)dv = \frac{8\pi h v^3 \mu^3}{c^3 [exp(hv/kT) - 1]},$$ (3.34)

where μ is the refractive index. Combining (2.7) and (2.9), we find,

$$g_1 B_{12} = g_2 B_{12},$$ (3.35)

And

$$\frac{A_{21}}{B_{21}} = \frac{8\pi h v^3 \mu^3}{c^3}.$$ (3.36)

Now, the stimulated emission rate B_{21} is given by,

$$B_{21} = \frac{\sigma_{21} c}{h v g(v) \mu}.$$ (3.37)

So,

$$\sigma_{21} = \frac{h v g(v) \mu B_{21}}{c}.$$ (3.38)

Combining (3.36) and (3.37), we derive the Fuchtbauer-Ladenburg equation for the emission cross-section,

$$\sigma_{21} = \frac{\lambda^2}{8\pi\mu^2} A_{21} g(v).$$ (3.39)

If we now consider absorption, we have,

$$B_{12} = \frac{\sigma_{12} c}{h v g'(v) \mu},$$ (3.40)

where the line shape for absorption $g'(v)$ is not necessarily the same as that for emission $g(v)$. The Fuchtbauer-Ladenburg equation for the absorption cross-section is thus,

$$\sigma_{12} = \frac{g_2}{g_1} \frac{\lambda^2}{8\pi\mu^2} A_{21} g'(v).$$ (3.41)

Let us consider a flux of photons ϕ, measured in quantity of photons per unit surface, per unit time interaction with a group of identical atoms with densities N_1 and N_2 per unit volume in the ground and excited state respectively. It can be shown that the rate of absorption of these photons is proportional to ϕ and N_1 as follows,

$$W_{12} = \sigma_{12}\phi(\nu)N_1 \tag{3.42}$$

Writing photon flux density as a function of intensity we obtain,

$$\phi(\nu) = \frac{I}{h\nu}, \tag{3.43}$$

where I is the incident intensity and $h\nu$ the energy of a unique photon. Substituting (3.43) into (3.42) we obtain,

$$W_{12} = \sigma_{12}\frac{I}{h\nu}N_1. \tag{3.44}$$

Similarly, the rate of stimulated emission can be written as,

$$W_{21} = \sigma_{21}\frac{I}{h\nu}N_2, \tag{3.45}$$

where σ_{21} is the emission cross-section.

We also know that the absorption and emission rates are related to the Einstein coefficients as follow,

$$W_{12} = B_{12}u(\nu)N_1, \tag{3.46}$$

and

$$W_{21} = B_{21}u(\nu)N_2, \tag{3.47}$$

where $u(\nu)$ is the energy density of per unit volume $[j/m^3]$ and is related to the intensity of light as,

$$u(\nu) = \frac{I}{c}n, \tag{3.48}$$

where n is the refractive index of the material of interest. Substituting (3.48) in (3.46) and (3.47) yields,

$$W_{12} = B_{12}N_1\frac{nI}{c}, \tag{3.49}$$

and

$$W_{21} = B_{21}N_2\frac{nI}{c}.$$ (3.50)

Comparing (3.50) with (3.47) yields,

$$A_{21} = \frac{\sigma_{21}8\pi\nu^2}{nc^2}$$ (3.51)

Substituting (3.51) into the Einstein relation (3.36) yields,

$$\frac{1}{\tau_{21}} = \frac{8\pi n^2}{c^2}\nu^2\sigma_{21}(\nu).$$ (3.52)

Equation (3.52) is known as the Fuchtbauer-Ladenburg equation and it relates radiative lifetime to the emission cross-section. Radiative lifetime is relatively easy to measure. Therefore, the emission cross-section can be deduced from the Radiative lifetime measurement.

Under the assumption that all sub-levels of an energy level are equally populated, the Einstein B_{12} and B_{21} coefficients are equal. Therefore, equation (3.52) can be written as,

$$\frac{1}{\tau_{21}} = \frac{8\pi n^2}{c^2}\nu^2\sigma_{21}(\nu).$$ (3.53)

In other words, absorption and emission cross-sections are also equal,

$$\sigma_{12} = \sigma_{21}.$$ (3.54)

3.6.6 McCumber Theory of Emission Cross Sections

In the previous discussion, we assumed that population density is equally distributed among the sub-levels of energy states. However, experiments have shown that this ideal case is far from reality. Therefore, McCumber derived a more general theory linking quantities like absorption and emission cross-section in degenerate energy levels and non-uniform population density distributions (McCumber 1964).

Let us consider the energy level system with sub-levels in each multiplet as described in Figure 3.5. The energy levels E_1 and E_2 correspond to lower and higher levels respectively and g_1 and g_2 are their respective degeneracies.

The emission and absorption cross-sections between two sub-levels are assumed equal. However, the overall absorption and emission cross sections for the levels 1 and 2 can be described as the summation of the individual cross-sections σ_{ij} between the levels,

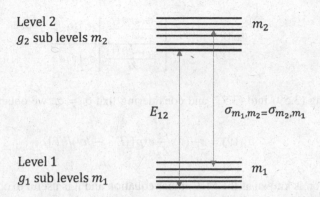

FIGURE 3.5 Energy levels of a 2 levels system showing sublevels in each multiplet.

$$\sigma_{21}(\nu) = \sum_{ij} d_j \left[\frac{exp(-E_j/kT)}{Z_2} \right] \sigma_{ij}(\nu) d_i, \tag{3.55}$$

and

$$\sigma_{12}(\nu) = \sum_{ij} d_i \left[\frac{exp(-E_i/kT)}{Z_1} \right] \sigma_{ij}(\nu) d_j. \tag{3.56}$$

The lineshape information is contained in the $\sigma(\nu)$ functions. Measurement of the energies E_i and E_j as well as the partition functions Z_1 and Z_2 are done from the lowest level of each one of the two multiplets. The factor in brackets represents the fractional thermal distribution of the jth and i^{th} levels of the upper (2) and lower (1) energy levels.

The Z_i partition is given by,

$$Z_i = \sum_{ij} e^{-E_i/KT}. \tag{3.57}$$

For any couple of lower and higher energy levels (Figure 3.5), the energy separation can be written as,

$$E_j - E_i = h\nu - E_{12}, \tag{3.58}$$

where E_{12} is the separation between the lowest components of states 1 and 2.

Let us try to find a relationship between $\sigma_{12}(\nu)$ and $\sigma_{21}(\nu)$ by dividing (3.55) by (3.56),

$$\frac{\sigma_{21}(\nu)}{\sigma_{12}(\nu)} = \frac{\sum_{ij} d_j \left[\frac{exp(-E_j/kT)}{Z_2} \right] \sigma_{ij}(\nu) d_i}{\sum_{ij} d_i \left[\frac{exp(-E_i/kT)}{Z_1} \right] \sigma_{ij}(\nu) d_j}. \tag{3.59}$$

Substituting (3.59) into (3.57) and considering that $\sigma_{ij} = \sigma_{ji}$ we obtain,

$$\sigma_{21}(\nu) = \sigma_{12}(\nu) \frac{Z_1}{Z_2} exp\left[(E_{12} - h\nu)/kT\right]. \tag{3.60}$$

This relation is known as the McCumber equation and it is useful to obtain emission cross-section knowing the absorption cross-section. In its commonly used form, the quantity $Z_1/Z_2 e^{E_{12}/kT}$ is replaced by the equivalent expression $e^{\varepsilon/kT}$ leading to the more compact form,

$$\sigma_{21}(\nu) = \sigma_{12}(\nu) e^{(\varepsilon - h\nu)/kT}. \tag{3.61}$$

Figure 3.6 demonstrates a good fit for experimental data obtained for the same glass sample. The McCumber theory provides not only absolute values of the cross-section but its spectral distribution as well, thus giving the complete picture of the cross-section. This theory allows one to determine the emission cross-section from the measured absorption cross-section and radiative lifetime. However, the McCumber theory requires knowledge of the electron structure of the ion.

Overall, the McCumber theory is a powerful instrument in the hands of researchers for calculating important spectroscopic parameters of laser-active optical centres and is often used in laser physics.

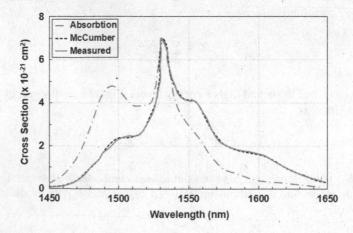

FIGURE 3.6 Comparison of the shape of the measured stimulated-emission cross-section with that calculated from the absorption cross-section using the McCumber theory.

3.6.7 LIFETIMES

The lifetime of a given energy level determines the time taken by an electron in the excited energy level before relaxing spontaneously to a lower energy level. The lifetime is inversely proportional to the transition probability to lower energy levels through both radiative and non-radiative processes. Radiative processes are always accompanied by the emission of photons whereas non-radiative processes are accompanied by the emission of phonons. The total lifetime combining radiative and non-radiative processes can be written as,

$$1/\tau = 1/\tau_{NR} + 1/\tau_R. \tag{3.62}$$

Radiative lifetime tends to be longer, in the order of microseconds to milliseconds, since they are forbidden (Laporte rules). Non-radiative lifetime, on the other hand, depends on the host. When an atom is relaxing from an excited state to a lower energy state, energy is released through phonons, which are units of lattice vibration. The radiative lifetimes of rare-earth ions are typically quite large since the electric dipole transition to lower excited states is forbidden. The non-radiative transition process occurs by the simultaneous emission of phonons. A longer lifetime corresponds to a bigger number of phonons involved in the bridging of the gap between the two energy levels involved in the transition. Therefore, the glass composition significantly influences the radiative and non-radiative lifetimes (Wu et al. 2003).

3.7 LINEWIDTH AND BROADENING

Broadening of laser transition is important because it affects the gain saturation of optical amplifiers and lasers. In a case of a homogeneously broadened gain medium, the gain saturation is equally homogeneous, whereas in the case of inhomogeneous broadening, gain saturation is also inhomogeneous i.e. gain saturation is stronger at certain wavelengths than others. This can be a serious problem in some regimes such as narrow-linewidth tunable lasers where efficiency will be dramatically low at these wavelengths.

3.7.1 HOMOGENEOUS BROADENING

Homogeneous broadening group effects expand the optical linewidth of an electron transition by affecting the radiating or absorbing atoms, in the same way. In other words, all the atoms of a homogeneously broadened sample interact identically with the radiation field so that each atom makes transitions with identical lineshape and widths. The spectral shape of the transition cross-sections of all involved atoms is therefore equal. The most important causes of homogeneous broadening include natural line broadening and collision broadening.

3.7.2 NATURAL BROADENING

Natural broadening is difficult to get rid of because it originates from the Heisenberg uncertainty principle,

$$\Delta x \Delta p \geq \frac{h}{4\pi}, \tag{3.63}$$

where Δp is the momentum and Δx the position of the particle. Using the laws of classical physics one can write:

$$\Delta x \times m \times v \geq \frac{h}{4\pi}, \tag{3.64}$$

where v is the velocity and m the mass of the particle. Dividing and multiplying by time, yields

$$\Delta x \times m \times a \times t \geq \frac{h}{4\pi}, \tag{3.65}$$

In this last expression $\Delta x \times m \times a = \Delta E$. With ΔE the energy. One can write,

$$\Delta E \times \Delta t \geq \frac{h}{4\pi}, \tag{3.66}$$

where Δt is the lifetime, therefore the uncertainty principle can be written as

$$\Delta E \times \tau \geq \frac{h}{4\pi}. \tag{3.67}$$

Using the Einstein-Planck formula, the equation can be written as,

$$\Delta v \geq \frac{h}{4\pi\tau}, \tag{3.68}$$

where v is the frequency. This expression relates the linewidth to the lifetime of the excited state. Because the system stays in the excited state only for a finite amount of time before relaxing to the ground state, the linewidth will always be broadened. We also know that the higher the energy gap between states involved in a transition, the greater will be the tendency to relax to the lower state. Therefore, the lifetime will be smaller and the line broader. So whenever the excited state is very close to the ground state, the system will have a narrow linewidth.

3.7.3 COLLISIONAL BROADENING

Rare-earth atoms are not isolated. Therefore, several identical atoms can undergo multiple collisions. When an excited and a non-excited atom are involved in a collision, the excited atom often transmits a portion of its excitation energy to the non-excited atom. In such an occurrence, its lifetime is reduced and for the same reason (uncertainty principle) as in the case of natural broadening, its linewidth will be broadened. Due to the nature of crystals (ordered solid), rare-earth ions only occupy the same type locations in the crystal lattice. Therefore, the interaction of these ions with the crystal lattice through phonons affect them identically.

The optical transition lineshape in the case of homogeneous broadening is represented by a Lorentzian function g_L expressed as follows (Koechner 2006):

$$g_L(\nu) = \frac{\Delta\nu_H/2\pi}{(\nu - \nu_0)^2 - \left(\frac{\Delta\nu_H}{2}\right)^2}, \tag{3.69}$$

where ν_0 is the central frequency of the optical transition and $\Delta\nu_H$ is the transition spectral line full wave at half maximum FWHM.

3.7.4 INHOMOGENEOUS BROADENING

Inhomogeneous broadening is the increase of an atomic transition linewidth resulting from effects that influence radiating or absorbing atoms differently. The consequence of this is that the absorption and emission cross-sections have different spectral shapes for different atoms. The emission spectrum from such a material is an average of the emission spectra of several atoms and can, therefore, be much broader than the ones observed for single atoms. Similarly, absorption spectra can be broadened. The most important cause of inhomogeneous broadening in solids is different surrounding crystal field and symmetry experienced by rare-earth ions in different locations of the host medium. This is particularly the case for glasses, but can also occur in crystalline materials. In inhomogeneously broadened transitions, different atoms experience different influences of the host medium according to their locations and this will result in different transitions. The overall shape of the transition can, therefore, be considered as the superposition of the individual homogeneously broadened lines corresponding to different locations. The lineshape of the inhomogeneously broadened optical transition is expressed by (3.70) as:

$$g_G(\nu) = \frac{1}{\Delta\nu_I}\sqrt{\frac{4ln2}{\pi}}\,exp\left[-4ln2\left(\frac{\nu - \nu_0}{\Delta\nu_I}\right)^2\right] \tag{3.70}$$

where $\Delta\nu_I$ is the FWHM of the non-homogeneously broadened line, determined by the influence of the surrounding fields. Unlike homogeneous broadening, the inhomogeneous broadening is independent of the temperature of the host material. Table 3.5 illustrate the values of homogeneous and inhomogeneous linewidth for

TABLE 3.5

Homogeneous and Inhomogeneous Linewidths for Few Rare-Earth Ions in Various Glasses

Rare-earth ion	Glass matrix	$\Delta \nu_H$ (cm^{-1})	$\Delta \nu_{INH}$ (cm^{-1})	Source
Nd^{3+}	Silicate	110	50	(Pellegrino et al. 1980)
Yb^{3+}	Phosphate	80	66	(Fournier and Bartram 1970)
Er^{3+}	Germano-silicate	17	30	(Desurvire et al. 1990; Sun et al. 2007; Bigot et al. 2AD)
Tm^{3+}	ZBLAN	32	450	(Roy et al. 2002)

different rare-earth ions in various glass materials measured at a temperature of 300°K.

At room temperature, it is difficult to discriminate between the effects of homogeneous and inhomogeneous contributions to the overall lineshape. This is particularly true in glasses (disordered solids). In such a case, the overall lineshape is described by a Voigt function, obtained as the convolution between the Gaussian and Lorentzian profiles,

$$g_V(\nu) = \int g_G(\mu) g_L(\nu - \mu) d\mu \qquad (3.71)$$

It is common, if one wants to get information regarding the inhomogeneous contribution, to lower the temperature of the sample (usually < 77 °K), to eliminate the homogeneous broadening contribution by "freezing" phonons. Using this technique, parameters such as the Stark structure can be precisely determined. One should note that the homogeneous and inhomogeneous linewidth contribution also depends on the nature of the glass matrix. For example, it has been shown that inhomogeneous broadening of the $^4I_{13/2} \rightarrow {}^4I_{15/2}$ in Er^{3+}-doped silica glass was greater in alumina-silicate glasses compared to Germano-silicate glass (MacFarlane and Shelby 1987; Thiel et al. 2011).

3.8 IONS-IONS INTERACTION

When many rare-earths are together inside a doped material several phenomena can take place. The most important of these processes are energy transfer, cross-relaxation and cooperative upconversion. These processes are at the origin of concentration quenching that limits strongly the quantum efficiency of rare-earth-doped devices. It is worth mentioning that most of these processes become more pronounced at high rare-earth concentrations.

3.8.1 ENERGY TRANSFER MECHANISMS

In energy transfer, one excited ion transfers its energy to a neighbouring ion. The ion losing energy is referred to as the donor whereas the one gaining energy is called the acceptor. Energy transfer occurs between ions of either the same species or ions of different species. For example, energy transfer between Ytterbium and Erbium ions have been used successfully used to improve the efficiency of Erbium-doped amplifiers and lasers (Nilsson et al. 2003; Sobon et al. 2012) whereas Erbium to Erbium energy transfer is one of the most important dissipative mechanisms in Erbium-doped lasers and amplifiers (Myslinski et al. 1997). When the energy lost by the donor is exactly equal to the energy gained by the acceptor the energy transfer is said to be resonant. When the lost and gained energies are not equals, the additional energy is not equal; the conservation of energy is achieved through absorption or emission of phonons (Moos and Riseberg 1968). Because lattice crystal vibration is involved (through phonons), energy transfer is an energy-dependent process. A detailed analysis of energy transfer that includes all calculations involved in energy transfer can be found in the literature (Pisarska et al. 2014; Wei et al. 2014; Sontakke and Kalyandurg 2012; Huang et al. 2014).

The resonant transfer means that their conservation of energies between the two ions or the energy lost by one ion is exactly the energy gained by the next ion. Energy transfer has been used advantageously in co-doped gain medium to increase the efficiency of lasers.

3.8.2 COOPERATIVE UP-CONVERSION

The cooperative up-conversion is illustrated in Figure 3.7(a) In this process, one excited ion transfers its energy to a neighbouring excited ion and relaxes to the ground state while the acceptor is promoted to a higher energy level, from where it can relax non-radiatively to its initial excited state. Losing one photon in the process or radiatively decay to the ground state releasing a photon with lower wavelength. In this regard, this process can be beneficial because it can be used for making up-conversion lasers. Cooperative up-conversion is a deleterious effect because every time it takes place one photon is lost. Its effect becomes more pronounced at a high concentration as the probability of ion-ion interactions increases. Cooperative up-conversion is the most deteriorating factor of Erbium-doped lasers and amplifiers (Myslinski et al. 1997) and the power limiting factor of short cavity fiber lasers. However, cooperative upconversion can also be exploited beneficiary to build up-conversion laser also known as infrared to visible lasers.

3.8.3 CROSS RELAXATION

In the cross-relaxation process as illustrated in Figure 3.7, an ion in the excited state transfers part of its energy to another ion in the ground state. In the process, the donor decay to a lower energy level whereas the acceptor is promoted to a higher energy level, both ions ending in an intermediate energy state from which the decay nonradiatively to the ground state. The result of cross-relaxation is the loss of one

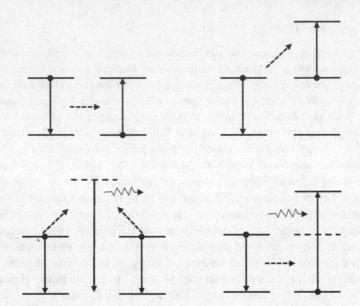

FIGURE 3.7 Ion-ion energy mechanisms. (a): Resonant energy transfer; (b): stepwise upconversion; (c): cooperative luminescence; (d): cooperative energy transfer and simultaneous photon absorption.

photon whose energy is converted to phonon energy. Cross relaxation is the dominant cause of loss efficiency in Nd^{3+} doped lasers and amplifiers (Arai et al. 1986). However, due to the lack of an intermediate energy level to accommodate the ions between the metastable and ground level, cross-relaxation does not take place in erbium-doped devices.

3.9 GENERAL CONSIDERATIONS ON FIBER LASER OPERATION

Concisely, a laser is an oscillator; therefore, it is made of a light amplifier, a feedback mechanism and a pumping system to provide energy. As the name says it, a fiber laser is a type of laser with an optical fiber doped with ions of rare-earth as its amplifying medium. Given the geometry of the fiber, various types of resonators are possible in fiber laser, as it will be discussed in this section. Many of the fundamental theories that sustain the working principle of rare-earth-doped fiber lasers have been discussed in previous sections of this chapter. This chapter puts everything together and describes fiber lasers in terms of their design and operation.

3.9.1 LASER GENERAL GAIN COEFFICIENT

Suppose a fiber laser is represented as in Figure 3.8. If I_z is the intensity of the incoming light, $I(z + dz)$ will be the intensity of the outgoing light after a distance dz.

The amplification condition inside the gain medium is given by,

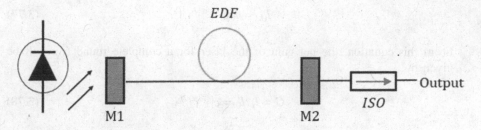

FIGURE 3.8 Linear cavity fiber laser.

$$I(z + dz) > I(z). \tag{3.72}$$

The following differential equation governs the intensity change as light passes through the active medium,

$$\frac{dI}{dz} = G(\omega)I(z), \tag{3.73}$$

where $G(\omega)$ is the gain coefficient. The equation tells us that the outgoing intensity can be larger or smaller than the incoming intensity after travelling a certain distance in the amplifying medium. If amplification has to occur, the outgoing intensity will be larger than the incoming intensity. G depends on the population difference between the two energy levels of interest. The higher the population difference, the higher will be the gain. Solutions to equation (3.72) take the form,

$$I_{(z)} = I_0 e^{Gz}. \tag{3.74}$$

Depending on whether G is positive or negative, amplification or attenuation can take place.

For an un-pumped fiber laser, let the loss per unit length at the laser wavelength due to absorption and scattering be α, then the light intensity across the length of the fiber decreases according to,

$$I_{(z)} = I_0 e^{-\alpha z}, \tag{3.75}$$

where z is the distance along with the fiber laser and I_0 the intensity at $z = 0$.

For pumped laser, let the gain per unit length produced by the inverted ions be given by g. Then the net gain (or loss) per unit length is given by $(g - \alpha)$,

$$I_{(z)} = I_0 e^{(g-\alpha)z}. \tag{3.76}$$

For a fiber of length L, the expression becomes,

$$I_{(L)} = I_0 e^{(g-\alpha)L}. \tag{3.77}$$

From this equation, the net gain of the laser for a complete round trip will be given by,

$$G = I_p/I_0 = e^{(g-\alpha)2L}. \tag{3.78}$$

Because g is the net amplification per unit length, for the entire length of the gain medium this coefficient will become,

$$G = \frac{I_{out}}{I_{in}} = exp\left[\int_0^L g(z)dz\right]. \tag{3.79}$$

The main source of power loss inside the laser is the absorption of a fraction of power by the mirror and the background loss of the fiber medium resulting from absorption and scattering. Absorption and scattering are called by the more generic term of "distributed losses". Laser action can only occur and be maintained if the total gain in the cavity overcome the losses. If the gain is equal to the losses, oscillation can take place internally but there will not be power coupled out of the laser because all the gain produced will be used to overcome the losses. The lasing condition to be satisfied can be written as,

If the reflectance of reflecting mirrors at the ends of the cavity is R, the laser's oscillation condition is for the net gain be greater than $\frac{1}{R}$, therefore, the threshold condition for oscillation will be,

$$I_p/I_0 = e^{(g-\alpha)2L} = \frac{1}{R}. \tag{3.80}$$

In a general case, the reflectance's of the two mirrors are different. If we note them R_1 and R_2 respectively the threshold oscillation condition becomes,

$$e^{(g-\alpha)2L} = \frac{1}{\sqrt{R_1 R_2}}. \tag{3.81}$$

The right-hand side of the equation represents the maximum gain and can be written as,

$$G_{max} = (R_1 R_2)^{-\frac{1}{2}}, \tag{3.82}$$

where $g(z)$ is the gain per unit length in the section of length $d(z)$.

The gain of the fiber laser can be found by considering the rate equation for the rare-earth ion population density distribution on the different energy level of the doped fiber as well as the propagation equations of the pump and laser field along the fiber length. For the sake of simplification, we are going to consider a two-level system and neglect the effects of amplified spontaneous emission (Barnes and Walsh 1999; Yılmaz et al. 2015). If the fiber laser is pumped at a pump power P_p the equations for the two-level system can be written as,

$$\frac{dN_1}{dt} = -\frac{dN_2}{dt} - N_1 \left(\frac{P_p(z)\sigma_a(\lambda_p)}{Ah\nu_p} + \frac{P_s(z)\sigma_a(\lambda_s)}{Ah\nu_s} \right)$$

$$+ N_2 \left(\frac{1}{\tau} + \frac{P_s(z)\sigma_e(\lambda_s)}{Ah\nu_s} + \frac{P_p(z)\sigma_e(\lambda_p)}{Ah\nu_p} \right), \qquad (3.83)$$

where, N_1, and N_2 are the population density of energy levels 1 and 2 respectively, P_s the signal power, σ_a and σ_e the absorption and emission cross-section respectively, A the fiber section, τ the lifetime of the metastable level and h the planck constant. These equations are completed with the conservation equation giving the total number of ions as,

$$N_t = N_1 + N_2. \qquad (3.84)$$

Under continuous-wave operation, all-time derivatives vanish; equations (3.83) and (3.84) form a system of an algebraic equation. Solving this system of equation for N_2 yields:

$$\frac{N_2(z)}{N_t} = \frac{\dfrac{P_p(z)\sigma_a(\lambda_p)}{h\nu_p A} + \dfrac{P_s(z)\sigma_a(\lambda_s)}{h\nu_s A}}{\dfrac{P_p(z)\left(\sigma_a(\lambda_p) + \sigma_e(\lambda_p)\right)}{h\nu_p A} + \dfrac{1}{\tau} + \dfrac{P_p(z)(\sigma_a(\lambda_s) + \sigma_e(\lambda_s))}{h\nu_p A}} \qquad (3.85)$$

These are the population densities at energy level 1 and 2 of our simplified system.

Inside the laser gain medium, optical field at pump and laser wavelengths propagate in the longitudinal z-direction. Therefore, for each of these fields, we can write differential equations governing the spatial variation of optical power. These equations are called propagation equations and can be written as,

$$\frac{dP_p(z)}{dz} = [(\sigma_a(\lambda_p) + \sigma_e(\lambda_p))N_2(z) - N\sigma_a(\lambda_p)]P_p(z) - \alpha_p P_p(z) \qquad (3.86)$$

Similarly, the propagation equation for the laser optical field can be written as,

$$\frac{dP_s(z)}{dz} = [(\sigma_a(\lambda_s) + \sigma_e(\lambda_s))N_2(z) - N\sigma_a(\lambda_s)]P_s(z) - \alpha_s P_s(z) \qquad (3.87)$$

Equations (3.86) and (3.87) are coupled differential equations that can be solved with appropriate boundary conditions given for a linear cavity by:

$$P_p(z = 0) = P_p$$
$$P_s^+(z = 0) = P_s^-(z = 0)R_1 \qquad (3.88)$$
$$P_s^-(z = L) = P_s^+(z = L)R_2$$

where L is the length of the cavity, R_1 and R_2 are the reflectivity of the mirrors and the superscripts $+$ and $^-$ represent the forward and backwards propagation respectively. For a ring cavity fiber laser, these boundary conditions will reduce to:

$$P_p(z = 0) = P_p$$
$$P_s(0) = RP_s(L) \qquad (3.89)$$

where R is the coupling ratio of the output coupler. Analytical solution of these equations is often cumbersome. Using some simplifications several authors proposed what is known as quasi analytical solutions (Elahi and Zare 2009; Kazemizadeh et al. 2016; Yahel et al. 2003). The weakness of the analytical solutions is that they cannot be used to solve laser equations when simplifications are no longer taken into account, for example, for more than two levels or structures like co-doped gain mediums. In such an instance, one often resort to numerical solutions, which will be described in more details in the next chapter of this book. The solution of the laser equations are P_p and P_s as a function of z. These functions are very important in determining characteristics of fiber laser-like optimum length, threshold power, slope efficiency, etc.

The solutions to equations (3.86) and (3.87) are functions describing the propagations of pump and laser power. These functions can be used to find critical parameters of the laser such as output power, threshold, optimum fiber length, slope efficiency, etc. In the following chapters of the book, we will see in detail how these characteristics can be obtained for a specific type of fiber laser.

3.9.2 RESONATORS: LINEAR AND RING CAVITY

The resonant cavity of a laser provides the feedback necessary to sustain the oscillation. It acts as a filter that selects only one or several almost monochromatic components in the stimulated emission spectrum. The most used resonators for fiber laser are the Fabry-Perot Cavity and the ring cavity described in the following section.

3.9.2.1 Fabry-Perot Cavity

A Fabry-Perot cavity illustrated in Figure 3.9 is a type of resonator formed by using two parallel mirrors. In this type of resonators, an optical wave enters the cavity and makes an infinite number of reflections interfering with itself. For wavelengths with constructive interference, a maximum intensity develops inside the cavity. Whereas for wavelength with destructive interference, the wave vanishes after several reflections, therefore, no energy at these wavelengths can be stored inside the cavity.

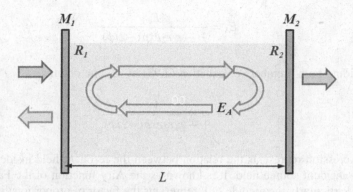

FIGURE 3.9 Schematic diagram of a Fabry-Perot cavity.

At any point in the cavity, there exists an electrical field E_A, with a frequency ν as indicated in Figure 3.9. At any time, this field will be the superimposition of the initial incident field E_s and the fields having successfully travelled 1, 2, 3..., n round trips in the cavity. Es can be an outside field in the case of a Fabry-Perot etalon or the spontaneous emission created by the medium itself in the case of a laser. For each round trip, the field Es undergo a phase difference 2ϕ which is the result of the propagation on a round trip,

$$\phi = \frac{2\pi n l}{\lambda} = \frac{2\pi n l \nu}{c}, \tag{3.90}$$

where l is the total cavity length, λ the wavelength of the optical field, ν the frequency, n the refractive index of the medium and c the speed of light. In addition to the 2ϕ phase difference, the field also undergoes a reduction of amplitude because of losses at the surface of the mirrors. The field, therefore, can be described by the following equation,

$$E_A = E_s + r_1 r_2 exp(-2i\phi)E_s + (r_1 r_2)^2 exp(-4i\phi)E_s + K$$
$$+ (r_1 r_2)^n exp(-2ni\phi)E_s. \tag{3.91}$$

This equation can be written in the more compact form as:

$$E_A = E_s(1 + r_1 r_2 exp(-2i\phi) + [r_1 r_2 exp(-2i\phi)]^2 + K$$
$$+ [r_1 r_2 exp(-2i\phi)]^n). \tag{3.92}$$

A pattern of a geometric sequence can be seen in the equation with the term $r_1 r_2 exp(-2i\phi)$ as the common ratio and the modulus is less than unity. The limit of expression (3.92) when n tends to infinity is given by:

$$E_A = \frac{E_s}{1 - r_1 r_2 \exp(-2i\phi)}. \tag{3.93}$$

By dividing both members of equation (3.93) by E_s one obtains,

$$F = \frac{1}{1 - r_1 r_2 \exp(-2i\phi)}. \tag{3.94}$$

This expression represents the relation between the resonant field inside the cavity and the incident source field. It is known as the Airy function of the Fabry-Perrot cavity. Particularly the module of F represents the factor of proportionality between the amplitudes of cavity field and incident source field. This factor is a function of the frequency ν of the electric field through the phase change, which is a function of frequency as seen in equation (3.90). The dependency of $|F|$ as a function of frequency ν is illustrated in Figure 3.10, where the peaks correspond to the frequencies for which the optical field is maximum. The frequencies correspond to those undergoing constructive interference which takes place when the phase shift at each end of the round trip is an integer multiple of 2π. From equation (3.90), we can write,

$$\nu_k = k\,(c/2nl). \tag{3.95}$$

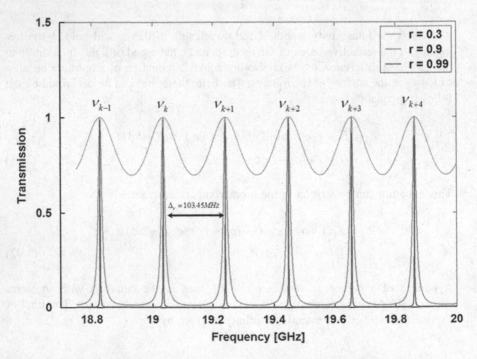

FIGURE 3.10   Irradiance as a function of wavelength (frequency) in the Fabry-Perot cavity.

In other words, this means that the optical length of the cavity must be an integer multiple of half wavelengths. The ν_k frequencies that are allowed to oscillate in the cavity are known as longitudinal modes of the cavity. The gap of frequencies between two successive longitudinal modes is given by:

$$\Delta \nu = \nu_{k+1} - \nu_k = c/2nl, \qquad (3.96)$$

where $\Delta \nu$ is called the Free Spectral Range (FSR) of the cavity.

The role of the cavity in a linear Fabry-Perot laser is crucial because it limits the number of longitudinal modes, which can oscillate. Taking into account the gain profile of the laser-amplifying medium, it is easy to see that two conditions need to be satisfied in laser action. First, the longitudinal mode must correspond to the ones allowed by the Fabry-Perot cavity, and secondly, these modes must fall under the gain profile of the amplifying medium as it can be seen in Figure 3.11. By increasing the space between the longitudinal modes, only one mode can fall under the gain profile of the medium. If that happens, a single longitudinal mode laser can be obtained. Single longitudinal mode fiber lasers will be treated in more details in Chapter 7.

There is another type of resonant cavity, which is particularly advantageous for the fiber laser. This cavity is the ring cavity presented in the following section.

3.9.2.2 Ring Cavity

One of the worst drawbacks of the linear cavity resonator such as the Fabry-Perot, is the presence of hole-burning due to development of standing wave patterns in the cavity. Eliminating standing waves can be made possible by making the optical field travel in only one direction by using an optical isolator inside a ring cavity. Similar to the case of the linear Fabry-Perot cavity, the E_A field is a superimposition of multiple electrical fields having travelled in the cavity several times. The ring

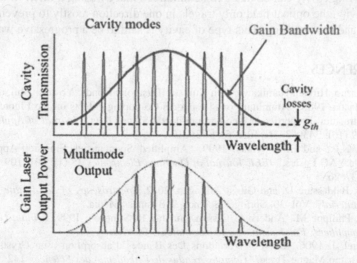

FIGURE 3.11   Gain profile and longitudinal modes.

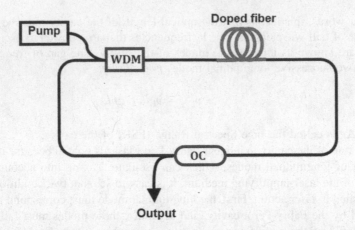

FIGURE 3.12   Ring cavity: bi-directional and unidirectional.

cavity does not have mirrors and light is coupled out of the laser by using an optical coupler as shown in Figure 3.12. Here, the loss of the cavity is given by the coupling coefficient (K) of the coupler. The mathematical treatment is similar to that of a Fabry-Perot cavity and F is also an Airy function. The Free Spectral Range (FSR) in the case of a ring cavity is given by,

$$\Delta \nu = \nu_{k+1} - \nu_k = c/nl. \tag{3.97}$$

The relation shows clearly that the FSR of a ring cavity is twice as large as for a Fabry-Perot cavity. This is an advantage as the space between longitudinal modes is larger and fewer modes can oscillate. Therefore, it is much easier to build a single longitudinal mode fiber laser with a ring cavity than with linear Fabry-Perot cavity.

The ring cavity can be made to be unidirectional by using a filter. In this type of ring cavity, the optical field only travels in one direction mostly to prevent standing waves and hole-burning. This type of cavity is said to be a progressive wave cavity.

REFERENCES

Arai, Kazuo, Hiroshi Namikawa, Ken Kumata, Tatsutoku Honda, Yoshiro Ishii, and Takashi Handa. 1986. "Aluminum or Phosphorus Co-doping Effects on the Fluorescence and Structural Properties of Neodymium-doped Silica Glass." *Journal of Applied Physics* 59 (10): 3430–3436. 10.1063/1.336810.

Barnes, N.P., and B.M. Walsh. 1999. "Amplified Spontaneous Emission-Application to Nd:YAG Lasers." *IEEE Journal of Quantum Electronics* 35 (1): 101–109. 10.1109/3.737626.

Bartolo, Baldassare Di and Guzin Armagan. 2012. *Spectroscopy of Solid-State Laser-Type Materials*. Vol. 30. Springer Science & Business Media.

Becker, Philippe M., Anders A. Olsson, and Jay R. Simpson. 1999. *Erbium-Doped Fiber Amplifiers: Fundamentals and Technology*. Elsevier.

Becquerel, J. 1906. "Sur Les Variations Des Bandes d'absorption d'un Crystal Dans Un Champ Magne Tique." *Comptes rendus de l'Académie des Sciences* 142: 775–779.

Becquerel, Jean. 1907. "Influence Des Variations de Température Sur l'absorption Dans Les Corps Solides."Radium (Paris), 1907, 4 (9), pp.328-339

Bigot, Laurent, Stephan Guy, Bernard Jacquier, Anne-marie Jurdyc, Dominique Bayart, Stéphanie Blanchandin, and Laurent Gasca. 2AD. "Homogeneous and Inhomogeneous Line Broadening in EDFA," Optical Amplifiers and Their Applications, May 2001, paper OTuB1.

Broer, L.J.F., C.J. Gorter, and J. Hoogschagen. 1945. "On the Intensities and the Multipole Character in the Spectra of the Rare-Earth Ions." *Physica* 11 (4): 231–250.

Desurvire, E., J.L. Zyskind, and J.R. Simpson. 1990. "Spectral Gain Hole-Burning at 1.53 Mu m in Erbium-Doped Fiber Amplifiers." *IEEE Photonics Technology Letters* 2 (4): 246–248. 10.1109/68.53251.

Elahi, Parviz and N. Zare. 2009. "The Analytical Solution of Rate Equations in End Pumped Fiber Lasers with Minimum Approximation and Temperature Distribution during the Laser Operation." *Acta Physica Polonica A - ACTA PHYS POL A* 116 (October). 10.12 693/APhysPolA.116.522.

Fournier, J.T. and R.H. Bartram. 1970. "Inhomogeneous Broadening of the Optical Spectra of Yb3+ in Phosphate Glass." *Journal of Physics and Chemistry of Solids* 31 (12): 2615–2624. 10.1016/0022-3697(70)90256-8.

Fowler, W. Beall and D.L. Dexter. 1962. "Relation between Absorption and Emission Probabilities in Luminescent Centers in Ionic Solids." *Physical Review* 128 (5): 2154–2165. 10.1103/PhysRev.128.2154.

Funabashi, Koichi and John L. Magee. 1957. "Central-Field Approximation for the Electronic Wave Functions of Simple Molecules." *The Journal of Chemical Physics* 26 (2): 407–411.

Griffiths, David J. and Darrell F. Schroeter. 2018. *Introduction to Quantum Mechanics*. Cambridge University Press.

Hartree, Douglas R. 1928. "The Wave Mechanics of an Atom with a Non-Coulomb Central Field. Part I. Theory and Methods." In *Mathematical Proceedings of the Cambridge Philosophical Society*, 24:89–110. Cambridge University Press.

Hartree, Douglas Rayne. 1928. "The Wave Mechanics of an Atom with a Non-Coulomb Central Field. Part II. Some Results and Discussion." In *Mathematical Proceedings of the Cambridge Philosophical Society*, 24:111–132. Cambridge University Press.

Huang, Feifei, Xueqiang Liu, Lili Hu, and Danping Chen. 2014. "Spectroscopic Properties and Energy Transfer Parameters of Er3+-Doped Fluorozirconate and Oxyfluoroaluminate Glasses." *Scientific Reports* 4 (May): 5053. 10.1038/srep05053.

Hyde, Kenneth E. 1975. "Methods for Obtaining Russell-Saunders Term Symbols from Electronic Configurations." *Journal of Chemical Education* 52 (2): 87.

Jammer, Max. 1966. *The Conceptual Development of Quantum Mechanics*. New York: McGraw-Hill.

Jørgensen, Chr Klixbüll and B.R. Judd. 1964. "Hypersensitive Pseudoquadrupole Transitions in Lanthanides." *Molecular Physics* 8 (3): 281–290.

Judd, Brian R. 1962. "Optical Absorption Intensities of Rare-Earth Ions." *Physical Review* 127 (3): 750.

Kazemizadeh, Fatemeh, Rasoul Malekfar, and F. Shahshahani. 2016. "An Analytical Model for Rare-Earth Doped Fiber Lasers Consisting of High Reflectivity Mirrors." *International Journal of Optics and Photonics* 10 (November): 101–110. 10.18869/acadpub.ijop.10.2.101.

Kenyon, Anthony. 2002. "Recent Developments in Rare-Earth Doped Materials for Optoelectronics Progress." *Progress in Quantum Electronics* 26 (December): 225–284. 10.1016/S0079-6727(02)00014-9.

Koechner, Walter. 2006. *Solid-State Laser Engineering*. 10.1007/0-387-29338-8.

Landau, L.D., and E.M. Lifshitz. 1980. "Chapter I - The Fundamental Principles of Statistical Physics." Vol 9. Pergamon Press Inc, New York, USA

Laporte, Otto. 1924. "Die Struktur Des Eisenspektrums." *Zeitschrift Für Physik* 23 (1): 135–175.

MacFarlane, R.M. and R.M. Shelby. 1987. "Homogeneous Line Broadening of Optical Transitions of Ions and Molecules in Glasses." *Journal of Luminescence* 36 (4–5): 179–207. 10.1016/0022-2313(87)90194-3.

Malta, Oscar L. and Luís D. Carlos. 2003. "Intensities of 4f-4f Transitions in Glass Materials." *Quimica Nova* 26 (6): 889–895. 10.1590/S0100-40422003000600018.

Marks, Tobin J. and R. Dieter Fischer. 1979. *Organometallics of the F-Elements*. Springer.

Mason, Stephen F., Robert D. Peacock, and Brian Stewart. 1975. "Ligand-Polarization Contributions to the Intensity of Hypersensitive Trivalent Lanthanide Transitions." *Molecular Physics* 30 (6): 1829–1841.

Mayer, M. Goeppert. 1941. "Rare-Earth and Transuranic Elements." *Physical Review* 60 (3): 184–187. 10.1103/PhysRev.60.184.

McCumber, D.E. 1964. "Einstein Relations Connecting Broadband Emission and Absorption Spectra." *Physical Review* 136 (4A): A954–A957. 10.1103/PhysRev.136.A954.

Moos, H. and L. Riseberg. 1968. "Multiphonon Orbit-Lattice Relaxation of Excited States of Rare-Earth Ions in Crystal." *Physical Review* 174 (November). 10.1103/PhysRev.174.429.

Myslinski, P., D. Nguyen, and J. Chrostowski. 1997. "Effects of Concentration on the Performance of Erbium-Doped Fiber Amplifiers." *Journal of Lightwave Technology* 15 (1): 112–120. 10.1109/50.552118.

Nazeeruddin, Md K., S.M. Zakeeruddin, and K. Kalyanasundaram. 1993. "Enhanced Intensities of the Ligand-to-Metal Charge-Transfer Transitions in Ruthenium (III) and Osmium (III) Complexes of Substituted Bipyridines." *The Journal of Physical Chemistry* 97 (38): 9607–9612.

Nielson, C.W., and George F. Koster. 1963. *Spectroscopic Coefficients for the Pn, Dn, and Fn Configurations*. MIT press.

Nilsson, J., S.- Alam, J.A. Alvarez-Chavez, P.W. Turner, W.A. Clarkson, and A.B. Grudinin. 2003. "High-Power and Tunable Operation of Erbium-Ytterbium Co-Doped Cladding-Pumped Fiber Lasers." *IEEE Journal of Quantum Electronics* 39 (8): 987–994. 10.11 09/JQE.2003.814373.

Ofelt, G.S. 1962. "Intensities of Crystal Spectra of Rare-Earth Ions." *The Journal of Chemical Physics* 37 (3): 511–520. 10.1063/1.1701366.

Peacock, R.D. 1975. "Structure and Bonding." *Berlin* 22: 83.

Peijzel, P S.s., A. Meijerink, R.T. Wegh, M.F. Reid, and Gary W. Burdick. 2005. "A Complete 4fn Energy Level Diagram for All Trivalent Lanthanide Ions." *Journal of Solid State Chemistry* 178 (2): 448–453.

Pellegrino, J.M., W.M. Yen, and M.J. Weber. 1980. "Composition Dependence of Nd3+ Homogeneous Linewidths in Glasses." *Journal of Applied Physics* 51 (12): 6332–6336. 10.1063/1.327621.

Pisarska, J., M. Sołtys, L. Żur, W.A. Pisarski, and C.K. Jayasankar. 2014. "Excitation and Luminescence of Rare-Earth-Doped Lead Phosphate Glasses." *Applied Physics B* 116 (4): 837–845. 10.1007/s00340-014-5770-9.

Reisfeld, R., G. Katz, C. Jacoboni, R. De Pape, M.G. Drexhage, R.N. Brown, and C.K. Jørgensen. 1983. "The Comparison of Calculated Transition Probabilities with Luminescence Characteristics of Erbium(III) in Fluoride Glasses and in the Mixed Yttrium-Zirconium Oxide Crystal." *Journal of Solid State Chemistry* 48 (3): 323–332. 10.1016/0022-4596(83)90089-0.

Roy, Fabien, Dominique Bayart, Céline Heerdt, André Le Sauze, and Pascal Baniel. 2002. "Spectral Hole Burning Measurement Thulium-Doped Fiber Amplifiers." *Optics Letters* 27 (1): 10–12. 10.1364/OL.27.000010.

Sobon, Grzegorz, Pawel Kaczmarek, and Krzysztof M Abramski. 2012. "Erbium–Ytterbium Co-Doped Fiber Amplifier Operating at 1550nm with Stimulated Lasing at 1064nm." *Optics Communications* 285 (7): 1929–1933. 10.1016/j.optcom.2011.12.080.

Sontakke, Atul D. and Annapurna Kalyandurg. 2012. "Energy Transfer Kinetics in Oxy-Fluoride Glass and Glass-Ceramics Doped with Rare-Earth Ions." *Journal of Applied Physics* 112 (July). 10.1063/1.4731732.

Sun, Yihao, Rufus Cone, Laurent Bigot, and B. Jacquier. 2007. "Exceptionally Narrow Homogeneous Linewidth in Erbium-Doped Glasses." *Optics Letters* 31 (January): 3453–3455. 10.1364/OL.31.003453.

Thiel, C.W., Thomas Böttger, and R.L. Cone. 2011. "Rare-Earth-Doped Materials for Applications in Quantum Information Storage and Signal Processing." *Journal of Luminescence* 131 (3): 353–361. 10.1016/j.jlumin.2010.12.015.

Vleck, J H van. 1937. "The Puzzle of Rare-Earth Spectra in Solids." *Journal of Physical Chemistry* 41 (1): 67–80.

Walsh, Brian M. 2006. "Judd-Ofelt Theory: Principles and Practices." In *Advances in Spectroscopy for Lasers and Sensing*, 403–433. Springer.

Wang, Xianfen, Guandong Liu, Tatsumi Iwao, Masashi Okubo, and Atsuo Yamada. 2014. "Role of Ligand-to-Metal Charge Transfer in O3-Type NaFeO2–NaNiO2 Solid Solution for Enhanced Electrochemical Properties." *The Journal of Physical Chemistry C* 118 (6): 2970–2976.

Wegh, René T., Andries Meijerink, Ralf-Johan Lamminmäki, and Jorma Hölsä. 2000. "Extending Dieke's Diagram." *Journal of Luminescence* 87: 1002–1004.

Wei, Tao, Ying Tian, Fangze Chen, Muzhi Cai, Junjie Zhang, Xufeng Jing, Fengchao Wang, Qinyuan Zhang, and Shiqing Xu. 2014. "Mid-Infrared Fluorescence, Energy Transfer Process and Rate Equation Analysis in Er3+ Doped Germanate Glass." *Scientific Reports* 4 (1): 6060. 10.1038/srep06060.

Weinberg, Steven and Steven Weinberg. 2015. "The Quantum Theory of Radiation." *Lectures on Quantum Mechanics* 1 (4): 361–391. 10.1017/cbo9781316276105.013.

Winkler, R., S. Papadakis, E. De Poortere, and M. Shayegan. 2003. *Spin-Orbit Coupling in Two-Dimensional Electron and Hole Systems*. Vol. 41. Springer.

Wu, Ruikun, John D. Myers, Michael Myers, and Charles F. Rapp. 2003. "Fluorescence Lifetime and 980nm Pump Energy Transfer Dynamics in Erbium and Ytterbium Co-Doped Phosphate Laser Glasses," June. 10.1117/12.478261.

Yahel, Eldad, Amos Hardy, and Abstract High-power Er. 2003. "Modeling High-Power Er3+ – Yb3+ Codoped Fiber Lasers." *Lightwave* 21 (9): 2044–2052.

Yılmaz, Saim, Parviz Elahi, Hamit Kalaycioglu, and Fatih Ilday. 2015. "Amplified Spontaneous Emission in High-Power Burst-Mode Fiber Lasers." *Journal of the Optical Society of America B* 32 (December): 2462. 10.1364/JOSAB.32.002462.

4 Mathematical Methods for Fiber Lasers

4.1 INTRODUCTION

Performance analysis and behaviour prediction of fiber lasers depend on several parameters. Therefore, developing fiber lasers relying on experimental work alone is inefficient. Such an approach often leads to tedious, costly, and time-consuming development methods. To alleviate such constraints an accurate model of rare-earth-doped lasers is desirable. Working with a numerical model is advantageous because it provides an understanding of the system under investigation and allows one to conduct a wide range of experiments to optimize the design of the device and foresee its limitations. Using a good model helps to predict the characteristics of the fiber laser without conducting experiments, which can be difficult to carry out in laboratory conditions. Experimenting is fundamentally a trial-and-error method and plagued by all drawbacks of such methods namely, waste of materials and dangerous test (material can be destroyed if not handled carefully).

The simulation of the fiber laser requires the solution of two groups of equations namely the rate equations of the active medium and the propagation equation for the optical field inside the cavity. Rate equations represent the multiple interactions between the propagating optical field and the rare-earth ions as well as ion-ion interactions (Strohhöfer and Polman 2001). These include absorption, spontaneous emission, stimulated emission, and nonradiative relaxation of excited ions through multi-phonon transitions. Ion-ion interaction includes cooperative up-conversion and energy transfer between two species of ions in the case of the co-doped gain medium, the most common being Erbium-Ytterbium. It is also possible to observe excited state absorption both at pump and laser wavelength. Depending on the level of accuracy one wants to achieve, some of the previous processes could be included or not in the final system of equations. On the other hand, the complexity of the resulting system must be taken into account. Ion-ions interactions, for example, will add nonlinear terms to the equations increasing the level of difficulty in finding the solution.

The propagation equations are also a system of differential equations describing the variation of the pump, laser, and amplified spontaneous emission fields as a function of position across the gain medium. Depending on the type of laser, these equations can be ordinary differential or partial differential equations of time and position.

Finally, to solve the equations, appropriate boundary conditions must be used. The boundary conditions depend on the working regime (continuous or pulsed) of the fiber laser.

DOI: 10.1201/9781003256380-4

The resulting systems of equations are complex and solutions whether analytical or numerical can quickly become impossible to find, therefore, approximations and simplifications are often inevitable. Neglecting some transitions and effects regarded as less important to the resulting characteristics of the fiber laser leads to systems of equations reported to yield results close to those obtained with experiments (Yahel and Hardy 2003; Yahel et al. 2003).

The other difficulty in fiber laser modelling is related to obtaining values of the parameters and coefficients in the equations. Parameters are often obtained indirectly through measurement of quantities like absorption coefficient (Barnard et al. 1994) which may result from the contribution of several processes. Under such circumstances, it is obvious that there are difficult to measure, and their values can only be obtained through assumptions and data fitting.

Any good simulation must work on a great number of situations to allow one to compare different designs to attain maximum efficiency. Finally, the computational time must be considered. Accurate simulations often take longer computational time whereas a simple model can be computed in fractions of seconds, but do not represent accurately the system under investigation. Therefore, the level of simplification must be such that a balance is found between accuracy and computation time.

There exists powerful commercial software to model and simulate fiber lasers. Nevertheless, a personal modelling method gives one flexibility to extend simulations in particular cases that is not always taken into account by more general commercial software and the insight on the internal working principle of the device is strongly improved when one design his tool. The process of designing the tool allows one to have a deep understanding of the internal mechanism of the device.

4.2   RATE EQUATIONS FOR THE GAIN MEDIUM

The rate equations for doped media are a system of ordinary differential equations derived from interactions between optical fields and atoms of the gain rare-earth-doped medium. These equations express the variation of population densities of rare-earth ions in different energy levels (Nguyen et al. 2007; Bao and Son 2007; Han et al. 2015; Wang and Po 2003). The variations of population densities can be easily quantified by their rates of variations which are functions of many parameters. These parameters can be grouped into two categories, namely the "fiber parameters" and the spectroscopic data. Under the fiber, parameters are its length, core diameter, its numerical aperture, and attenuation coefficient. The majority of these parameters can be found in the datasheet provided by the fiber manufacturer. The spectroscopic data, on the other hand, includes rare-earth ions concentration, emission and absorption cross-sections, up-conversion coefficients, energy-transfer coefficients, and excited-state absorption (ESA) coefficients. These parameters are most of the time obtained by measurement performed by the designer or obtained from published data. The complexity of the rate equations depends on the number of processes considered. For example, a model which takes into account cooperative up-conversion will result in a more complex system of rate equations because this process introduces nonlinearity in the equations (Hakeim and I. 2011; Valery Rudyak and Minakov 2009; Yu et al. 2008; Sorbello et al. 2001; "Theoretical Modeling of Er Doped Fiber Amplifiers with

Excited State Absorption.Pdf," n.d.). The above implies that a model of the fiber laser including all possible processes will result in a complex system of differential equations that can quickly become intractable. Therefore, assumptions and simplifications are inevitable. This is the subject of the next section of this chapter.

4.2.1 TWO ENERGY LEVELS SYSTEMS

As said in the previous section, simplifications may be necessary for modelling the fiber gain medium of fiber lasers. One of the simplification schemes is the two-energy level formalism. This simplification is valid for 3 energy levels systems with a short lifetime in the uppermost energy level such as the $^{4}I_{11/2}$ manifold of Erbium (Träger 2012). In such a case, all the excited ions are considered to repopulate the metastable level with a much longer lifetime. Also, because there are no other energy levels above level 2, cooperative up-conversion and excited-state absorption factors will not appear in the equations. The resulting system of equations (4.1) and (4.2) is simple and its solution can be found easily using well-established algorithms like 4th order Runge-Kutta methods (Esfandiari 2017). In the continuous-wave regime, the time derivatives become zero and the system reduces to a system of algebraic equations which can be solved analytically by simple substitutions. The solution of the equations provides the distribution of the rare-earth ions population at different energy levels for given pump and laser powers (Thorlabs 2018; "Theoretical Modeling of Er Doped Fiber Amplifiers with Excited State Absorption.Pdf," n.d.). The system of rate equations for a two-level system is written as:

$$\frac{dN_1(z, t)}{dt} = -\left[W_{sa}(z) + W_p(z) \right] N_1(z, t) + \left[W_{se}(z) + \frac{1}{\tau_{21}} \right] N_2(z, t) \quad (4.1)$$

$$\frac{dN_2(z, t)}{dt} = -\left[W_{se}(z) + \frac{1}{\tau_{21}} \right] N_2(z, t) + [W_{sa}(z) + W_p(z)]N_1(z, t) \quad (4.2)$$

where τ_{21} is the lifetime of the metastable level. W_{sa}, and W_p are the stimulated absorption, stimulated emission, and pumping rates, respectively,

$$W_{sa}(z) = \frac{1}{A_{eff}} \int_0^\infty \frac{\sigma_{12}(\nu)}{h\nu} [P_{ASE}^+(z, \nu) + P_{ASE}^-(z, \nu)]\Gamma(\nu)d\nu \quad (4.3)$$

$$W_{se}(z) = \sigma_{21}(\nu_p) \frac{P_s^+(z) + P_s^-(z)}{h\nu_p A_{eff}}\Gamma_s + \frac{1}{A_{eff}} \int_0^\infty \frac{\sigma_{21}(\nu)}{h\nu} [P_{ASE}^+(z, \nu) + P_{ASE}^-(z, \nu)]\Gamma(\nu)d\nu \quad (4.4)$$

$$W_p(z) = \sigma_{12}(\nu_p) \frac{P_p^+(z) + P_p^-(z)}{h\nu_p A_{eff}}\Gamma_p \quad (4.5)$$

In equations (4.1)–(4.5), σ_{ij} represents the emission or absorption cross-sections for the $i \rightarrow j$ transition, h is Planck's constant. The pump and amplified spontaneous emission powers are represented by $P_p^\pm(z)$ and $P_{ASE}^\pm(z, \nu)$ respectively. It can also easily be seen that there is a dependency on amplified spontaneous emission on frequency. Such a dependency does not exist for pump power because pumping is done at a known specific frequency. To include this fact in numerical modelling, amplified spontaneous emission, (ASE) distribution is often represented in slots of frequencies of width $\delta\nu$. Generally, 30 intervals in the ASE spectra are accepted in most simulations but when higher precision is required, the intervals may be increased to 60 or more. Increasing intervals are done at the expense of computation time and must only be considered when high accuracy is required. A trade-off must always be found between the level of precision required and the computation time.

A_{eff} is the effective doped fiber core area and $\Gamma(\nu)$ the overlap integral between the intensity distribution of the fundamental linearly polarized mode (LP$_{01}$) and the core rare-earth doping density transverse distribution. The effective doped fiber core area is expressed as (Barnard et al. 1994):

$$A_{eff} = \int \frac{N_t(r, \phi, z)dA}{N_t(0)} \tag{4.6}$$

where $N_t(0)$ is the total rare-earth concentration at radius $r = 0$, and $N_t(r, \phi, z)$ the concentration of rare-earth ions inside the core, given in cylindrical coordinates. The overlap integral or confinement factor can be expressed as (Giles and Desurvire 1991):

$$\Gamma_{k,i}(z) = \frac{\int_0^{2\pi} \int_0^\infty i_k(r, \phi)N_i(r, \phi, z)rdrd\phi}{\bar{N}_i} \tag{4.7}$$

where the indices k and i represent the kth signal (a wavelength dependency) and the energy levels respectively. $\bar{N}_i$ is the averaged population density at level i given by (Giles and Desurvire 1991):

$$\bar{N}_i(z) = \frac{\int_0^{2\pi} \int_0^\infty N_i(r, \phi, z)rdrd\phi}{\pi b_{eff}^2} \tag{4.8}$$

where b_{eff} is the effective radius given by (Giles and Desurvire 1991):

$$b_{eff} = \left[\frac{1}{2} \int_0^\infty \frac{N_i(r)}{N_i(0)} rdrd\phi \right]^{1/2} \tag{4.9}$$

For the fundamental mode approximated by a Gaussian profile, the overlap factor is given by

$$\Gamma_t = 1 - e^{-2b^2/w^2} \tag{4.10}$$

where w is the fundamental mode radius. In the Gaussian profile approximation, Marcuse proposed an empirical relation to calculate the mode radius as (Marcuse 1977):

$$w = a\left(0.65 + \frac{1.65}{V^{3/2}} + \frac{2.879}{V^6}\right) \tag{4.11}$$

where a is the fiber's core radius and V the normalized frequency (V number of the fiber). It is worth mentioning that the Marcuse relation (4.11) holds only for $1.4 < V < 3$. Marcuse's approximation is not unique, other empirical approximations for mode radius have also been proposed namely by Mylinski (Myslinski and Chrostowski 1996) or Desurvire (Giles and Desurvire 1991).

4.2.2 Systems with More than 3 Energy Levels

The simplified two levels system described above fails to include processes that become important in some cases. For example, in short cavity Erbium-doped fiber lasers, the concentration of rare-earth ions must be high. Therefore, the effect of ion-ion interaction becomes important and cannot be neglected in an accurate model. Other cases of more than three energy levels include thulium and Neodymium ions which cannot be described as a two levels system. Also because of up-conversion and energy transfer, systems with three energy levels or more often result in nonlinear systems of equation. Analytical solutions of such systems of equations are difficult to find and one rely often on numerical solutions. The system of equations (4.12)–(4.17) represents rate equations for an Erbium-Ytterbium co-doped gain medium. In the equation, the squared variable represents up-conversion, and terms representing the energy transfer can also be easily seen.

$$\frac{dN_2(z, t)}{dt} = W_{12}N_1 - W_{21}N_2 - W_{23}N_2 - A_{21}N_2 + A_{32}N_3 - 2C_{22}N_2^2 \tag{4.12}$$

$$\frac{dN_3(z, t)}{dt} = W_{13}N_1 - W_{31}N_3 - W_{23}N_2 - W_{34}N_3 - A_{32}N_3 - A_{31}N_3$$
$$+ A_{43}N_4 + C_{22}N_2^2 - 2C_{33}N_3^2 + K_{61}N_1N_6 - K_{35}N_3N_5 \tag{4.13}$$

$$\frac{dN_6(z, t)}{dt} = W_{56}N_5 - W_{65}N_6 - A_{65}N_6 + K_{35}N_3N_5 - K_{61}N_1N_6 \tag{4.14}$$

$$\frac{dN_4(z, t)}{dt} = W_{34}N_3 - A_{41}N_4 - A_{43}N_4 + C_{33}N_2^2 \tag{4.15}$$

$$N_{er} = N_1 + N_2 + N_3 + N_4 \tag{4.16}$$

TABLE 4.1
Coefficients Used in Equations (4.12)–(4.17)

Symbol	Parameter
A_{21}	Spontaneous emission rate of Er^{3+} from energy level 2 (reverse of lifetime)
A_{32}	Nonradiative decay rate of Er^{3+} from energy level 3 to level 2
A_{31}	Spontaneous emission rate of Er^{3+} from energy level 3
A_{41}	Spontaneous emission rate of Er^{3+} from energy level 1 (green light)
A_{43}	Nonradiative decay rate of Er^{3+} from energy level 4 to level 3
A_{65}	Spontaneous emission rate from Yb^{3+} level 6
C_{22}	Cooperative up-conversion coefficient of Er^{3+} ions at energy level 2
C_{33}	Cooperative up-conversion coefficient of Er^{3+} ions at energy level 3
K_{35}	Energy transfer coefficient from Er^{3+} to Yb^{3+}
K_{61}	Energy transfer coefficient from Yb^{3+} to Er^{3+}

$$N_{Yb} = N_5 + N_6 \tag{4.17}$$

This system of equations represents the 4 energy levels (N_1 to N_4) for Erbium in addition to the two energy levels of Ytterbium (N_5 and N_6). Various coefficients used in the equations are defined in Table 4.1:

In the system of equation (4.12)–(4.17), W_{ij} represent the rate of stimulated transition between energy levels i and j respectively given by:

$$W_{ij} = \frac{\Gamma \sigma_{ij}(\lambda) P}{A_{eff} h\nu} \tag{4.18}$$

where P is power (pump, signal, or amplified spontaneous emission), σ_{ij} is the absorption or emission cross-section between energy levels i and, h the Planck constant and ν the frequency of the radiation.

Similar to the case of a two levels energy system, in the continuous-wave regime, the time derivatives become zero and the system is reduced to an algebraic system of equations. However, due to square terms resulting from up-conversion, the system is nonlinear and cannot be easily solved by simple substitution. To find the solution in this case numerical root-finding methods Amongst the most popular methods are the Secant method (Press et al. 1992b) and the Newton-Raphson methods (Press et al. 1992a). The Secant method is well suited for small systems of equations. When the system of equations becomes larger, the Secant method either fail to converge or becomes very slow. In such a case using a Newton-Raphson method is usually preferred because it inevitably converges after a few iterations in most of the cases provided that a good initial guess is given.

Any system of algebraic equations can be written in matrix form as:

$$f(x) = \begin{bmatrix} f_1(x) \\ f_2(x) \\ \vdots \\ f_I(x) \end{bmatrix} \tag{4.19}$$

Newton-Raphson method shows that the solution to such a system of equations can be approximated by successive iterations given by:

$$x_{k+1} = x_k - J_k^{-1} f(x_k) \tag{4.20}$$

where J_k is the Jacobian matrix given by:

$$J_k = f'(x) = \begin{bmatrix} \frac{\partial f_1(x)}{\partial x_1} & \frac{\partial f_1(x)}{\partial x_2} & \cdots & \frac{\partial f_1(x)}{\partial x_n} \\ \frac{\partial f_2(x)}{\partial x_1} & \ddots & & \frac{\partial f_2(x)}{\partial x_n} \\ \vdots & & \ddots & \vdots \\ \frac{\partial f_n(x)}{\partial x_1} & \frac{\partial f_n(x)}{\partial x_2} & \cdots & \frac{\partial f_n(x)}{\partial x_n} \end{bmatrix} \tag{4.21}$$

In equation (4.20), k is the iteration counter. The convergence condition is given by:

$$|x_{k-1} - x_k| < \varepsilon \tag{4.22}$$

where ε is the user-defined error value. The convergence condition is checked after each iteration and the algorithm stops when the user-defined error or the maximum number of iteration is reached. As we have already mentioned, the Newton-Raphson method major drawback when solving multidimensional nonlinear equations remain the dependence of convergence on the initial guess. In fact, if the initial guess is not sufficiently close to the root, the algorithm will diverge in the majority of cases. For a system with rate equations of fiber lasers, the initial guess may be extremely difficult to find. Therefore alternative powerful methods like W4 method (Okawa et al. 2018), Muller's methods (Mathews and Fink 2004), homotopy algorithms (Watson 1990), or quasi-Newton methods (Broyden 1969; Johnson 1988; Saheya et al. 2016). Finally, several other algorithms have been reported to solve a specific system of equations (Peterka et al. 2011; Yelen et al. 2005)

4.3 COUPLED PROPAGATION EQUATIONS

We consider the general case of bi-directionally pumped linear cavity fiber laser with pump powers launched at $z = 0$ and $z = L$ and propagating in the $+z$ and

— z directions, respectively. Pumping induces along the fiber a gain factor that depends on z. Photons released by stimulated emission are guided by the fiber and travel in both forward and backward directions and constitute "signal power" P_s. Excited ions can make a spontaneous transition to a lower energy state releasing a photon. A fraction of these spontaneously released photons is guided by the fiber and undergoes amplification as it propagates down the core of the fiber. This parasitic process is known as amplified spontaneous emission. Photons released by spontaneous emission are amplified in both the forward (+z) and backward (-z) directions and generate two ASE wave, one in each direction with powers P_{ASE}^+ and P_{ASE}^- for the forward and backward directions respectively. ASE is an important source of noise and poor performance of rare-earth-doped fiber amplifiers and lasers.

This physical system is described by coupled equations involving the pump power, the signal power and the forward and backward ASE powers.

The noise generated by ASE spread across the whole emission spectrum; therefore, to model the evolution of the pump, laser and spontaneous emission radiations along the active fiber, the wavelength spectrum is divided into slots of width Δ_i centred at frequency ν_i, were $i = 1, 2, 3 \ldots n$. In each slot, the following equations are considered

To describe the evolution of the pump, laser and amplified spontaneous emission radiations along with the active fiber, the wavelength spectrum is divided into slots and in each slot, the evolution of these quantities along the z-direction is determined by a set of coupled first-order differential equations (King et al. 2003) as follows:

$$\frac{dP_p^\pm(z)}{dz} = \mp \Gamma_p \left[\sigma_a(\nu_p)N_1(z) - \sigma_e(\nu_p)N_2(z) \right] P_p(z, t) - \alpha_p P_p(z, t) \quad (4.23)$$

$$\frac{dP_s^\pm(z)}{dz} = \pm \Gamma_s [\sigma_{es}N_2 - \sigma_{as}N_1] P_s(z, t) \mp \alpha_s P_s(z, t) \quad (4.24)$$

$$\frac{dP_{ASE}^\pm(z, \nu_i)}{dz} = \pm \Gamma_{ASE} [\sigma_e(\nu_i)N_2(z) - \sigma_a(\nu_i)N_1(z)] P_{ASE}(z, \nu_i) \mp \alpha_{ASE} P_{ASE}(z, \nu_i)$$

$$\pm 2\sigma_e(\nu_i)h\nu_i\Delta\nu_i N_2(z) \quad (4.25)$$

where $P_{ASE}^\pm(z, \nu_i)$ is the amplified spontaneous power at position z within the interval of frequencies $(\nu_i, \nu_i + \Delta_i)$ in the forward and backward directions. The factor "2" before the last term of equation (4.25) account for the number of transverse modes propagating in the fiber. In the case of LP_{01} mode, this factor represents the two orthogonal polarization of the guided mode. Equations (4.23)–(4.25) are completed by the boundary conditions derived from the resonator configuration. For the linear cavity, the boundary conditions are written as:

$$P_p^+\left(0, \nu_p\right) = P_{p0}$$

$$P_p^-\left(L, \nu_p\right) = P_{pL}$$

$$R_1 P_s^-(0, \nu_s) = P_s^+(0, \nu_s)$$

$$R_2 P_s^+(L, \nu_s) = P_s^-(L, \nu_s)$$

$$R_1(\nu_i) P_{ASE}^-(0, \nu_i) = P_{ASE}^+(0, \nu_i)$$

$$R_2(\nu_i) P_{ASE}^+(L, \nu_i) = P_{ASE}^-(L, \nu_i)$$

$$i = 1, 2 \ldots n$$

(4.26)

The number of equations to be solved depends on the width and hence the number of slots used to discretize the spectrum. If the spectrum is divided into n intervals, the number of equations to be solved will be $(2n + 4)$. The coefficient 2 accounts for the forward and backward amplified spontaneous emission frequencies in each slot. For example, the spectrum of Erbium ions expands from approximately 1500 nm to 1650 nm and can be divided into 100 slots of 1.5 nm each resulting in 204 equations to be solved. It is, however, worth mentioning that the computational speed is related to the number of equations hence methods to optimize the computation must be found. For example, the number of equations can be reduced to a minimum acceptable without compromising accuracy. Another technique consists in dividing the spectrum into slots of different width, the shorter slots being attributed to spectral components one intends to give special attention.

Equations (4.23)–(4.25) is a set of first-order differential equations with a unique independent variable z. Therefore, these type of equations can be solved by a Runge-Kutta method. The Runge-Kutta method solves initial value problems by approaching its solutions through multiple approximations. If one supposes a given differential equation is:

$$\frac{dy}{dx} = f(x, y)$$

(4.27)

With initial conditions x_0 and y_0. To apply the Runge-Kutta method one must first choose a step interval h which is the next point at which the equation must be solved $x + h$ and find the corresponding value of the function at this point using the relation:

$$x_{n+1} = x_n + h$$

(4.28)

$$y_{n+1} = y_n + \Delta_y$$

(4.29)

where,

$$k_1 = h. f(x_n, y_n)$$

(4.30)

$$k_2 = h.f\left(x_n + \frac{h}{2}, y_n + \frac{k_1}{2}\right) \qquad (4.31)$$

$$k_3 = h.f\left(x_n + \frac{h}{2}, y_n + \frac{k_2}{2}\right) \qquad (4.32)$$

$$k_4 = h.f\left(x_n + h, y_n + k_3\right) \qquad (4.33)$$

$$\Delta_y = \frac{1}{6}(k_1 + 2k_2 + 2k_3 + k_4) \qquad (4.34)$$

4.4 SOLUTIONS ALGORITHMS

The solutions of equations (4.23)–(4.25) provide pump, signal, and ASE powers distribution along the fiber length. Therefore, to find these quantities the gain fiber is divided into several sections as illustrated in Figure 4.1. In each section, the pump power, laser power and ASE are assumed to be constant in each propagating direction. To simplify the analysis, we are going to assume the forward co-propagating waves, i.e. signal and pump wave propagate in the same direction from $z = 0$ towards $z = L$, where L is the length of the cavity also we also assume unidirectional pumping. For bidirectional pumping and counter-propagating waves, the analysis still stands with slight modifications namely in boundary conditions.

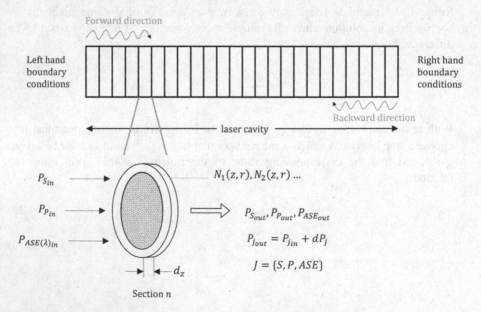

FIGURE 4.1 Fiber laser cavity divided into multiple sections Δz.

In the case of an amplifier, the incident pump power and signal power at $z = 0$ is known, thus the steady-state population density of the metastable level $N_2(z = 0)$ can easily be calculated from (4.1) and (4.2). The value of $N_2(z = 0)$ is then used to propagate the pump, signal and ASE from $z = 0$ to $z = \Delta z$ to find the incident powers of the next section using (4.23)–(4.25). These new incident powers are used to calculate $N_2(z = \Delta z)$ which is used to further propagate pump, laser, and ASE. The process is repeated until one reaches the end of the fiber at $z = L$. It is worth mentioning that the propagation equations in each section are solved with the 4th order Runge-Kutta scheme described in the previous section. The case of fiber amplifier described here is, clearly, an initial value problem, because pump and signal powers at one end of the fiber are known. In the case of fiber laser, the problem to solve is a boundary value problem with boundary condition provided by equation (4.26). The boundary conditions show that pump power is known at $z = 0$. The power of the forward propagating ASE at $z = 0$ is equal to zeros for the entire ASE spectrum, whereas the corresponding backwards propagating ASE at the same point is maximum and vice versa. Finally, the signal there is not incident signal power since oscillation is initiated inside the cavity. For modelling purpose, an equivalent input noise power must be included to start the oscillation process. There are two main ways of solving such boundary value problems which are shooting method and relaxation method (Flannery et al. 1992). These methods will be described in the below section.

For fiber lasers, in general, the field is propagated from section to section as described in the case of fiber amplifiers until the end of the gain medium where a criterion of convergence is tested. If the criteria of convergence satisfied, the algorithm ends otherwise, the process is repeated until convergence is achieved.

4.5   SHOOTING METHODS

The shooting method is a trial and error scheme in which the unknown boundary value at $z = 0$ is guessed. The system of equations is then integrated from $z = 0$ to $z = L$, where the boundary condition at $z = L$ is applied to the field. The system is then integrated from $z = L$ to $z = 0$. The calculated boundary value at $z = 0$ is compared with the guessed value and the difference is used to correct the guess in subsequent trials. Guess correction can be done using a classical one-dimensional root-finding technique such as the secant method or Newton-Raphson method. Below is the shooting method algorithm for a linear cavity Fabry-Perot fiber laser. The major drawback of shooting method is its implementation complexity when the number of equations to integrate becomes large. In a fiber laser, three types of optical fields are presents namely, pump wave, signal wave, and forward and backward ASE waves (see Figure 4.2 and Figure 4.3), therefore is not uncommon to end up with a system of hundreds of equations. Since one does not know the shape the spectrum, it is common to assume some arbitrary spectral distribution. Using the arbitrary spectral distribution at $z = 0$, one integrates the equations in the forward direction and the shooting algorithm is applied as before. Experience shows that using such a method results in very slow convergence towards the solution.

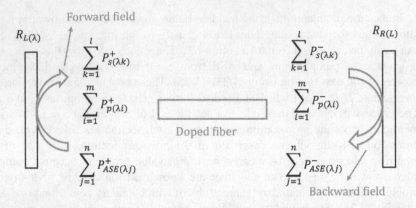

FIGURE 4.2 Example of a linear cavity with all the propagating waves represented.

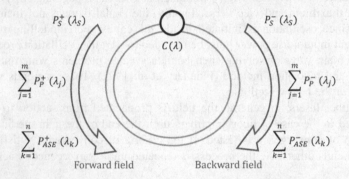

FIGURE 4.3 Ring cavity representation with all the propagating fields.

Sophisticated algorithms have been proposed (Anzueto-Sánchez et al. 2005; Han et al. 2005; Lali-Dastjerdi et al. 2008; Liu et al. 2010; Karimi and Farahbod 2014) to improve the guess process and make it decrease convergence time but these methods remain tedious to implement and in some case, the solution depends on the chosen algorithm. For this reason, the shooting method is often not preferred, and the more robust relaxation method is widely used:

- Provide the initial values for the forward pump power $P_p^+(0)$, signal power $P_s^+(0)$, set forward amplified spontaneous emission $P_{ASE}^+(0) = 0$ and provide a guess value for backward amplified spontaneous;
- Set the longitudinal section size Δz;
- Integrate propagation equations in the forward direction using the RK4 scheme in each section until $z = L$;
- Compare the value of P_{ASE}^- at $z = L$ with the boundary condition $(P_{ASE}^-(L) = 0)$ and use the difference to refine the guess with the secant method;

- If the value of $P_{ASE}^- \leq \varepsilon$ (convergence criteria) stop the algorithm, else go to step 3.

A variant of the shooting method can be used, where signal power instead of amplified spontaneous emission is used as a comparison value as shown in the following algorithm:

- Provide the initial values for the forward pump power $P_p^+(0)$, and provide a guess for signal power $P_s^+(0)$, set forward amplified spontaneous emission $P_{ASE}^+(0) = 0$ and provide a guess value for backward amplified spontaneous;
- Set the longitudinal section size Δz;
- Integrate propagation equations in the forward direction using the RK4 scheme in each section until $z = L$;
- Use the boundary conditions to calculate $P_s^-(L)$ and $P_{ASE}^-(L)$ given by: $P_s^-(L) = P_s^+(L)R_2$ and $P_{ASE}^-(L) = P_{ASE}^+(L)R_2$;
- Integrate propagation equations in the backward direction using the RK4 scheme in each section until $z = 0$;
- Use the boundary conditions to calculate $P_s^+(0)$ and $P_{ASE}^+(0)$ given by: $P_s^+(0) = P_s^-(0)R_1$ and $P_{ASE}^+(0) = P_{ASE}^-(0)R_1$;
- Compare the new value of $P_s^+(0)$ to the guessed value. If the difference between the two values is larger than a residual value ε, use this difference to refine the guess with a standard secant method.

4.6   RELAXATION METHODS

The relaxation method has the advantage of a high speed of convergence and robustness. In this method, the pump power, signal power, and ASE are integrated back and forth using the boundary conditions at each end until the change of population density distribution from one iteration to the next is below a fixed value. The relaxation method algorithm is presented in the following. Again, for clarity, we assume unidirectional pumping at $z = 0$.

- Provide the initial values for the forward pump power $P_p^+(0)$, signal power $P_s^+(0)$ and forward amplified spontaneous emission $P_{ASE}^+(0)$ and setting backward values to zero $P_s^-(0) = 0$, $P_{ASE}^- = 0$;
- Integrate propagation equations in the forward direction using the RK4 scheme in each section until $z = L$;
- At $z = L$ apply boundary conditions to all fields such that $P_s^-(L) = P_s^+(L)R_2$ and $P_{ASE}^-(L) = P_{ASE}^+(L)R_2$;
- With $P_s^-(L)$ and $P_{ASE}^-(L)$ as new initial values, integrate propagation equations from $z = L$ to $z = 0$ using RK4 scheme in each section;
- At $z = 0$ apply boundary conditions to all incident fields such that $P_s^+(0) = P_s^-(0)R_1$ and $P_{ASE}^+(0) = P_{ASE}^-(0)R_1$;
- Check the convergence criteria: $\frac{N_i(z=0)_{step1} - N_i(z=0)_{step3}}{N_i(z=0)_{etape1}} \leq \varepsilon$;

- Compare the calculated values from step 6 at $z = 0$ with the initial guess. If the difference is bigger than the allowed tolerance, take the values obtained at 6 as initial values and go to step 3;
- Stop.

4.7 FINITE DIFFERENCE METHODS

The rate and propagation equations we have presented in the previous section are suitable to model continuous wave fiber lasers. However, when it comes to pulsed fiber lasers or strongly time-dependent phenomena such as transient states, they are no longer useful because the temporal dependence of population densities and other parameters become very significant. In such a case a travelling wave model is then preferred to be used.

The travelling wave model consists of a system of coupled partial differential equations with time and space as variables. In the simple case of a two-level fiber laser system, travelling wave equations can be written as follow:

$$N = N_1 + N_2 \tag{4.35}$$

$$\frac{\partial N_2}{\partial t} + \frac{N_2}{\tau} = \frac{\Gamma_p \lambda_p}{hcA} [\sigma_a(\lambda_p) N_1 - \sigma_e(\lambda_p) N_2](P_p^+ + P_p^-)$$

$$+ \sum_k \frac{\Gamma_k \lambda_k}{hcA} [\sigma_a(\lambda_k) N_1 - \sigma_e(\lambda_k) N_1](P_k^+ + P_k^-) \tag{4.36}$$

$$\pm \frac{\partial P_p^\pm}{\partial z} + \frac{1}{\nu_p} \frac{\partial P_p^\pm}{\partial t} = -\Gamma_p [\sigma_e(\lambda_p) N_2 - \sigma_a N_1] P_p^\pm(z, t) - \alpha_p(\lambda_p) P_p^\pm(z, t) \tag{4.37}$$

$$\pm \frac{\partial P_k^\pm}{\partial z} + \frac{1}{\nu_k} \frac{\partial P_k^\pm}{\partial t} = -\Gamma_k [\sigma_e(\lambda_k) N_2 - \sigma_a(\lambda_k) N_1]$$

$$P_k^\pm(z, t) - \alpha_a(\lambda_k) P_k^\pm(z, t) + 2\sigma_e(\lambda_k) N_2 \frac{hc^2}{\lambda_k^3} \Delta \lambda_k \tag{4.38}$$

$$k = 1, \ldots, K$$

In the equations (4.35)–(4.38), N is the dopant concentration and is for simplicity assumed to be constant inside the core. N_1 and N_2 represent the ground and upper-level ion population densities respectively as described in the previous section. The parameters ν_k and ν_p represent the group velocities of the amplified spontaneous emission and pump power in the doped fiber respectively and c is the speed of light in free space. These equations are also completed by boundary condition that depends on the specific laser configuration. Below we provide the boundary conditions for a linear cavity Q-switched fiber laser. In the particular case of pulsed lasers such as Q-switched fiber lasers, the boundary conditions are time-dependent and written as:

$$P_p^+(0) = \eta_p P_p \tag{4.39}$$

$$P_k^+(0, t) = R_R(\lambda_k, t) P_k^-(0, t) \tag{4.40}$$

$$P_k^-(L, t) = R_L(\lambda_k, t) P_k^+(L, t) \tag{4.41}$$

$$k = 1, \ldots, K$$

In equations (4.40) and (4.41)(4.40), (4.41)R_L and R_R are left and right coupler reflectivity respectively. Depending on the type of fiber laser, these reflectivities can also be time dependent. In the case of an active Q-switch fiber laser the total reflectivity of the output coupler can be thought of as the product of the constant reflectivity and a time dependency given by an acousto-optic modulator or as a filter.

The time-dependent equation of output reflectivity can be written as:

$$R_R(tot) = R_R(\lambda_k)T(t) \tag{4.42}$$

In equation (4.42)(4.42)$T(t)$ is the switching element transmission.

Taking into consideration the time-dependent boundary condition, the coupled propagation equations must be solved numerically using the finite difference method (Zwillinger 1997). The finite difference method consists of replacing all the partial derivatives in the equations by finite differences. The space axis and time axis are divided into small pieces Δ_z and Δ_t respectively. In this way, the differential equation is transformed into an algebraic equation that can be usually be solved numerically. One obtains in this way a grid of points that give at every point of the gain medium and every time the value of the pump power, forward and backward output power. Generally, the finite difference methods can be divided into three groups, namely, explicit finite difference, implicit finite difference, and Crank-Nicholson (Zwillinger 1997). Amongst all these, the explicit method is the easiest to implement. However, it is only numerically stable and convergent when the relation $\Delta_t = \frac{\Delta_z}{\nu}$ known as the courant condition (Zwillinger 1997) is satisfied. In relation, Δ_t is the section in the time axis and Δ_z a section in the space axis. The other mentioned methods namely, implicit and Crank-Nicholson, are always stable but more computationally intensive. The Crank-Nicholson is better in terms of stability. Below an explicit scheme is used to discretize the rate and propagation equations for a Q-switch fiber laser. This particular method is called the forwards time, forwards space method.

$$N_2(z, t + \Delta t) = N_2(z, t)\left[1 - \frac{\Delta t}{\tau}\right]\frac{\Gamma_p\lambda_p}{hcA}[\sigma_a(\lambda_p)N_1 - \sigma_e(\lambda_p)N_2](P_p^+ + P_p^-)$$

$$+ \sum_k \frac{\Gamma_k\lambda_k}{hcA}[\sigma_a(\lambda_k)N_1 - \sigma_e(\lambda_k)N_1](P_k^+ + P_k^-) \qquad (4.43)$$

$$\frac{P_p(z + \Delta z, t + \Delta t) - P_p(z + \Delta t)}{\Delta z} + \frac{1}{\nu_p}\frac{P_p(z + \Delta z, t + \Delta t) - P_p(z + \Delta z, t)}{\Delta t}$$

$$= -\Gamma_p[\sigma_e(\lambda_p)N_2 - \sigma_a N_1]P_p^\pm(z, t) - \alpha_p(\lambda_p)P_p^\pm(z, t) \qquad (4.44)$$

$$\frac{P_s^+(z + \Delta z, t + \Delta t) - P_s^+(z + \Delta t)}{\Delta z} + \frac{1}{\nu_s}\frac{P_s^+(z + \Delta z, t + \Delta t) - P_s^+(z + \Delta z, t)}{\Delta t}$$

$$= \Gamma_s[\sigma_e(\lambda_s)N_2 - \sigma_a(\lambda_s)N_1]P_s^+(z, t) - \alpha_a(\lambda_s)P_s^+(z, t) + 2\sigma_e(\lambda_s)N_2\frac{hc^2}{\lambda_s^3}\Delta\lambda_s$$

$$(4.45)$$

$$\frac{P_s^-(z - \Delta z, t + \Delta t) - P_s^-(z + \Delta t)}{\Delta z} + \frac{1}{\nu_s}\frac{P_s^-(z - \Delta z, t + \Delta t) - P_s^-(z - \Delta z, t)}{\Delta t}$$

$$= \Gamma_s[\sigma_e(\lambda_s)N_2 - \sigma_a(\lambda_s)N_1]P_s^-(z, t) - \alpha_a(\lambda_s)P_s^-(z, t) + 2\sigma_e(\lambda_s)N_2\frac{hc^2}{\lambda_s^3}\Delta\lambda_s$$

$$(4.46)$$

These difference equations are solved together with the boundary conditions. The spatial steps range from a minimum of hundreds and time-space can be obtained through the courant condition. Stable convergence for 200 or greater spatial steps has been reported in the literature (Yang et al. 2012). The solution to the equation gives values for the evolution of output pulse as a function of time. Transient conditions like when starting on the fiber laser can also be found using the same method.

4.8   CONCLUSION

In this chapter numerical methods for solving fiber laser rate and propagation, equations have been investigated. These equations are generally systems of coupled differential equation whose analytical solutions are most of the time very difficult to find. One has to resort to a numerical solution to solve these equations. Depending on the gain medium, (2 or more level, co-doped fiber) and laser configuration (continuous wave fiber laser, linear or ring cavity, or pulsed fiber laser) several algorithms can be successfully implemented to solve these equations. The numerical methods and algorithms described in this chapter will be used in the subsequent chapter to simulated various types of rare-earth-doped fiber lasers.

REFERENCES

Anzueto-Sánchez, G., A. Martínez-Rios, R. Selvas Aguilar, I. Torres-Gómez, J.A. Álvarez-Chávez, J. Sánchez-Mondragón, and D. May-Arrioja. 2005. "Simple Numerical Modeling of Yb-Doped Fiber Lasers." In *Proc. SPIE.* Vol. 5970. 10.1117/12.628646.

Barnard, G., P. Myslinski, J. Chrostowski, and M. Kavehrad. 1994. "Analytical Model for Rare-Earth-Doped Fiber Amplifiers and Lasers." *IEEE Journal of Quantum Electronics* 30 (8): 1817–1830. 10.1109/3.301646.

Broyden, C.G. 1969. "A New Method of Solving Nonlinear Simultaneous Equations." *The Computer Journal* 12 (1): 94–99. 10.1093/comjnl/12.1.94.

Esfandiari, Ramin S. 2017. *Numerical Methods for Engineers and Scientists Using MATLAB®.* CRC Press.

Flannery, Brian P., William H. Press, Saul A. Teukolsky, and William Vetterling. 1992. "Numerical Recipes in C." *Press Syndicate of the University of Cambridge, New York* 24: 78.

Giles, C. Randy and Emmanuel Desurvire. 1991. "Modeling Erbium-Doped Fiber Amplifiers." *Journal of Lightwave Technology* 9 (2): 271–283.

Hakeim, Abdel and Fady I. 2011. "Model of Temperature Dependence Shape of Ytterbium-Doped Fiber Amplifier Operating at 915 Nm Pumping Configuration." *International Journal of Advanced Computer Science and Applications* 2 (10): 10–13. 10.14569/IJACSA.2011.021002.

Han, Qun, Tiegen Liu, Xiaoying Lü, and Kun Ren. 2015. "Numerical Methods for High-Power Er/Yb-Codoped Fiber Amplifiers." *Optical and Quantum Electronics* 47 (7): 2199–2212. 10.1007/s11082-014-0096-8.

Han, Qun, Jiping Ning, Zhiqiang Chen, Lianju Shang, and Guofang Fan. 2005. "An Efficient Shooting Method for Fiber Raman Amplifier Design." *Journal of Optics A: Pure and Applied Optics* 7 (8): 386–390. 10.1088/1464-4258/7/8/006.

Johnson, D.D. 1988. "Modified Broyden's Method for Accelerating Convergence in Self-Consistent Calculations." *Physical Review B* 38 (18): 12807–12813. 10.1103/PhysRevB.3 8.12807.

Karimi, Maryam and Amir Hossein Farahbod. 2014. "Improved Shooting Algorithm Using Answer Ranges Definition to Design Doped Optical Fiber Laser." *Optics Communications* 324: 212–220. 10.1016/j.optcom.2014.03.013.

King, A.C., J. Billingham, and S.R. Otto. 2003. *Differential Equations: Linear, Nonlinear, Ordinary, Partial.* Cambridge University Press.

Lali-Dastjerdi, Zohreh, Feisal Kroushawi, and Mohammad Rahmani. 2008. "An Efficient Shooting Method for Fiber Amplifiers and Lasers." *Optics & Laser Technology* 40 (November): 1041–1046. 10.1016/j.optlastec.2008.02.006.

Liu, Jinglin, Chujun Zhao, Shuangchun Wen, Dianyuan Fan, and Cijun Shuai. 2010. "An Improved Shooting Algorithm and Its Application to High-Power Fiber Lasers." *Optics Communications* 283 (October): 3764–3767. 10.1016/j.optcom.2010.05.060.

Marcuse, D. 1977. "Loss Analysis of Single-Mode Fiber Splices." *The Bell System Technical Journal* 56 (5): 703–718. 10.1002/j.1538-7305.1977.tb00534.x.

Mathews, John H. and Kurtis D. Fink. 2004. *Numerical Methods Using MATLAB.* Vol. 4. NJ: Pearson Prentice Hall Upper Saddle River.

Myslinski, P. and J. Chrostowski. 1996. "Gaussian-Mode Radius Polynomials for Modeling Doped Fiber Amplifiers and Lasers." *Microwave and Optical Technology Letters* 11 (2): 61–64. 10.1002/(SICI)1098-2760(19960205)11:2<61::AID-MOP3>3.0.CO;2-M.

Nguyen, Dan T., Arturo Chavez-Pirson, Shibin Jiang, and Nasser Peyghambarian. 2007. "A Novel Approach of Modeling Cladding-Pumped Highly Er-Yb Co-Doped Fiber Amplifiers." *IEEE Journal of Quantum Electronics* 43 (11): 1018–1027. 10.1109/JQE.2007.905010.

Okawa, Hirotada, Kotaro Fujisawa, Yu Yamamoto, Ryosuke Hirai, Nobutoshi Yasutake, Hiroki Nagakura, and Shoichi Yamada. *The W4 Method: A New Multi-Dimensional Root- Finding Scheme for Nonlinear Systems of Equations, ArXiv abs/1809.04495 (2018)*.

Peterka, Pavel, Ivan Kasik, Anirban Dhar, Bernard Dussardier, and Wilfried Blanc. 2011. "Theoretical Modeling of Fiber Laser at 810 Nm Based on Thulium-Doped Silica Fibers with Enhanced 3H4 Level Lifetime." *Optics Express* 19 (3): 2773–2781. 10.13 64/OE.19.002773.

Phung, B.Q. and Le Hong Son. 2007. "Gain and Noise in Erbium-Doped Fiber Amplifier (Edfa)-a Rate Equation Approach (Rea)." *Communications in Physics* 14 (1): 1–6. 10.3125/cip.v14i1.348.

Press, William H., Saul A. Teukolsky, William T. Vettereling, and Brain P. Flannery. *Numerical Recipes in C: The Art of Scientific Computing, 2nd Ed. Cambridge University Press, NY: USA,* 1992

Press, William H., Saul A. Teukolsky, William T. Vettereling, and Brain P. Flannery. *Numerical Recipes in C: The Art of Scientific Computing, 2nd Ed*. Cambridge University Press, NY: USA, 1992.

Rudyak, Valery and Andrey Minakov. 2009. "Modeling and Optimization of An." *Micromachines* 33 (1): 75–88. 10.3390/mi5040886.

Saheya, B., Guo-qing Chen, Yun-kang Sui, and Cai-ying Wu. 2016. "A New Newton-like Method for Solving Nonlinear Equations." *SpringerPlus* 5 (1): 1269. 10.1186/s40064-016-2909-7.

Sorbello, G., S. Taccheo, and P. Laporta. 2001. "Numerical Modelling and Experimental Investigation of Double-Cladding Erbium- Ytterbium-Doped Fiber Amplirers." *Optical and Quantum Electronics* 33: 599–619.

Strohhöfer, Christof and Albert Polman. 2001. "Relationship between Gain and Yb 3+ Concentration in Er 3+–Yb 3+ Doped Waveguide Amplifiers." *Journal of Applied Physics* 90 (9): 4314–4320.

"Theoretical Modeling of Er Doped Fiber Amplifiers with Excited State Absorption.Pdf." n.d.

Thorlabs. 2018. "Ytterbium Doped Fiber."

Träger, Frank. 2012. *Springer Handbook of Lasers and Optics*. Springer Science & Business Media.

Wang, Yong and Hong Po. 2003. "Dynamic Characteristics of Double-Clad Fiber Amplifiers for High-Power Pulse Amplification." *Journal of Lightwave Technology* 21 (10): 2262–2270. 10.1109/JLT.2003.818166.

Watson, Layne T. 1990. "Globally Convergent Homotopy Algorithms for Nonlinear Systems of Equations." *Nonlinear Dynamics* 1 (2): 143–191.

Yahel, Eldad and Amos A. Hardy. 2003. "Modeling and Optimization of Short Er 3 + – Yb 3 + Codoped Fiber Lasers." *Journal of Lightwave Technology* 24 (3): 1601 - 1609.

Yahel, Eldad, Amos Hardy, and Abstract High-power Er. 2003. "Modeling High-Power Er 3 + – Yb 3 + Codoped Fiber Lasers." *Lightwave* 21 (9): 2044–2052.

Yang, Jianlong, Yulong Tang, Rui Zhang, and Jianqiu Xu. 2012. "Modeling and Characteristics of Gain-Switched Diode-Pumped Er-Yb Codoped Fiber Lasers." *IEEE Journal of Quantum Electronics* 48 (12): 1560–1567. 10.1109/JQE.2012.2225416.

Yelen, Kuthan, Louise M.B. Hickey, and Mikhail N. Zervas. 2005. "Experimentally Verified Modeling of Erbium-Ytterbium Co-Doped DFB Fiber Lasers." *Journal of Lightwave Technology* 23 (3): 1380.

Yu, Xiaochen, Shibin Jiang, Zhenzhou Cheng, Ruiyuan Su, Jianguo Tian, Weiping Zang, Changguang Zou, and Feng Song. 2008. "Numerical Investigation of Gain Characteristics of Er3+/Yb3+ Co-Doped Fiber Amplifiers." *Optical and Quantum Electronics* 40 (13): 1021–1031. 10.1007/s11082-009-9295-0.

Zwillinger, D. 1997. *Handbook of Differential Equations*. Edited by Academic Press. Third edit.

5 Continuous-Wave Silica Fiber Lasers

5.1 INTRODUCTION

This chapter reviews one of the two main types of fiber lasers: namely continuous wave fiber lasers. Continuous operation of a fiber laser (and any other laser of this type) means that the laser continuously emits light under continuous pumping conditions. The fiber lasers we are reviewing in this chapter are typically silica-based. We shall focus our attention on points such as rare-earth ion spectroscopy, resonator configurations, theoretical performance as well as laser characteristics including output power, lasing threshold, slope efficiency, and the influence of detrimental effects. The architecture of the laser is dictated by the type of optical resonator. The two main types of optical resonators are linear and ring cavity. We shall thoroughly describe each one of these cavities, highlighting their respective advantages and weaknesses. We shall derive a mathematical model of every one of these laser configurations using Erbium and Ytterbium-doped silica fibers as amplifying medium. The mathematical model is based on the rate equation derived from the ion's light interaction and propagation equations in the fiber medium with appropriate boundary conditions. First, a rigorous mathematical treatment is applied to the resulting system of partial differential equations and next a simplified analytical solution is derived while ignoring parameters assumed to be less influential to the behaviour of the laser. This last solution is then compared with the true numerical solution to assess accuracy. Finally, in the last section different types of fiber laser are presented namely, high power fiber lasers, narrow-linewidth fiber lasers, and up-conversion fiber lasers. In each case, the results of the simulation obtained from the mathematical model are compared with experimental work published to assess the validity of the model.

5.2 ARCHITECTURE AND THEORY OF OPERATION

As with any other laser, rare-earth doped fiber lasers are formed by using an optical light amplifier combined with an appropriate feedback mechanism. The amplifying medium in the case of a fiber laser is a fiber with its core doped with rare-earth ions. The most common doping material is Erbium (Er^{3+}), Ytterbium (Yb^{3+}), Neodymium (Nd^{3+}), Thulium (Tm^{3+}), Holmium (Ho^{3+}), Samarium (Sm^{3+}), and Praseodymium (Pr^{3+}). The feedback, on the other hand, is obtained differently depending on the cavity configuration. In the case of a linear cavity, the feedback is obtained by using mirrors outside of the gain medium (Yamamoto et al. 1994; Shimizu et al. 1987). In recent years, the mirrors are often replaced by Bragg gratings printed in the fiber core (Ball et al. 1991). Ring cavity fiber lasers, on the other hand, are obtained by making

DOI: 10.1201/9781003256380-5

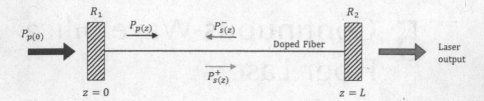

FIGURE 5.1   Schematic diagram of a linear cavity fiber laser.

a loop with the rare-earth doped optical fiber combined with an optical coupler. The whole system, doped fiber and feedback is optically pumped using a laser diode. The pump wavelength is often in the range of 980 nm for Erbium and 1480 nm for Ytterbium. Other rare-earth ions use different wavelengths, typically 650 nm for Thulium. Continuous-wave fiber optic lasers offer key advantages when compared to other solid-state bulk lasers. The active medium can theoretically be unlimited leading to excellent control of lasing modes and linewidth with the added advantage of the fiber coiled into a small volume. Fiber lasers are known to have a large gain bandwidth, which is the result of the strongly broadened rare-earth ions transitions in glass, allowing for wide wavelength range tunability as shown in Figure 5.1.

5.2.1   AMPLIFYING MEDIUM

The amplifying medium for fiber laser is an optical fiber doped with rare-earth ions. The description of rare-earth ions in this chapter is brief as a detailed description has been provided in Chapter 3. The reader interested in this aspect is advised to review Chapter 2 of this book. Here we shall describe mainly Erbium- and Ytterbium-doped fiber the other rare-earth ions are described in a purely informative purpose as they are beyond the scope of this book.

5.2.1.1   *Erbium*

Erbium-doped fibers are the most popular type of rare-earth-doped fiber, mainly due to their importance in the telecommunication industry where they are mainly used as amplifiers (Parekhan and Banaz 2008; Giles and Desurvire 1991; Semmalar and Malarkkan 2013). Erbium is often modelled as a three levels system (Urquhart, n.d.). Its emission wavelength ranges from 1520 nm to about 1565 nm (Urquhart, n.d.). Erbium accepts pumping wavelengths of 510, 532, 665, 810, 980 nm and 1480 nm. In practice, 980 nm and 1480 nm pumping wavelengths are preferred over the other wavelengths because the influence of excited state absorption is minimal at these wavelengths (Laming 1989) and absorption is higher. Figure 5.2 illustrates the absorption spectra of Alumino-Silicate Erbium-doped fiber at room temperature. The figure shows two regions of high absorption, namely around 1000 nm and between 1400 nm and 1600 nm. The region around 1000 nm is particularly interesting because it corresponds to the emission wavelength of efficient laser diodes for pumping the lasers (Thyagarajan 2006).

In addition to absorption and emission cross-sections, the lifetime of the excited energy level is an important characteristic of rare-earth-doped fibers. The glass host

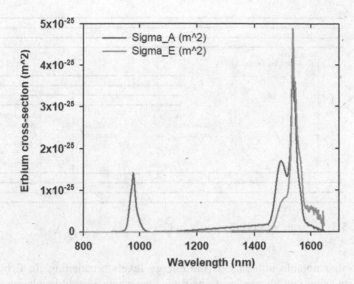

FIGURE 5.2 Absorption characteristic of the alumino-silicate fiber.

TABLE 5.1
Lifetime of Erbium in Various Glass

Host glass	Lifetime (ms)	Reference
Al-P silica	10.8	(Miniscalco 1991)
Al-Ge silica	9.5–10.0	(Giles and Desurvire 1991)
Phosphate	10.7	(Thyagarajan 2006)
Fluorophosphates	8.0	(Thyagarajan 2006)

composition has a notable influence on the lifetime of the excited ions, as a lifetime of the same ion can vary greatly with the nature of the host (Miniscalco 1991). The lifetime of Er^{3+} $^4I_{13/2}$ energy level in different types of glass is provided in Table 5.1 (Thyagarajan 2006). It can be seen that with 8.0 ms, fluorophosphate glass possesses the shortest lifetime for the $^4I_{13/2}$ energy level, and the alumino-silicate, the longest(10 ms). A longer lifetime of the metastable level is desirable because it allows us to achieve a larger population inversion.

The host glass also plays an important role on the Erbium ion energy level known as Stark splitting. Stark splitting is the splitting of the energy levels of the Erbium ions due to the influence of the electric field induced by the host electrical field. A consequence of this is that substructures can be observed in the electronic transition (Urquhart, n.d.). The substructures are called Stark components of the parent energy manifold (Figure 5.3).

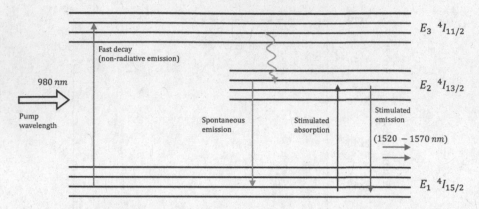

FIGURE 5.3 Erbium ions energy levels in glass.

The other notable influence is the energy levels broadening. In Erbium-doped media, emission and absorption transitions occur between sublevels resulting from Stark levels. The emission and absorption lines are therefore non-continuous but made of distinct transitions. The homogeneous and inhomogeneous broadening of the sublevels determines the transition widths. Inhomogeneous broadening occurs because of the host electrical field's variation from site to site. Homogeneous broadening, on the other hand, is produced by internal mechanisms that are not a function of the ion site such as thermal fluctuations or intrinsic level lifetimes. Figure 5.4 shows the emission and absorption cross-section as a function of wavelength near 1550 nm for the Al-Ge silica glass. At high concentration,

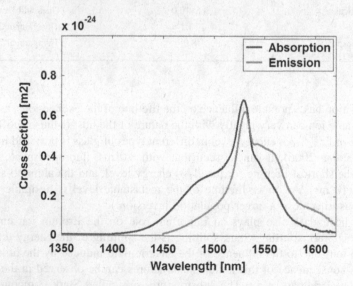

FIGURE 5.4 Emission and absorption cross-sections of Erbium-doped Aluminosilicate glass in the region of 1350–1650 nm.

Erbium-doped glasses are plagued by detrimental effects due to ion-ion interactions. The most notable are energy transfer, excited-state absorption, and up-conversion. These are mainly due to the tendency of Erbium ions to cluster in glass hosts. The consequence of these effects is luminosity quenching which reduces fiber laser efficiency and increases the lasing threshold. One solution to address this issue is to use phosphate glass instead of simple silica glass because Erbium ions are more soluble in phosphate glass than silica glass.

5.2.1.2 Ytterbium

The interest of using Erbium-doped fiber lasers results from its ability to obtain high gains around 1550 nm which corresponds to the third telecommunication window. However, outside of the telecommunication domain, that unique advantage of Erbium-doped glass fibers does not stand anymore. For fiber lasers emitting at wavelengths other than 1550 nm, other types of rare-earth ions are better candidates. One of the most used rare-earth elements (especially for high power fiber lasers applications) is the Ytterbium. With only two energy levels, namely the $^2F_{7/2}$ ground level and the $^2F_{5/2}$ excited level, Ytterbium possesses a simple energy levels structure compared to other rare-earth materials. Ytterbium can provide amplification from 975 nm to almost 1200 nm (Strohhöfer and Polman 2003). Also, Ytterbium-doped fiber lasers can exhibit higher output laser power and better conversion efficiency. Because of its energy levels structure, deleterious effects that affect Erbium-doped fibers like excited state absorption and clustering at high concentration are suppressed in Ytterbium, making it possible to increase doping concentration beyond value unacceptable for Erbium ions for example. Therefore, achieving a high gain in a relatively short length of fiber becomes possible. The other advantage of Ytterbium-doped fiber is its wide absorption band, hence, the possibility to use pump wavelengths covering a wide range. Figure 5.5 illustrates the energy level diagram of ytterbium with all possible transitions and their corresponding wavelengths.

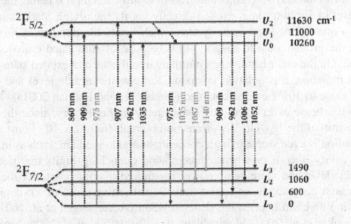

FIGURE 5.5 Ytterbium ions energy levels.

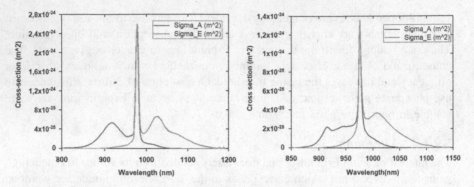

FIGURE 5.6 Emission (red) and absorption (black) cross-section of Yb3+ in a) Aluminosilicate and b) Phosphosilicate fiber.

It is worth mentioning that since Ytterbium is a two-level system, transitions at a different wavelength, which make population inversion possible is only due to Stark splitting in the host. Absorption and emission cross-sections are functions of the various possible transition. Figure 5.6 illustrates this absorption and emission cross-section as a function of wavelength for Ytterbium ions in Aluminosilicate and Phosphosilicate fibers.

The shape of the emission spectrum is very similar for the two fibers, but Aluminosilicate fiber has a cross-section almost two times larger than Germanosilicate in the region around 1000 nm. Absorption spectrum, on the other hand, is almost constant between 900 nm and close to 960 nm for Phosphosilicate fiber, whereas it has a peak around 950 nm, falls to a minimum around 950 to reach the maximum at around 980 nm for Aluminosilicate fiber.

5.2.1.3 *Erbium-Ytterbium Co-doping*

Two different pump wavelengths can be used to pump Er^{3+} doped fiber lasers, namely 1480 nm and 980 nm. Under the 1480 nm excitation scheme, the metastable $^4I_{13/2}$ level of Erbium ion is excited directly via its high-lying Stark states. The ion acts as a quasi-three-level laser system which limits the population inversion concerning the ground state to roughly 40% because of stimulated emission by pump radiation. On the other hand, when pumping into the second excited state $^4I_{11/2}$ using 980 nm radiation, a population inversion between metastable level and the ground state of close to 100% can be obtained (Strohhöfer and Polman 2001a). This second excitation scheme of Er^{3+} requires considerably higher energy, since the absorption cross-section of the $^4I_{15/2}$ to $^4I_{11/2}$ transition is small (only 1×10^{-21} cm^2 at 975 nm). The situation is even worse in high concentration gain medium, such as in the case of a short cavity fiber laser where concentration quenching limits the lifetime of the metastable level and more pump power is required to compensate for the loss of the ions through cooperative up-conversion. To solve this problem, co-doping Erbium ions with Ytterbium is often the better solution (Achtenhagen et al. 2001; Bai et al. 2015; Denker et al. 2007; Moghaddam et al. 2011; Nie et al. 2006; Strohhöfer and Polman 2001b; Prajzler et al. 2008). The absorption cross-section of Yb^{3+} at 980 nm

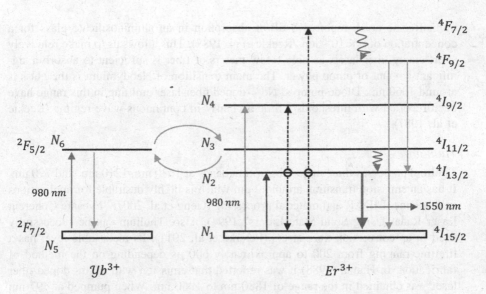

FIGURE 5.7 Energy level diagram of Erbium-Ytterbium co-doped fiber.

is about one order of magnitude larger than that of Er^{3+}. Also, Yb^{3+} has the advantage of the good spectral overlap of its emission transition with Er^{3+} absorption transition leading to an efficient energy transfer from Yb^{3+} to Er^{3+}. Like Er^{3+}, Yb^{3+} also tends to form clusters. Since the ions are similar, clusters, if they can take place at all, will not be between two or more Er^{3+} ions but among Er^{3+} ions and several Yb^{3+} ions, thus reducing up-conversion between Er^{3+} ions (Federighi et al. 1995). The energy level diagram of Erbium-Ytterbium co-doped medium and possible transitions are represented in Figure 5.7.

Most of the possible transitions are represented in Figure 5.7. Some of these transitions like cooperative up-conversion among Erbium ions and energy transfer between Erbium and Ytterbium ions result from the ion-ions interactions, whereas the others like absorption at the ground state, stimulated emission, and absorption at excited state is the result of ion-light interaction. In addition to these transitions, spontaneous emission and non-radiative transitions are also possible. Because of the large number of transitions, the system of rate equations of an Erbium-Ytterbium medium is always more complex than that of a single rare-earth system as it will be seen in subsequent sections of this chapter.

5.2.1.4 *Other Rare-earth Ions*

Neodymium

Neodymium was the first rare-earth ion to be used in fiber lasers. Snitzer and Koester reported the first Neodymium- doped fiber laser as early as 1964 (Pátek 1970). Nd^{3+}-doped fiber lasers can be pumped at 514.5 nm, and 752 nm using gas lasers such as Ar-ion or Kr-ion lasers. The most practical pumping wavelength, however, is 810 nm because the AlGaAs semiconductor laser emits in this band. A critical advantage of Nd^{3+}-doped glass fiber is that its absorption coefficient is quite high. Reekie et al. reported 8 dB/m absorption in an aluminosilicate glass for a concentration of 9×10^{24} ions/Reekie et al. 1987). This allows us to make relatively short cavity fiber lasers as only a few meters of fiber is sufficient to absorb a significant amount of pump power. The main transition of Neodymium is the glass is around 1060 nm. Diode-pumped Nd^{3+}-doped fiber laser emitting in this range have been reported with thresholds a little as 1 mW in continuous-wave regime (Reekie et al. 1987).

Thulium

Thulium possesses three major absorption bands at 1630 nm, 1210 nm, and 790 nm. It has an emission transition around 2 μm which is highly desirable for applications in medicine, LIDAR and material processing (Geng et al. 2007; "Eyesafe Coherent Laser Radar Using Solid State Lasers" 1994). Also, Thulium can be successfully used in up-conversion fiber lasers (Peterka et al. 2011). Its metastable level has a lifetime ranging from 200 to approximately 600 μs depending on the method of fabrication. In Hanna (1988) it was reported that emission with a Tm- doped fiber laser was obtained in the range of 1880 nm to 1960 nm. When pumped at 797 nm with a dye laser, a threshold of 21 mW and slope efficiency of 13% were obtained for a laser oscillation around 1900 nm. Hanna et al. reported an improved version of Thulium doped fiber laser. Pumped at 810 nm, it oscillated at 1900 nm with 44 mW output power for 167 mW pump power. The slope efficiency was reported to be 36% (Hanna et al. 1990). Other works have reported high power continuous wave Thulium-doped fiber lasers (Percival et al. 1992; Jeong et al. 2007; Peterka et al. 2011).

Praseodymium

The most interesting characteristic of Praseodymium is that it possesses multiple metastable levels. Therefore, laser emitting several wavelengths, as well as down-conversion operation of fiber lasers, is possible. Also, it possesses a large absorption range allowing large flexibility in pumping. Other rare-earth used fiber lasers are Holmium- and Samarium-doped fiber lasers. They find applications in lasers but also as optical amplifiers (Kim et al. 2009; N Simakov et al. 2015; Ng et al. 2016; Nikita Simakov et al. 2013; Hemming et al. 2012; 2014).

5.2.2 OPTICAL RESONATORS AND FEEDBACK

Several types of optical resonators have been used in rare-earth-doped fiber lasers, each with its advantages and disadvantages. As already mentioned in Chapter 3, there are two main types of resonators used in fiber lasers, namely the linear cavity with reflector mirror on both ends of the doped fiber and ring cavity fiber lasers. Technically, in the linear cavity configuration, the dielectric mirrors are butted to the right-angled cleaved fiber at each end. This technique has been used in numerous reported works and demonstrated good results (Freeman et al. 1988; Mears et al. 1985; Yamamoto et al. 1994; Ball et al. 1991; Shimizu et al. 1987). Although this approach is good for laboratory conditions, it is cumbersome and not convenient for mass production.

The Fresnel reflection from a cleaved tip of the fiber of interest can also be taken advantage of. However, the Fresnel reflection at a fiber glass-air interface is only around 4% and the gain of the fiber gain medium must be high enough to maintain oscillation.

The majority of fiber lasers in modern days, however, use fiber Bragg gratings printed either in the doped fiber or in an undoped fiber, which is spliced onto the doped fiber. The biggest advantage of fiber Bragg grating over traditional mirrors is the ability to select the laser wavelength, thus eliminating mode-hopping and achieving narrow linewidth operation.

5.2.2.1 *Fiber Bragg Grating*

As mentioned in the last paragraph, the majority of fiber lasers use fiber Bragg grating (FBG) as a reflective mirror for their feedback. A fiber Bragg grating is a photonic structure obtained by periodically varying the refractive index of the core of an optical fiber. Such structures have the property of reflecting a fraction of the incident light at positions where the refractive index changes. If the reflected light waves have the same phase, the amplitudes add up due to constructive interference. This situation is referred to as the Bragg condition and occurs at a specific wavelength called Bragg wavelength. A fiber Bragg grating is, therefore, a wavelength selective mirror, which reflects only light waves at frequencies that fall within its reflection bandwidth and transmits the rest. Using FBG as feedback mirrors, narrow linewidth operation of fiber lasers can be achieved (Spiegelberg et al. 2004; Ball et al. 1991). The fiber Bragg grating can either be spliced onto the fiber end or directly printed into the active fiber core, the latter case reduces the number of splices hence the cavity loss, resulting in improved efficiency. In other words, the Bragg condition is simply the requirement that both energy and momentum conservations must be satisfied. Energy conservation requires that the frequency of incident and reflected light wave does not change ($\lambda f_i = \lambda f_r$), whereas momentum conservation requires that the sum of incident wavevector and grating wavevector is equal to the wavevector of the scattered radiation as in equation 5.1 below:

$$k_i + K = k_r \tag{5.1}$$

The grating wave vector K has a magnitude $2\pi/\Lambda$, where Λ is the grating period and a direction normal to the grating plane. The incident and diffracted wavevectors are equal in magnitude but have opposite directions. Therefore, the momentum conservation condition can be written as:

$$2\left(\frac{2\pi n_{eff}}{\lambda_B}\right) = \frac{2\pi}{\Lambda} \tag{5.2}$$

Which can be written as:

$$\lambda_B = 2n_{eff}\Lambda \tag{5.3}$$

Equation (5.3) represents the Bragg condition, where λ_B is the Bragg wavelength and n_{eff} the effective refractive index of the fiber.

Optical field propagation in fiber Bragg grating is described by the coupled mode wave theory (McCall 2000; Liau et al. 2015) which is succinctly described in this section. The reader interested in a detailed analysis of the theory can find it in references (Erdogan 1997; Hill and Meltz 1997; McCall 2000). Mathematical description of coupled wave theory is done with coupled-mode equations. Solving the coupled-mode equations allows the analysis of fiber Bragg grating spectral characteristics like reflection and transmission spectra, spectral width and side lobes etc.

The coupled-mode equations are derived from the general electromagnetic wave equation in the matter. In the "synchronous approximation" (Kogelnik 1988)these equations can be written as:

$$\frac{dR(z)}{dz} = i\hat{\sigma}R(z) + ikS(z) \tag{5.4}$$

$$\frac{dS(z)}{dz} = -i\hat{\sigma}S(z) - ik^*R(z) \tag{5.5}$$

where $S(z)$ and $R(z)$ are the complex amplitudes of the forward propagating radiation A and backward propagating optical radiation B given by:

$$R(z) = A(z)exp(i\delta z - \phi/2) \tag{5.6}$$

and

$$S(z) = B(z)exp(-i\delta z + \phi/2) \tag{5.7}$$

where k is the "AC" coupling coefficient and $\hat{\sigma}$ is the general "DC" self-coupling coefficient expressed as:

$$\hat{\sigma} \equiv \delta + \sigma - \frac{1}{2}\frac{d\phi}{dz} \tag{5.8}$$

where δ is the frequency detuning parameter defined as:

$$\begin{aligned} \delta &= \beta - \frac{\pi}{\Lambda} \\ &= \beta - \beta_D \\ &= 2\pi n_{eff}\left(\frac{1}{\lambda} - \frac{1}{\lambda_D}\right) \end{aligned} \tag{5.9}$$

where $\lambda_D = 2n_{eff}\Lambda$ represents the "design wavelength" for an infinitesimally weak grating ($\delta n_{eff} \rightarrow 0$) with a period Λ. When $\delta = 0$ equation (5.34) reduces to

$\lambda = 2n_{eff}\Lambda$, the Bragg condition obtained with equation (5.3), the complex coefficient σ represents a factor on which depend absorption loss in the grating given by $\alpha = 2Im(\sigma)$ and the derivative $(1/2)d\phi/dz$ represents a possible chirp of the grating period, where $\phi(z)$ is the phase.

For the simple case of single-mode fiber Bragg grating, the expressions of "DC" and "AC" coupling coefficients reduces to:

$$\sigma = \frac{2\pi}{\lambda}\overline{\delta n_{eff}} \tag{5.10}$$

$$k = \frac{\pi}{\lambda}v\overline{\delta n_{eff}} \tag{5.11}$$

For uniform gratings, $d\phi/dz = 0$, $\overline{\delta n_{eff}}$ is constant, therefore k, σ and $\hat{\sigma}$ are also constants and equations (5.4) and (5.5) reduce to coupled first-order differential equations with constant coefficients and can be solved analytically with appropriate boundary conditions. The boundary conditions are $S(0) = 1$, which means the forward incident wave on the grating is constant and $R(L) = 0$, which means, there is no backward wave entering the grating its end. By solving the equations, one can obtain the amplitude ρ and power R reflection coefficients given by:

$$\rho(L, \lambda) = \frac{S(0)}{R(0)} = \frac{-ksinh(\gamma L)}{\hat{\sigma}sinh(\gamma L) + i\gamma cosh(\gamma L)} \tag{5.12}$$

and

$$R(L, \lambda) = |\rho|^2 = \frac{sinh^2(\gamma L)}{cosh^2(\gamma L) - \frac{\hat{\sigma}^2}{k^2}} \tag{5.13}$$

where $\gamma = \sqrt{k^2 - \hat{\sigma}^2}$.

From equation (5.13) the maximum reflectivity is found for $\hat{\sigma} = 0$ as:

$$R_{max} = tanh^2(kL) \tag{5.14}$$

This maximum reflectivity occurs at the wavelength

$$\lambda_{max} = \left(1 + \frac{\overline{\delta n_{eff}}}{n_{eff}}\right)\lambda_D \tag{5.15}$$

Equation (5.15) shows that the wavelength λ_{max} corresponding to the maximum reflectivity drifts from the design wavelength λ_D by a factor $\frac{\overline{\delta n_{eff}}}{n_{eff}}\lambda_D$. When the wavelength of the incident radiation approaches, a bandgap is formed as a result of the periodic modulation of the refractive index. The Bandwidth of the fiber Bragg grating is then defined as

$$\Delta\lambda_{FW} = v\frac{\delta\overline{n_{eff}}}{n_{eff}}\lambda_D \tag{5.16}$$

For uniform fiber Bragg grating, it is useful to define the bandwidth between the first zeros as:

$$\Delta\lambda_0 = v\frac{\delta\overline{n_{eff}}}{n_{eff}}\sqrt{1 + \left(\frac{\lambda_D}{v\delta\overline{n_{eff}}}\right)^2}\lambda_D \tag{5.17}$$

Equations (5.12)–(5.17) plays an important role in designing a uniform fiber Bragg grating and calculate its reflectivity spectrum. Figure 5.8 illustrates the reflection spectrum of a 10 mm uniform fiber Bragg grating with an index change $\delta_{neff} = 1e - 3$.

It can be seen from the theory of fiber Bragg grating developed in the previous section that the most important parameters in designing a uniform fiber Bragg grating are the length and the index change. This can be observed from equation (5.14) where the maximum reflectivity of the grating is a function of the length L and the coupling coefficient k which in terms is a function of index change. Equation (5.14) shows that the maximum reflectivity increases if the product kL referred to in the literature as "grating strength" increases. For a uniform grating, the visibility $v = 1$, equation (5.42), therefore reduces to:

$$\Delta\lambda_0 = \frac{1}{n_{eff}}\sqrt{\left(\delta\overline{n_{eff}}\right)^2 + \left(\frac{\lambda_D}{L}\right)^2}\lambda_D \tag{5.18}$$

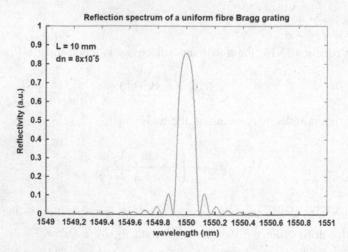

FIGURE 5.8   Reflection spectra of a uniform fiber Bragg grating.

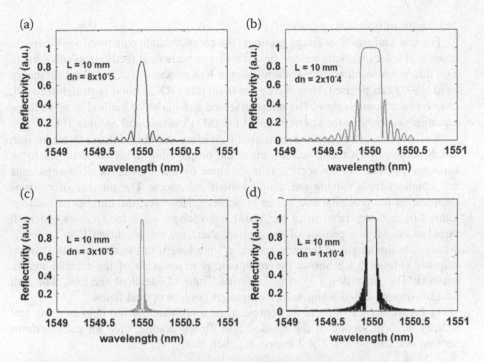

FIGURE 5.9 Fiber Bragg grating spectra for various lengths and index modulation (a) L = 10 mm, $\delta n = 8*10^{-5}$; (b) L = 10 mm, $\delta n = 2*10^{-4}$; (c) L = 50 mm, $\delta n = 3*10^{-5}$; (d) L = 50 mm, $\delta n = 1*10^{-4}$.

Equation (5.18) shows that the bandwidth of any non-chirped grating is wider for larger index changes and narrower for longer fiber Bragg gratings. Figure 5.9

Substituting equation (5.11) in (5.18) yields:

$$\Delta\lambda_0 = \frac{\lambda_D^2}{n_{eff}L}\sqrt{1 + \left(\frac{kL}{\pi}\right)^2} \tag{5.19}$$

which relates the grating bandwidth with the grating strength. For weak gratings, equation (5.19) reduces to:

$$\Delta\lambda_0 \frac{2\lambda_D}{N} \tag{5.20}$$

where N is the number of periods inside the grating. On the other hand, for strong gratings (large value of kL) equation (5.19) becomes:

$$\Delta\lambda_0 = \frac{\delta\overline{n}_{eff}}{n_{eff}}\lambda_D \tag{5.21}$$

Non-uniform fiber Bragg gratings:

For non-uniform fiber Bragg gratings, the coupled-mode equations don't have an analytical solution, therefore one has to resort to numerical methods to solve them. For this purpose, direct integration using a Runge-Kutta (RK) method (Flannery et al. 1992) can be used. However, even though the RK method is straightforward, convergence is often slow. The alternative and most popular method to solve these equations is the transfer matrix method (TMM) (Yamada and Sakuda 1987).

The essence of the method consists of dividing the non-uniform structure into small sections such that in each section the propagating fields are assumed to be uniform. Each one of the section is represented by a 2×2 matrix whose elements are obtained from solving the coupled-mode equations. The number of sections depends on the precision one desire to achieve. However, the number of sections cannot be arbitrary large since the TMM is no longer valid for sections of length equal to only a few periods of the grating, therefore the condition $L \gg L^m \gg \Lambda$, where L is the length of the grating and L^m the length of the mth section is often required. Figure 5.10, Shows the schematic representation of the transfer matrix method. The fiber Bragg grating is divided into M identical sections. The mth section is represented along with its input and output optical fields.

In Figure 5.10, if R_m and S_m represent the amplitudes of the field entering and exiting the Mth section of the divided fiber Bragg grating, then for each uniform section one can write a 2×2 matrix T_m such that:

$$\begin{pmatrix} S_m \\ R_m \end{pmatrix} = T_m \begin{pmatrix} S_{m-1} \\ R_{m-1} \end{pmatrix} \qquad (5.22)$$

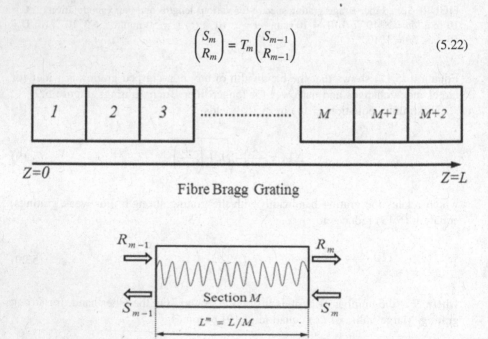

FIGURE 5.10 Schematic representation of the transfer matrix method. The fiber Bragg grating is divided into M identical sections. The mth section is represented with its input and output optical fields.

where the elements of the matrix T_m describing the Mth section is given by:

$$T_{11}^m = \left[\cosh(\gamma^m L^m) + i\frac{\hat{\sigma}^m}{\gamma^m} \sinh(\gamma^m L^m) \right]$$

$$T_{12}^m = -\frac{k^m}{\gamma^m} \sinh(\gamma^m L^m)$$

$$T_{21}^m = -\frac{k^m}{\gamma^m} \sinh(\gamma^m L^m)$$ (5.23)

$$T_{22}^m = \left[\cosh(\gamma^m L^m) - i\frac{\hat{\sigma}^m}{\gamma^m} \sinh(\gamma^m L^m) \right]$$

where γ^m, $\hat{\sigma}^m$, and ϕ^m are defined as:

$$\gamma^m = \sqrt{(k^m)^2 - (\hat{\sigma}^m)^2},$$

$$\phi^m = \phi_{initial} + \Sigma_{m=1}^{M-1} 2\pi L^m / \Lambda^m,$$ (5.24)

$$\hat{\sigma}^m = 2\pi n_{eff}(1/\lambda_L - 1/\lambda_B^m),$$

In equation (5.24), Λ^m is the grating period in the mth section, ϕ^m its phase and $\phi_{initial}$ its initial phase. $\hat{\sigma}^m$ represents the deviation of the propagation constant at the Bragg wavelength. In the computations, the initial phase is often given the value zero for simplicity.

The transfer matrix of the Mth section of the grating is written:

$$T_m = \begin{bmatrix} T_{11}^m & T_{12}^m \\ T_{21}^m & T_{22}^m \end{bmatrix}$$ (5.25)

and the entire grating is described by a matrix T written as:

$$T = T_M T_{M-1} \dots T_\phi \dots T_3 T_2 T_1$$ (5.26)

Finally, the relation between the input and output fields of the grating will be given by:

$$\begin{bmatrix} S(L) \\ R(L) \end{bmatrix} = T \begin{bmatrix} S(0) \\ R(0) \end{bmatrix}$$ (5.27)

where $R(L)$ is the reflectivity of the fiber Bragg grating.

5.2.2.2 Fabry-Perot Resonator

A Fabry-Perot resonator is formed by placing miniature planar dielectric resonator in intimate contact with the end of the doped fiber which is either polished or cleaved perpendicular to the fiber axis. The pump beam is usually focused into the

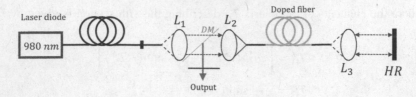

FIGURE 5.11 Basic setup of a Fabry-Perot fiber laser. The pump power is coupled into the doped fiber through a dichroic mirror and the laser light is extracted through the same dichroic mirror after a full round trip in the laser cavity.

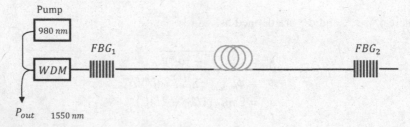

FIGURE 5.12 Fabry-Perot Fiber laser using fiber Bragg grating as reflector mirrors.

fiber through the high reflecting mirror which must be dichroic to transmit only the pump light as shown in Figure 5.11. This first approach has been used successfully in laboratory conditions; however, the approach is cumbersome and not suitable for mass production. The most popular approach to make a Fabry- Perot cavity is to use fiber Bragg gratings as mirrors as illustrated in Figure 5.12 (Wang 1974; Pradhan et al. 2006; Li et al. 2015; Wong et al. 2010b; Canning et al. 2003).

5.2.2.3 *Distributed Bragg Reflector Fiber Laser (DBR)*

For many applications, the CW laser frequency mustn't change randomly. Such a laser can be designed by shortening the cavity enough so that the longitudinal mode spacing is comparable with the Bragg grating bandwidth. For a Bragg grating bandwidth below 0.2 nm, it is desirable to have mode spacing of the order of 10 GHz i.e. cavity length in the order of 1 cm. This type of configuration is called Distributed Bragg Reflector (DBR) fiber laser (Lin et al. 2012; Shi, n.d.; Canning et al. 2003). The difference between a Fabry-Perot cavity using fiber Bragg gratings and a DBR lies in the fact that the gain medium of a DBR fiber laser is comparable in size with the length of the fiber Bragg grating. DBR fiber lasers are mostly used as high-resolution sensors (Lyu et al. 2013; Beregovski et al. 1998; Quintela and Laarossi 2014; Wong et al. 2010a; Zhang et al. 2016; Rodríguez-Cobo et al. 2014).

5.2.2.4 *Distributed Feedback*

A distributed feedback (DFB) cavity is made of one grating printed throughout the amplifying medium to provide required feedback (V. C. Lauridsen et al. 1998; Vibeke Claudia Lauridsen et al. 1997). The feedback of a DFB fiber laser is therefore distributed along the length with a phase shift printed into the middle of

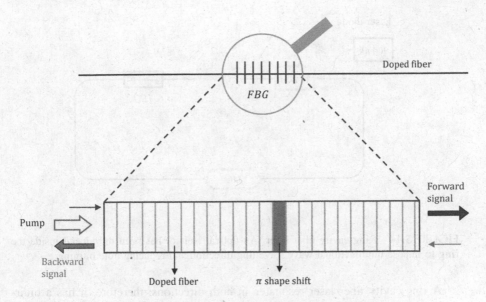

FIGURE 5.13 Schematic representation of the distributed feedback fiber laser.

the device as illustrated in Figure 5.13 to ensure single-frequency operation. For this reason, DFB fiber lasers are relatively short. They are usually no longer than 10 cm because beyond this length fiber Bragg gratings become very difficult to manufacture. The operation of a DFB fiber laser is based on the coupling of two counter-propagating waves via backward Bragg reflection from the periodic perturbation of the refractive index of the fiber. The theoretical analysis of DFB laser operation was first proposed in 1972 by Kogelnik and Shank for semiconductor lasers (Kogelnik and Shank 1972).

Later in the mid-1990s, Kringlebotn et al. demonstrated the concept in erbium-doped fiber (Kringlebotn et al. 1998). Later, the same group reported numerical models of DFB fiber lasers in Erbium (V. C. Lauridsen et al. 1998), Erbium-Ytterbium (Vibeke Claudia Lauridsen et al. 1997) and a method for optimising the output power (V. C. Lauridsen et al. 1999). Soon after another type of DFB fiber lasers were reported in photonic crystal fibers (Ndergaard 2000), with apodized fiber Bragg grating (A. I. Azmi and Peng 2008) or high power output (Ibsen et al. 1999). Since then, many research has been reported focusing mostly on improving efficiency (Kuthan Yelen et al. 2004; Horak et al. 2005; K Yelen et al. 2005) as well as applications in telecommunication (Varming et al. 1997) and sensing (Sorin et al. 2010; Foster et al. 2017). Theoretical analysis and characteristics of the DFB fiber laser are discussed in detail in section 5.3.4 of this chapter.

5.2.2.5 Ring Cavities

A ring cavity is easily obtained by forming a loop with a doped fiber and an optical coupler as illustrated in Figure 5.14. The pump light is injected into the resonator through a wavelength division multiplexer (WDM) coupler placed in the loop.

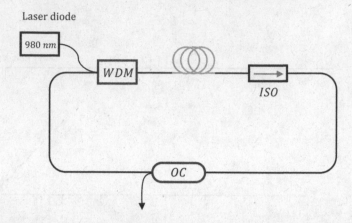

FIGURE 5.14 Ring cavity fiber laser. An optical isolator has been introduced inside the ring to impose unidirectional wave travelling direction and reducing hole burning.

A ring cavity fiber laser resonates in both directions; therefore, it has a bidirectional output. Its conversion efficiency is only half as high as that of a Fabry-Perrot cavity. This limitation can be removed by introducing an optical isolator inside the loop which forces unidirectional operation (see Figure 5.14). The unidirectional operation also permits to suppress spatial hole burning and prevent mode hopping.

Another advantage of a ring cavity fiber laser is the ability to deliver narrow-linewidth output. According to Schawlow-Townes theory, the linewidth of a single longitudinal mode laser is given by:

$$\Delta \nu_{laser} = \frac{\pi h \nu (\Delta \nu_c)^2}{P_{out}} \tag{5.28}$$

where $\Delta \nu_c$ is the FWHM of the Lorentzian linewidth of the passive resonator mode, P_{out} the output power of the laser, h the Planck constant and ν the frequency of the oscillating mode. The linewidth of the resonator mode of the ring cavity fiber laser is given by:

$$\Delta \nu_c = \frac{1}{2\pi \tau_c} \tag{5.29}$$

where τ_c is the photon cavity decay time given by:

$$\tau_c = \frac{nL}{C(kL - lnR_{oc})} \tag{5.30}$$

where L is the length of the cavity, C the speed of light in free space, k the attenuation coefficient per unit length and R_{oc} the reflectivity of the output coupler.

Equations (5.29) and (5.33) show that the linewidth of the resonator mode $\Delta \nu_c$ is inversely proportional to the photon cavity decay time τ_c which in term is proportional to the cavity length. It is then obvious that a longer laser cavity corresponds to a narrow linewidth operation of the fiber laser.

5.3 CONTINUOUS WAVE FIBER LASER MODELING

5.3.1 Formalism

The operation of a fiber laser involves numerous phenomena which influence each other in a very sophisticated manner. To understand the impact of each one of these effects on the characteristics of the fiber laser, one needs to find a tool for manipulating each effect. A model of a fiber laser is an excellent tool in the process of designing the device of interest as it allows one to conduct tests and experiments which would be both difficult and costly in a laboratory environment. To optimize the performance and to make a judicious choice of laser components, a model is required. A mathematical model is usually a set of mathematical equations governing the behaviour of the device. These equations are derived from well-known phenomena translated into mathematical equations. Fiber lasers are usually modelled by rate equations which are a set of time-dependent differential equations obtained from the energy level of the rare-earth ions inside a glass host. These equations result from the interactions of pump and laser field intensity with the doped ions as well as interaction energy exchange between ions of the same species or different species. Parameters appearing in the equations are pump and signal powers, absorption and emission cross-section, overlap factors for pump and signal intensities, radiative and non-radiative lifetimes, often obtained by measurement. After solving the equations, one can vary parameters to optimize the fiber laser performance. Interesting characteristics to look at include pump threshold, output power as a function of the pump power, output power as a function of doped fiber length, output power as a function of output coupling ratio, to name a few. When modelling a fiber laser, one must find a trade-off between oversimplification and a too sophisticated model. An oversimplified model is usually easy to solve and excellent in terms of computational time but it doesn't give a good insight into the device under investigation. On the other hand, too sophisticated models give rise to complicated systems of equations that are not easy to solve and perform poorly in terms of computational time.

Numerous theoretical models of rare-earth doped fiber lasers have been investigated and results presented in the literature. These include models for Fabry-Perot, ring cavity, distributed Bragg reflectors, and distributed feedback fiber laser. The general approach used is the same as for bulk lasers, combining the rate equations of the laser which describes stimulated and spontaneous transitions taking place inside the laser gain medium, with the propagation equations for laser and pump optical field along with the gain medium. If we consider unidirectional pumping, the pump is injected at one end of the laser. There exist two counter-propagating fields inside the laser cavity, which with the pump power make a system of three equations. This set of equations is completed by boundary conditions dictated by the reflections at mirrors both ends of the laser. At the left end of the laser, the pump power is equal to the launched power, the forward propagating laser power is equal to the backward propagating power times the reflectivity of the mirror, and the backward propagating signal at the right end of the cavity is equal to the forward propagating laser power times the right end mirror reflectivity. Modelling of other cavities is similar to that of the Fabry-Perot. The only difference comes from the boundary conditions that change from one type of laser to another. The resulting system of coupled differential equations can be solved by many numerical methods such as Runge-Kutta, relaxation methods, finite difference method, etc. There are several commercial software packages for modelling fiber lasers, but to derive a model always provide one with unique insight because every model is unique in terms of its parameters and goals.

Most of the models presented here are done for Erbium and Ytterbium-doped fiber laser, but the methods used are perfectly suitable for other types of rare-earth-doped species with more or less complexities of rate equations.

5.3.2   Linear Cavity Fiber Laser

Modelling linear cavity fiber laser is similar to the ring cavity except that the boundary conditions are different. This type of fiber lasers uses a piece of doped fiber between two mirrors as illustrated in Figure 5.15.

The rate equation for linear cavity fiber laser is similar to the one described for the ring laser or any other rare-earth doped medium. For the sake of simplification, a 2-energy level system is adopted to conduct the computation. This approach is valid because of the short lifetime of $^4I_{11/2}$ manifold of erbium. All the ions accumulated by pumping in the energy level $^4I_{11/2}$ are considered to repopulate the metastable level $^4I_{13/2}$. The resulting rate equations are given as:

$$\frac{dN_2(z,t)}{dt}$$

$$= \frac{\Gamma_p \lambda_p}{hCA}\left(\sigma_{ap}(\lambda_p)N_1 - \sigma_{ep}(\lambda_p)N_2\right)P_p + \frac{\Gamma_s \lambda_s}{hCA}(\sigma_{as}(\lambda_s)N_1 \tag{5.31}$$

$$- \sigma_{es}(\lambda_s)N_2)(P_s^+ + P_s^-) - \frac{N_2}{\tau}$$

$$\frac{dN_1(z,t)}{dt} = -\frac{dN_2(z,t)}{dt} \tag{5.32}$$

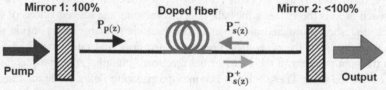

FIGURE 5.15   Schematic representation of a Fabry-Perot fiber laser.

$$N_t = N_1 + N_2 \tag{5.33}$$

where N_1 and N_2 are the population densities of the ground state and upper-level metastable level respectively, σ_{ap} the absorption cross-section at pump power, σ_{ep} the emission cross-section at the pump wavelength, σ_{as} the absorption cross-section at the laser wavelength and σ_{es} the emission cross-section at the signal wavelength. Γ_p and Γ_s are the overlap factors at pump and laser wavelength respectively. τ is the lifetime of the metastable level $^4I_{13/2}$, A the area of the doped core, C the speed of light in free space and h the Planck's constant.

For continuous wave operation the laser reaches its steady state, the derivatives in (5.31) and (5.33) become zero and the system of equations (5.31)–(5.33) has a solution given by:

$$N_2(z) = \frac{\dfrac{[P_p^+(z) + P_p^-(z)]\sigma_{ap}\Gamma_p\lambda_p}{hCA} + \dfrac{[P_s^+(z) + P_s^-(z)]\sigma_{as}\Gamma_s\lambda_s}{hCA}}{\dfrac{[P_p^+(z) + P_p^-(z)](\sigma_{ap} + \sigma_{ep})\Gamma_p\lambda_p}{hCA} + \dfrac{[P_s^+(z) + P_s^-(z)](\sigma_{as} + \sigma_{es})\Gamma_s\lambda_s}{hCA} + \dfrac{1}{\tau}} N_t \tag{5.34}$$

Equation (5.38) is solved together with the following propagation equations, describing the propagation of the optical fields inside the cavity:

$$\pm \frac{dP_p^\pm(z)}{dz} = -\Gamma_p[\sigma_{ap}N - (\sigma_{ap} + \sigma_{ep})N_2(z)]P_p^\pm(z) - \alpha_p P_p^\pm(z) \tag{5.35}$$

$$\pm \frac{dP_s^\pm(z)}{dz} = -\Gamma_s[\sigma_{as}N - (\sigma_{as} + \sigma_{es})N_2(z)]P_s^\pm(z) - \alpha_s P_s^\pm(z) \tag{5.36}$$

where the sign $\pm$ represents the forward and backward propagating fields respectively. σ_{as}, σ_{es}, σ_{ap}, and σ_{ep} represent the absorption and emission cross-section at the laser and pump wavelength respectively. Finally, Γ_p, Γ_s, α_p, and α_s represent the overlap factors and background losses at pump and laser wavelength respectively and N the population density of rare-earth ions inside the glass matrix. In the simple ideal case of a two-level system like Erbium pumped at 980 nm, N_2 can be obtained by solving the rate equations analytically. The propagating equations (5.35) and (5.36) are completed by the boundary conditions for forward (5.37) and backward (5.38) pumping given as:

$$
\begin{aligned}
P_p^+(0) &= P_{pF} \\
P_p^-(L) &= P_p^+(L) \\
P_s^+(0) &= R_1 P_s^-(0) \\
P_s^-(L) &= R_2 P_s^+(L)
\end{aligned}
\tag{5.37}
$$

$$P_p^-(L) = P_{pB}$$
$$P_p^+(0) = P_p^-(0)$$
$$P_s^+(0) = R_1 P_s^-(0)$$
$$P_s^-(L) = R_2 P_s^+(L)$$

$$(5.38)$$

where P_{pF} and P_{pB} represent the launched pump powers at the left and right end of the cavity respectively and R_1 and R_2 the reflectivities of mirror 1 and mirror 2 respectively.

For illustration, the equations (5.34)–(5.36) are solved using a relaxation method with a 4th order Runge-Kutta algorithm. The parameters used in the equations are the same as the ones described in Table 5.2.

Additional Parameters

Parameter	Value	Unit
Erbium radiative lifetime	10	ms
Core diameter	4	μm
Core numerical aperture (NA)	0.2	–
Normalized frequency (V number)	2.56	–
Mode Field diameter (MFD)	4.22	μm
Overlap factor	0.83	–
Erbium concentration	9.8×10^{24}	Ions/m^3
Speed of light in vacuum	3×10^8	m/s
Planck constant	6.626×10^{-34}	m^{2}kg/s

Three pumping configurations are possible with this type of lasers namely, forward pumping, where the output mirror is situated opposite to the pumping side, backward pumping, where the output and pump are on the same end of the cavity and dual pumping. The details of the algorithms for forwarding and backward pumping are described below.

TABLE 5.2
Integration Parameter

Wavelength (nm)	Absorption cross-section (m^2)	Emission cross-section (m^2)
975	2.06×10^{-25}	0
980	1.87×10^{-25}	0
1529	6.80×10^{-25}	6.13×10^{-25}
1530	6.79×10^{-25}	6.15×10^{-25}
1531	6.72×10^{-25}	6.22×10^{-25}
1550	3.17×10^{-25}	4.44×10^{-25}

5.3.2.1 Backward Pumping

The pump power is launched at $z = L$, therefore the pump power, forward and backward laser powers are not known at $z = 0$. To implement the algorithm, the laser cavity is first divided into several uniform sections.

The algorithm starts by setting initial values for the backward pump, forward and backward signal as $P_p^-(0)_{guess}$, $P_s^+(0)_{guess}$ and $P_s^-(0)_{guess}$.

The fieldare then propagated through the cavity section by section and the fields at the end of each section is computed using a 4th order Runge-Kutta method to obtain $P_s^+(L)$, $P_s^-(L)$ and $P_p^-(L)$ at $z = L$.

At $z = L$, apply the boundary condition and check if the desired tolerance is achieved. If yes, the algorithm ends with convergence.

If the error exceeds the tolerance value, $P_s^- = R_2 P_s^+(L)$, $P_p^+(L) = P_{pB}$ and $P_s^+(L)$ are taken as the initial values and the equations are integrated backwards using a 4th order Runge-Kutta method to obtain $P_s^+(0)$, $P_s^-(0)$, and $P_p^-(0)$

At $z = 0$, apply the boundary conditions and check if the desired tolerance is achieved. If yes, the algorithm ends with convergence.

If the error exceeds the tolerance value, $P_s^+(0) = R_1 P_s^-(0)$, $P_s^-(0)$, and $P_p^-(0)$ are taken as initial values and the equations are integrated from $z = 0.$ to $z = L$ using the 4th order Runge-Kutta method to obtain $P_s^+(L)$, $P_s^-(L)$ and $P_p^-(L)$.

Steps 3 to 6 are repeated until convergence is achieved, more precisely, this means until the change of the value of the fields don't change significantly from one iteration to the next or until one of the quantities $|P_s^-(L) - R_2 P_s^+(L)|$, $\left| P_s^-(L) - P_{pB} \right|$ or $|P_s^+(0) - R_1 P_s^-(0)|$ falls below the tolerance value.

The same algorithm will be used for the case of the forward pumping, with the initial guesses taken from the right end of the fiber cavity, as well as for the dual pumping.

5.3.2.2 Pump Absorption for Various Lengths and Explanation

Pump absorption in a 1.5 m long fiber decays almost linearly, the slope of decay is bigger at high pump powers than low powers. This situation can be explained by the saturation of Erbium absorption. Because of the higher intensity developed in the core of fiber lasers, which result from the small cross-sections even at moderate pump powers, therefore the pump power is many orders of magnitude higher than the saturation power of the fiber. At these levels of pumping, the rate of absorption and stimulated emission are almost equal, the medium becomes transparent and no more power can be absorbed. The observed absorption is only possible due to spontaneous emission, which is the only possible way to dissipate the energy accumulated in the medium. As the rate of spontaneous energy per unit of length is almost constant, the resulting pump power will be linear. It can also be seen that the decay of pump power is much pronounced at high pump power than low pump powers to confirm the observation just discussed.

For the longer fiber of 10 metres, the same trend is observed as illustrated in Figure 5.16. Observation also shows that the rate of absorption of pump power in

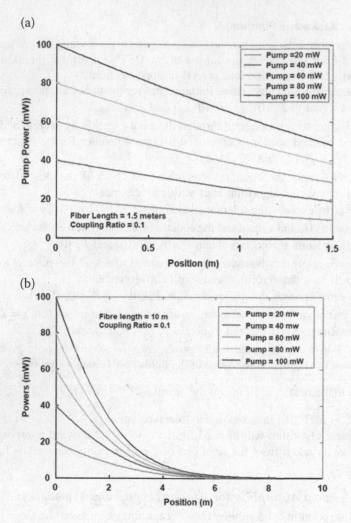

(a)

(b)

FIGURE 5.16 Pump Power Distribution for (a) a 1.5 m long fiber and (b) 10 m long fiber.

the first portion of the fiber is much higher for high pump power than for lower pump powers. As an illustration, for a pump power of 100 mW, 80% of the pump is absorbed in the 2.8 metres. This value falls to 70% within the same distance for the pump power of 40 mW.

This behaviour is easily explained as the high pump power creates higher population inversion and higher gain. Therefore, the laser radiation will grow quickly because the rate of stimulated emission is directly related to population inversion. The large stimulated emission will quickly depopulated the excited laser, repopulate the ground state and increases the rate of absorption that reduces the population inversion resulting in higher pump absorption. At a distance of about 6 metres, the pump is depleted, and laser radiation will start to be absorbed from that point. The interactions between pump and laser distribution as well as population density distribution inside the cavity are discussed in section 5.3.2.3 below.

5.3.2.3 *Internal Cavity Quantities Distribution*

Figure 5.17 illustrates the distribution of various quantities within the cavity of the linear cavity fiber laser. For the simulation, a 6 metres long fiber was considered, the doping concentration was 1.2×10^{25} ions/m^3 and the reflectivity of the output mirror was 10%. The reflectivity value of 10% was chosen arbitrarily as it will be shown in a subsequent section that the output coupler reflectivity does not have much influence on the output of the laser.

Figure 5.17 (a) and (b) show typical pump and signal light power distribution along the fiber laser cavity for forward and backward pump respectively. The pump power was 100 mW for a fiber length of 6 m. This value corresponds to the point where launched power becomes completely exhausted inside the cavity. As expected, the pump power attenuated exponentially along the propagation direction. Because of the rather low absorption and leakage loss (0.005), it takes few meters to absorb the pump. This value can quickly fall in the range of centimetres for highly doped fiber with high absorption fiber such as in the case of short cavity fiber lasers.

For forward pump, (Figure 5.17 (a)) the forward laser power increases rapidly first. Then its rate of increment gradually decreases, reaches a maximum and start decreasing. The intensity of the radiation decreases because of pump exhaustion and laser radiation reabsorption. For backward pump, the laser power first increases

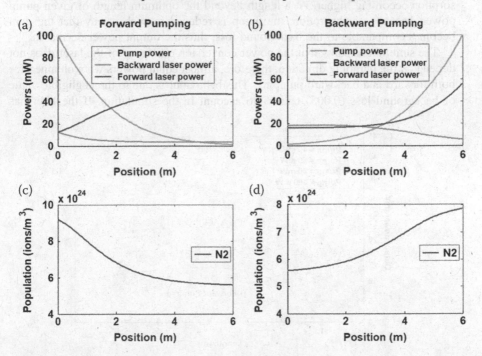

FIGURE 5.17 Distributions for 10 m and 100 mW pump.

slowly, then increase rapidly towards the end of the fiber. This behaviour is explained by the fact that the gain is higher towards the left end of the fiber for forward pumping whereas it is higher towards the right end of the fiber for backward pumping configuration. This can be seen in Figure 5.17 (c) and (d) which illustrates the distributions of upper energy level population density $N_2(z)$ along the fiber for forward and backward pumps respectively.

5.3.2.4   *Output Power Versus Fiber Length for Different Pump Powers*

The output laser power versus the fiber length for four pump powers of 40, 60, 80, and 100 mW and the output coupling ratio is 10% is illustrated in Figure 5.18. for a forward pumping scheme. This simulation is essential because it allows one to determine the optimum length of the doped fiber. The output power is defined as $P_{out} = (1 - R_{oc})P_s^+(L)$. P_{out} increases first with fiber length then start decreasing after reaching a maximum. The optimum fiber length is found around 3.9 m for the minimum pump power of 40 mW and increases with pump power reaching a value of around 4.8 m for a pump power of 100 mW. The existence of the optimum length is explained by the rare-earth. For a doped fiber shorter than the optimum length, the total number of erbium ions is small, therefore only a fraction of the pump power is absorbed, and the pump becomes saturated. As a result, the pump will pump all the ions from the ground to the excited state, and this condition will be maintained throughout the length of the fiber. If the fiber length is increased the quantity of active ions increases and the pump power decreases at the same time due to absorption becoming higher. At a length beyond the optimum length of given pump power, because of absorption, the pump is reduced in such a way that the gain becomes comparable to the background loss, thus the output power decreases.

The simulations show that the power conversion efficiency of the laser does not depend on the pumping direction, therefore, almost equal powers were obtained for both forward and backward pumping. This behaviour is due to the negligible value of background loss (0.005) taken into account in the simulation. If the fiber has

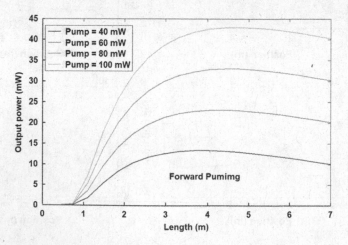

FIGURE 5.18   Output power versus fiber length.

substantial background loss, backward pumping is more efficient. This is because the signal is strong near the pump input end holding the excitation density down in that region so that the pump light is efficiently absorbed by the ions before parasitic losses start absorbing it.

Figure 5.19 illustrate the dependence between output power and pump power of the laser for 3 different pump powers namely 3, 6, and 9 metres in a forward pumping configuration. The output power monotonically increases with continuously increasing pump power, in a good linear relationship. However, there is a clear dependence between the length of the doped fiber, the pump threshold, and the slope efficiency. The laser with the shorter length of doped fiber (3 m) has the lowest pump threshold of around 10 mW, which is increased to 11.5 mW for 6 metres and almost 22 mW for the longest length of 9 m. Again this high absorption for longer fiber is explained by the higher background loss of the long fiber compared to shorter ones. The slope efficiency, on the other hand, was 67% for 3 metres fiber, increased to 77% for 6 metres and decreased back to 74% for 9 metres. The variation of slope efficiency also confirms the existence of the optimum fiber length.

It is worth mentioning that the previous results were obtained for an output coupling ratio of 10%. In the next section 5.3.2.5 the dependency of the output power as a function of coupling ratio is studied.

5.3.2.5 Output Power Versus Coupling Coefficient for Different Pump Powers and Optimum Length

The dependence of the output power on reflectivity for the forward pumping scheme is illustrated in Figure 5.20. The maximum power was obtained for a coupling ratio of 38% which corresponds to the optimum value. For low values of the output coupling ratio, the power of the signal reflected is too small, therefore, even though the gain can be large enough, the output will still be low. On the other

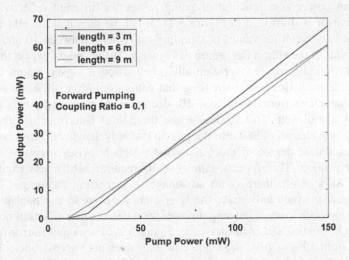

FIGURE 5.19 Output power as a function of pump power for different lengths and a coupling coefficient of 10%.

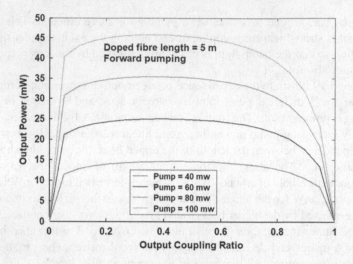

FIGURE 5.20 Output power versus output coupling ratio.

hand, a high reflectivity mirror at the output will couple only a tiny fraction of the laser radiation generated out of the resonator. Another parameter for optimizing the output power is the pumping wavelength. In the case of Erbium and Ytterbium-doped fiber laser, experience shows that the absorption cross-sections in the region of 970–980 nm are large enough so that even with only cleaved fiber with a reflection of 4% at the right fiber end and 100% at the left fiber end, one can still achieve lasing, though the threshold may be relatively high. It is also clear from Figure 5.20 that the optimum value of output coupling coefficient is independent of the pump power.

Output power as a function of pump power for different reflectivities of the output mirror is illustrated in Figure 5.21. It can be seen that the slope efficiency does not vary much for values of coupling coefficients between 10 and 70%. This corresponds to the almost flat region of Figure 5.20. One might expect that the laser threshold will be reduced substantially when using a higher output coupler reflectivity is used. However, increasing that value from 4% to 40%, thus decreasing the resonator loss from 14 dB to 4 dB, decreases the threshold power only to the value of a milliwatt. This is due to the three-level behaviour of erbium at the pumping wavelength of 980 nm: the gain at that wavelength becomes positive only for an excitation degree of more than about 50% but rises very fast for slightly larger excitation. Therefore, a gain of 4 dB requires hardly less excitation than 14 dB. As a result, there is no advantage in increasing the output coupler reflectivity; that would only make the laser more sensitive to any parasitic losses.

The slope efficiency of the output power for different reflectivities of output mirror can also be studied and is illustrated in Figure 5.21. For output mirror reflectivities varying from 10% to 70%, the slope efficiency does not vary significantly.

For Backward pumping slope efficiencies of 60%, 58%, and 62% were obtained for mirror reflectivities of 10%, 40%, and 60% respectively as illustrated in Figure 5.22. A huge contrast was observed when the reflectivity was changed from 60% to 80%.

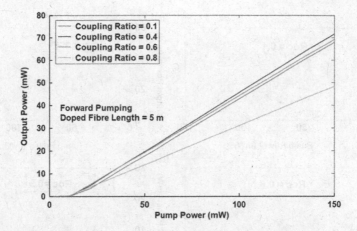

FIGURE 5.21 Output power as a function of pump power for different reflectivities of the output mirror.

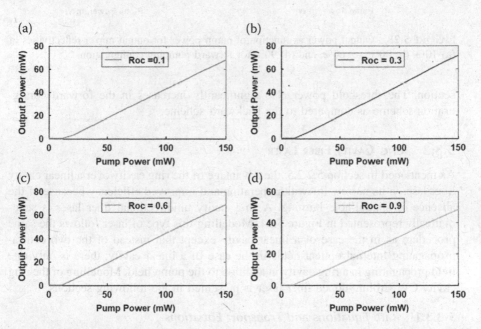

FIGURE 5.22 Output power as function of pump power for output mirror reflectivities of (a) 10%, (b) 30%, (c) 60%, and (d) 90% for backward pumping configuration.

Figure 5.23 illustrates the output power as a function of pump power for different reflectivities of output mirror in a forward pumping configuration. It can be observed that the output power of the forward pumping configuration is slightly lower than that of backward pumping configuration for the reasons described above in this

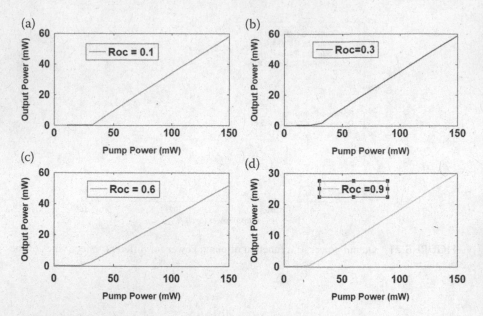

FIGURE 5.23 Output power as function of pump power for output mirror reflectivities of (a) 10%, (b) 30%, (c) 60%, and (d) 90% for forward pumping configuration.

section. The threshold power also significantly increases in the forward configuration scheme as compared to the backward scheme.

5.3.3 Ring Cavity Fiber Laser

As mentioned in section 5.2.2.5, the advantage of the ring cavity over a linear cavity fiber laser is its narrow linewidth operation and improved efficiency because of the absence of spatial-hole burning. A ring cavity unidirectional fiber laser is schematically represented in Figure 5.24. Modelling this type of laser follows the same procedure as in the case of a linear cavity except that instead of the two contra-propagating internal optical fields in the case of a linear cavity, there is only one field propagating in a ring cavity in addition to the pump field. Modelling of the ring cavity CW Erbium-doped fiber laser is presented in the following sections.

5.3.3.1 Rate Equations and Transport Equations

The continuous wave rare-earth-doped fiber laser is modelled by a set of ordinary differential equations, called rate equations, combined with propagations equations. The rate equations are derived from the transition between energy levels of the rare-earth doped fiber due to ion-ion and ion-light interaction. The propagation equations are derived from the pump and laser fields propagating inside the gain medium. The rate equations from the Erbium ions are assuming a two-level system:

$$N_0 = N_1 + N_2 \tag{5.39}$$

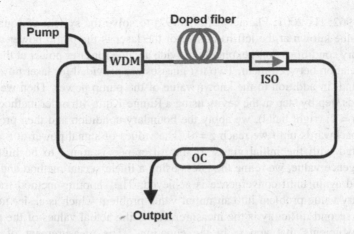

FIGURE 5.24 Schematic diagram of the ring cavity fiber laser used in the simulations.

$$\frac{\partial N_2}{\partial t} = \frac{\Gamma_p \lambda_p}{hCA}\left(\sigma_a(\lambda_p)N_1 - \sigma_e(\lambda_p)N_2\right)P_p + \frac{\Gamma_s \lambda_p}{hCA}(\sigma_a(\lambda_s)N_1$$
$$- \sigma_e(\lambda_s)N_2)(P_s^+ + P_s^-) - \frac{N_2}{\tau} \qquad (5.40)$$

These equations must be completed by the propagation equations for laser and pump filed intensity:

$$\frac{\partial P_p}{\partial z} = \Gamma_p\left(\sigma_e\left(\lambda_p\right)N_2 - \sigma_a\left(\lambda_p\right)N_1\right)P_p - \alpha_p P_p \qquad (5.41)$$

$$\frac{\partial P_s}{\partial z} = \Gamma_s(\sigma_e(\lambda_s)N_2 - \sigma_a(\lambda_s)N_1)P_s - \alpha_s P_s + 2\sigma_e(\lambda_s)N_2\frac{hC^2}{\lambda_s^3}\Delta\lambda_s \qquad (5.42)$$

In the equations (5.6) to (5.9), N_0 is the total Erbium ion population of the gain medium, N_1 and N_2 are the population densities of the ground state and excited state respectively. Γ_p and Γ_s are overlap factors for the signal and pump respectively. Only the part of the portion of the optical mode which overlaps with the Erbium ion distribution will stimulate absorption or emission from the Er^{3+} transitions. Since the Erbium is confined to the core of the optical fiber the mode intensity can more readily invert the Erbium ions.

5.3.3.2 Numerical Solutions

For the sake of simplicity, Erbium-doped gain medium is considered as a two levels system and unidirectional propagation is considered. The set of equations describing the fiber laser is a two-point boundary value problem. We used numerical methods described in the previous chapter to solve it. Here we use an algorithm involving a Runge-Kutta 4th order scheme along with a shooting method (Morrison et al. 1962; Ha 2001; Flannery et al. 1992) to solve the system of equations. The only value known at the left-hand side of the laser is the pump power as the laser boundary conditions don't explicitly provide a value for laser power at the boundary but a relation between them. Two fist guesses are provided for laser power at $z = 0$ (left hand) in addition to the known value of the pump power. Then we propagate the fields step by step in the cavity using a Runge-Kutta 4th order method. Once we reach $z = L$, (right hand), we apply the boundary condition and then propagate the fields backwards until we reach $z = 0$. The values of signal power at $z = 0$ is then compared with the initial guess. If the difference is found to be higher than a convergence value, we refine the guess using a linear secant method and propagate the field again until convergence is achieved. The shooting method transforms a boundary value problem into an initial value problem which is easier to solve.

The second difficulty is the measurement of the actual values of the parameters and coefficients that appear in the equations. The measurement of the active medium properties requires detailed spectroscopic investigation (Poole et al. 1989; Vienne et al. 1998) and data usually represent the combined effect of various parameters. Therefore, certain assumptions and fitting of some parameters cannot be avoided.

We used the absorption and emission cross-sections parameters provided by LIEKKI and illustrated in Figure 5.25. Table 5.3 values of the cross-sections for the relevant wavelength for Erbium-doped fiber lasers. A complete table with all the cross-sections at various wavelengths is available in Appendix A.

Figure 5.25 shows that Erbium has two absorption regions, around 1000 nm and around 1400–1600 nm, and a unique emission region in the range of 1400 to 1600 nm. The peak absorption and emission were measured at the wavelengths of 1529 and 1531 nm respectively. However, we assume pumping at 980 nm and lasing at

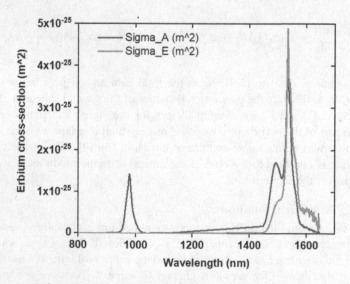

FIGURE 5.25 Erbium ions cross-section as a function of wavelength.

TABLE 5.3

Absorption and Emission Cross Sections for Erbium at Wavelengths Used in the Simulation

Wavelength (nm)	Absorption cross-section (m^2)	Emission cross-section (m^2)
975	2.06×10^{-25}	0
980	1.87×10^{-25}	0
1529	6.80×10^{-25}	6.13×10^{-25}
1530	6.79×10^{-25}	6.15×10^{-25}
1531	6.72×10^{-25}	6.22×10^{-25}
1550	3.17×10^{-25}	4.44×10^{-25}

TABLE 5.4

Additional Parameters Used in the Modelling of the Laser

Parameter	Value	Unit
Erbium radiative lifetime	10	ms
Core diameter	4	µm
Core numerical aperture (NA)	0.2	–
Normalized frequency (V number)	2.56	
Mode Field diameter (MFD)	4.22	µm
Overlap factor	0.83	
Erbium concentration	9.8×10^{24}	Ions/m^3

1550 nm, therefore the cross-sections at these wavelengths will be used as shown in Table 5.3.

Other parameters used in the modelling are also provided by LIEKKI as indicated in Table 5.4.

Figure 5.26 shows pump power absorption inside the cavity for pump powers ranging from 40 to 120 mW and a length of the doped fiber of 10 meters. The pump wave is attenuated because absorption cross-section at 980 nm is much higher than the emission cross-section. Also, the short lifetime of the $^4I_{11/2}$ manifold of Erbium reduces the population density of this level. It can be seen that the rate of absorption is almost the same for the different values of launched pump powers.

Figure 5.27 illustrates the population density distribution and internal laser power inside the cavity of the fiber laser for pump powers ranging from 40 mW to 100 mW. The simulations were conducted for an output coupling ratio of 10% and an Erbium-doped fiber length of 15 metres. The distribution of the population of metastable level N_2 is shown in Figure 5.27, A. At position $z = 0$, the population is inverted and N_2 reaches its maximum value for all pump powers. As z increases, the populations vary differently with different values of pump powers. For 40 mW, the

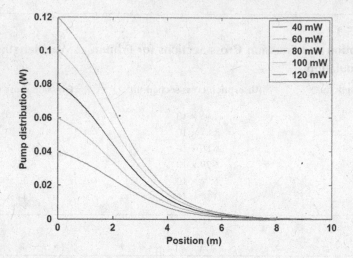

FIGURE 5.26 Different values of pump power absorption inside the cavity.

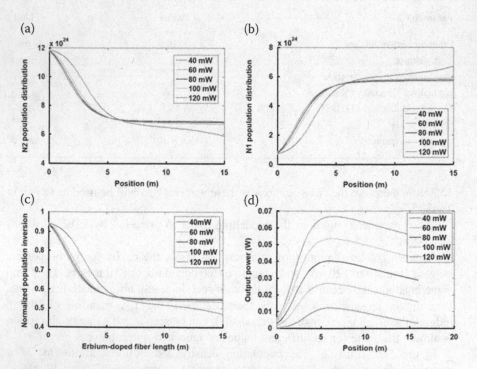

FIGURE 5.27 A) N_2 metastable level population density distribution as a function of position z, B) N_1 ground state population density distribution, C) Normalised N_2 population density distribution and D) laser power distribution.

rate of reduction of N_2 is slower than for the other values 60 mW through 120 mW. This situation is easily explained by the weak internal laser power corresponding to this value of pump power and which depopulate the metastable level at a lower rate as it can be seen in Figure 5.27, B illustrating the laser power distribution inside the cavity. It can also be observed that there is a point where the field reaches its maximum before it starts decreasing. The peak corresponds to the position where the pump power falls below a certain threshold such that it can no longer compensate the rate of stimulated emission. The distribution of N_1 population is illustrated in Figure 5.27, B and the profiles are the perfect reverses of N_2 profile. This is easily explained by the fact that the energy system was modelled as a two-level system in which the absorption and emission rate are identical. Population inversion is represented in Figure 5.27, C, at weak pumping population inversion, keep decreasing along the fiber length, whereas for higher pumps a constant value is reached beyond 5 metres for all the pump powers.

The variation of pump power and laser power inside the cavity is illustrated in Figure 5.28 for a pump power of 100 mW and a coupling coefficient of 10%. More than 60% of pump power is absorbed in the first 3 metres of the erbium-doped fiber which corresponds to 30% of the total length of 10 metres.

In this portion of the fiber, laser power also start increasing rapidly because of the strong pumping. Beyond 3 metres, 70% of the launched power has been absorbed population inversion is reduced. The slope efficiency begins to decrease, stabilize and become negative beyond 6 meters of length because not only the pump power is insufficient to repopulate the metastable level, but the signal starts to be reabsorbed by the Erbium ions in the ground state.

The output power as a function of pump power is illustrated in Figure 5.29. For an Erbium-doped fiber length of 6 metres and an output coupling ratio of 90%, the threshold was about 20 mW.

After reaching the threshold, output power grows almost steadily with a slope efficiency close to 12%. This is the typical behaviour of long cavity fiber lasers. Output powers as a function of pump power also vary with the length of the doped

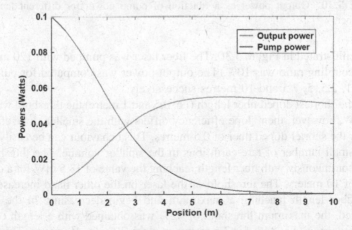

FIGURE 5.28 Pump and laser power distribution along the fiber laser cavity.

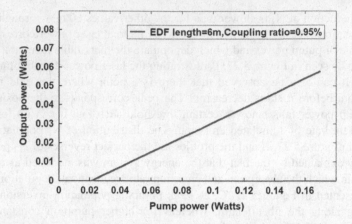

FIGURE 5.29 Pump power as a function of the output power of a ring cavity Erbium-doped fiber laser.

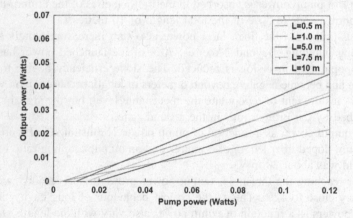

FIGURE 5.30 Output power as a function of pump power for different lengths of the doped fiber.

fiber as illustrated in Figure 5.30. The fiber laser was pumped with 120 mW and the output coupling ratio was 10%. The output power was computed for pump powers of 0.5, 1, 2.5, 5, 7.5 and 10 metres successively.

For the shortest doped fiber length (i.e. 0.5 and 1 metre) the threshold was found to be 3 mW, however, their slope efficiency differs with the smaller slope efficiency of 16% for the shorter doped fiber of 0.5 metres. This behaviour can be easily explained by the small number of rare-earth ions in the smaller volume. The threshold grows almost continuously with fiber length reaching the value of 18.5 mW for a doped fiber length of 10 meters. The threshold of the laser on the other hand increases with the doped fiber length reaching a maximum and begin decreasing. In the simulation conducted, the maximum threshold of 42% was obtained with a length of 5 metres and then decreased at 34% for 7.5 metres to reach 30% for 10 metres. This behaviour

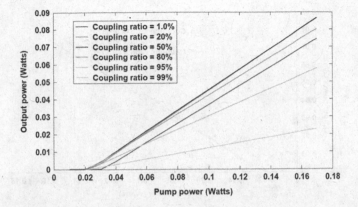

FIGURE 5.31   Output power as a function of pump power for different values of coupling ratio.

of the laser is an indication that there is an optimum length of the doped fiber which result in the maximum efficiency as is going to be shown in coming sections. Output power as a function of pump power was also computed for different output coupling ratio as illustrated in Figure 5.31. Simulations were performed for coupling ratio varying from 1% to 99% with an interval of 20 in between completed with a value of 95%. This choice of values was motivated by trying to target values where big changes can be observed. For the simulations, a 6 metres long Erbium-doped fiber was pumped with 150 mW at 980 nm.

Variations of laser threshold and slope efficiency are indicated in Figure 5.31 whereas Figure 5.32 and Figure 5.33 illustrate the same values with fewer curves for the sake of clarity. The smallest coupling ratio of 1% has the highest threshold of 32 mW because 99% of the power developed in the cavity constitutes the output, therefore, higher pump power is required to compensate for that loss and reach the threshold of the laser.

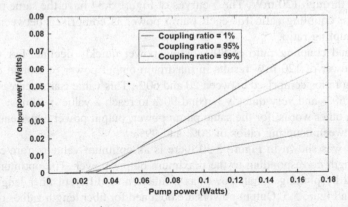

FIGURE 5.32   Output power as a function of pump power for different output coupling ratios.

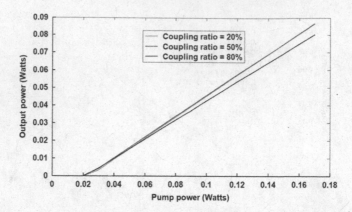

FIGURE 5.33 Output power as a function of pump power for different coupling ratios.

After reaching the threshold a slope efficiency of close to 64% is observed. When coupling ratio is increased, an increase in slope efficiency is also observed along with a reduction in the threshold. The output coupling values of 20%, 50%, and 80% have the same threshold of about 20 mW as illustrated in Figure 5.33 the slope efficiencies are around 74% with the lowest value of 71% for 20%.

On the other hand, the highest values of 95% and 99%, have the same threshold of 21 mW but the difference in slope efficiency is more pronounced as indicated in Figure 5.32. However, contrary to previous values of 1%, 20%, and 50% where slope efficiency was increasing with the coupling coefficient, here the slope efficiency is inversely proportional to the coupling coefficient. The coupling ratio of 95% corresponds to a slope efficiency of 48% whereas the slope efficiency falls to only 18% for a coupling ratio of 99%. This shows that there is an optimum value of coupling coefficient that provides the highest slope efficiency and the lowest threshold as illustrated in Figure 5.34 where output power is plotted against coupling coefficient for values of pump power ranging from 40 mW through 120 mW. The 5 curves of Figure 5.34 have the same profile. The optimum coupling ratio for each pump power is comprised between 20% and 60% coupling ratio.

Beyond coupling ratio of 80% output power quickly decline. For example, a pump power of 120 mW results in maximum output power of around 65 mW for coupling ratios comprised between 20 and 60%. This value starts decreasing slowly beyond 60% and very quickly beyond 90% to reach a value of almost 10 mW for 99%. In other words, for the same pump power, output power experiences an 80% drop between coupling ratios of 80% and 99%.

As it was shown in Figure 5.30 there is an optimum value of rare-earth-doped fiber length corresponding to the maximum output power. The optimum length is obtained by plotting output power as a function of different fiber lengths as illustrated in Figure 5.5. Output powers are obtained for fiber length values of 3 metres, 10 metres, 18 metres, and 20 metres respectively with a pump power of 100 mW and coupling ratio of 10% (Figure 5.35).

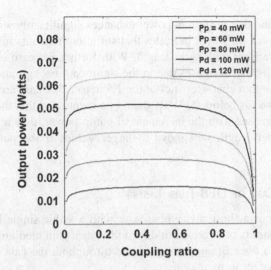

FIGURE 5.34 Output power as a function of coupling ratio for various pump powers of a 5 m Erbium-doped fiber.

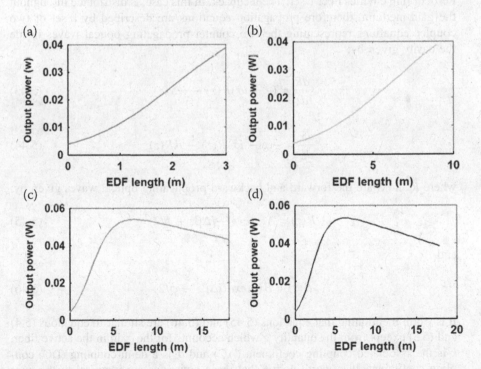

FIGURE 5.35 Output power as a function of fiber length, (a) 3 meters, (b) 5 metre, (c) 18 meters, and (c) 20 meters.

It can be found that the output power enhances significantly with increasing fiber length when $L < 1$ m, which indicates that sufficient amounts of pump powers are absorbed by the fiber with short length. With further increase of fiber length, the output power slightly decreases due to the significant background loss in the fiber. Therefore, the output characteristics of the 1.5 m to 2.5 m case are optimal when the input pump power is below 20 W. It should be mentioned that the optimal value of fiber length increases with the increment of pump power, thus it can speculate that the optimal fiber length will move to larger values if the pump power further increases.

5.3.4 The Case of DFB Fiber Lasers

DFB fiber laser is a short cavity fiber laser with a stable single longitudinal mode operation. It consists of a rare-earth-doped fiber as a gain medium and the feedback is achieved by a fiber Bragg grating printed throughout the gain medium with a pi phase-shifted defect in its middle.

In the case of a distributed feedback fiber laser, the rate equations describing processes in the gain medium are identical to the one described for standard Fabry-Perot or ring cavities fiber laser. The feedback, in this case, is distributed throughout the gain medium, therefore propagation equations are described by a set of two coupled equations representing the two counter-propagating optical waves inside the cavity given by:

$$\frac{dR}{dz} = (g - i\hat{\sigma})R(z) - ikS(z) \tag{5.43}$$

$$\frac{dS}{dz} = -(g - i\hat{\sigma})R(z) - ikS(z) \tag{5.44}$$

where R and S are the forward and backward propagating optical waves given by:

$$R(z) = A(z)exp(-i\Delta\beta z + \phi/2) \tag{5.45}$$

and

$$S(z) = B(z)exp(i\Delta\beta z - \phi/2) \tag{5.46}$$

It is worth mentioning that equations (5.45) and (5.46) are similar to equations (5.4) and (5.5) except from the quantity g which accounts for the gain in the active fiber. k is the associated coupling coefficient (AC) and $\hat{\sigma}$ is a demi-coupling (DC) coupling coefficient. It is worth noting that these equations are identical to the ones governing a passive fiber Bragg grating if the gain becomes zero. The DC coupling coefficient is given by:

$$\hat{\sigma} \equiv \Delta\beta + \sigma - \frac{1}{2}\frac{d\phi}{dz} \tag{5.47}$$

where $\Delta\beta$ represents the detuning from the Bragg wavelength of a weak grating, given by:

$$\begin{aligned}
\Delta\beta &= \beta - \frac{\pi}{\Lambda} \\
&= \beta - \beta_D \\
&= 2\pi n_{eff}\left(\frac{1}{\lambda} - \frac{1}{\lambda_D}\right)
\end{aligned} \tag{5.48}$$

In equation (5.48), λ_D is the design wavelength given by $\lambda_D = 2n_{eff}\Lambda$, where n_{eff} is the effective refractive index of the fiber core and Λ the period of the fiber Bragg grating. Finally, the derivative of the phase of the wave with respect to the position $\frac{1}{2}\frac{d\phi}{dz}$ represents a possible chirp of the Bragg grating period. In the simple case of a single transverse mode propagation, the coupling coefficients can be written as:

$$\sigma = \frac{2\pi}{\lambda}\overline{\delta n_{eff}} \tag{5.49}$$

And

$$k = \frac{\pi}{\lambda}\overline{\delta n_{eff}} \tag{5.50}$$

Because of the gain existing inside the doped fiber and the presence of the phase-shift, the grating of a DFB fiber laser cannot be treated as a uniform grating; therefore, the coupled-mode equations cannot be solved analytically. For the reasons given in section 5.2.2.1, the transfer matrix method (TMM) will be used to solve the equations. The DFB fiber laser is divided into several sections as illustrated in Figure 5.36.

In the case of a DFB fiber laser, if R_m and S_m represent the amplitudes of the field entering and exiting the Mth section of the divided fiber Bragg grating, then for each uniform section one can write a 2×2 matrix T_m such that:

$$\begin{pmatrix} S_m \\ R_m \end{pmatrix} = T_m \begin{pmatrix} S_{m-1} \\ R_{m-1} \end{pmatrix} \tag{5.51}$$

where the elements of the matrix T_m describing the Mth section are given by:

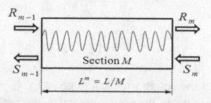

FIGURE 5.36 Schematic representation of the transfer matrix method. The fiber Bragg grating is divided into M identical sections. The mth section is represented with its input and output optical fields.

$$
\begin{aligned}
T_{11}^m &= \left[\cosh(\gamma^m L^m) + i \frac{\Delta\beta'^m}{\gamma^m} \sinh(\gamma^m L^m) \right] e^{i\beta_B L^m}, \\
T_{12}^m &= -\frac{k^m}{\gamma^m} \sinh(\gamma^m L^m) e^{-i(\beta_B L^m + \phi^m)}, \\
T_{21}^m &= -\frac{k^m}{\gamma^m} \sinh(\gamma^m L^m) e^{i(\beta_B L^m + \phi^m)}, \\
T_{22}^m &= \left[\cosh(\gamma^m L^m) - i \frac{\Delta\beta'^m}{\gamma^m} \sinh(\gamma^m L^m) \right] e^{-i\beta_B L^m},
\end{aligned}
\tag{5.52}
$$

where γ^m, $\Delta\beta'^m$ and ϕ^m are defined as:

$$
\begin{aligned}
\gamma^m &= \sqrt{(k^m)^2 - (\Delta\beta'^m)^2}, \\
\Delta\beta'^m &= i\Delta\beta^m - g^m, \\
\phi^m &= \phi_{initial} + \Sigma_{m=1}^{M-1} 2\pi L^m / \Lambda^m, \\
\Delta\beta^m &= 2\pi n_{eff} (1/\lambda_L - 1/\lambda_B^m),
\end{aligned}
\tag{5.53}
$$

In equation (5.53), Λ^m is the grating period, g^m the gain experienced by the propagating field in the mth section, ϕ^m its phase and $\phi_{initial}$ its initial phase. $\Delta\beta^m$ represents the deviation of the propagation constant at the Bragg wavelength. In the computations, the initial phase is often given the value zero for simplicity. Finally, the phase-shift in inside the grating is represented by the matrix:

$$
T_\phi = \begin{bmatrix} \exp(-i\phi/2) & 0 \\ 0 & \exp(i\phi/2) \end{bmatrix}
\tag{5.54}
$$

where ϕ is the value of the phase shift.

The transfer matrix of the Mth section of the grating is written:

$$T_m = \begin{bmatrix} T_{11}^m & T_{12}^m \\ T_{21}^m & T_{22}^m \end{bmatrix} \tag{5.55}$$

And the entire grating is described by a matrix T written as:

$$T = T_M T_{M-1} \dots T_\phi \dots T_3 T_2 T_1 \tag{5.56}$$

Finally, the relation between the input and output fields of the grating will be given by:

$$\begin{bmatrix} S(L) \\ R(L) \end{bmatrix} = T \begin{bmatrix} S(0) \\ R(0) \end{bmatrix} \tag{5.57}$$

The reflection and transmission spectra of a π phase-shifted fiber Bragg grating obtained using the transfer matrix method is illustrated in Figure 5.37.

5.3.5 DFB FIBER LASER OUTPUT POWER COMPUTATION

The length of a DFB fiber laser often does not exceed 10 cm, therefore the maximum achievable gain in this type of lasers is low, resulting in weak output power. To counteract this drawback, the concentration of rare-earth ions is often increased, which in terms the case of Erbium results in ion-ion interaction such as cooperative up-conversion. It has been proved that the interaction between ions of the same species is favoured by their poor solubility in silica glass. To the end of reducing these effects, phosphosilicate glasses are often preferred. But the best way of reducing cooperative up-conversion between Erbium ions remains using Ytterbium as a co-dopant to increase absorption and prevent interaction between ions. Therefore, in the present modelling of DFB fiber laser, Erbium-Ytterbium doped gain medium is used. The same principle is valid for Erbium or any other rare-earth element. The energy system of the Erbium-Ytterbium co-doped medium is illustrated in Figure 5.38.

The system of rate equation resulting from this energy system is complex and can be difficult to solve. By neglecting some of the transition reported to have less influence on the characteristics of the laser, another the energy system illustrated in Figure 5.39 is obtained.

By neglecting some of the transitions of less influence to the characteristic of the laser, one gains in computation time and effort but the accuracy of the model is not particularly affected, thus the assumption does not significantly affect the results. When the gain medium is pumped at 980 nm, the system of rate equations resulting from the simplified energy system of Figure 5.39 can be written as:

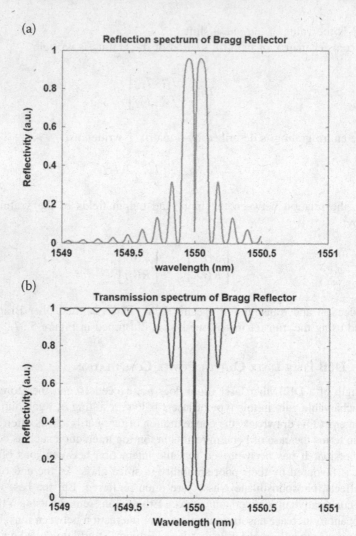

FIGURE 5.37 Reflection (a) and transmission (b) spectra of a 15 mm long πphase-shifted fiber Bragg grating with a Bragg wavelength of 1550 nm and $\Delta n = 1.447$. The phase shift can be seen as a narrow transmission band at the Bragg wavelength in the FBG stopband.

$$\frac{dN_2}{dt} = W_{12}N_1 - W_{21}N_2 - N_2A_2 + N_3A_3 - 2C_{up}^2N_2^2 \tag{5.58}$$

$$\frac{dN_3}{dt} = R_{13}(N_1 - N_3) - N_3A_3 + K_{tr}N_1N_6 + C_{up}N_2^2 - K_{tr}N_3N_6 \tag{5.59}$$

$$\frac{dN_6}{dt} = R_{56}N_5 - R_{65}N_6 - N_6A_6 - K_{tr}N_1N_6 - K_{tr}N_3N_6 \tag{5.60}$$

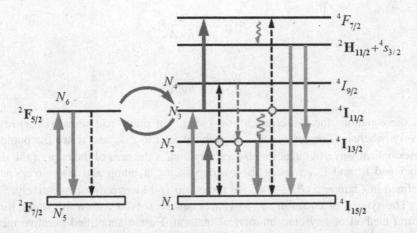

FIGURE 5.38 Energy levels of an Erbium-Ytterbium co-doped medium.

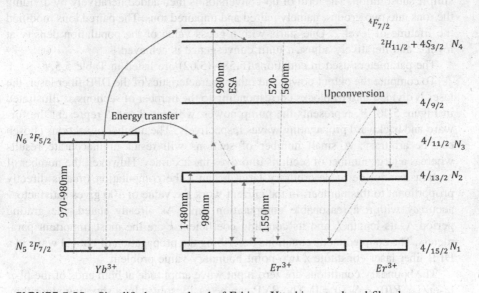

FIGURE 5.39 Simplified energy level of Erbium-Ytterbium co-doped fiber laser.

$$N_1 + N_2 + N_3 = N_{Er} \qquad (5.61)$$

$$N_5 + N_6 = N_{Yb} \qquad (5.62)$$

where:

$$R_{ij} = \Gamma_p \frac{\sigma_{ij}}{h\nu_p A} P_p \qquad (5.63)$$

$$W_{12} = \Gamma_p \frac{\sigma_{12}}{h\nu_s A} P_s \qquad (5.64)$$

$$W_{21} = \Gamma_s \frac{\sigma_{21}}{h\nu_s A} P_s \qquad (5.65)$$

In the equations the subscripts p and s stand for pump and signal (laser) respectively, whereas ij represents the transition from i to j. ν_p and ν_s are the pump and laser optical wave frequencies respectively and, A the area of the core of the doped fiber and Γ_p and Γ_s represent the overlap factor at pump and laser frequency as defined in Chapter 5. P_p and P_s are the pump and laser power respectively.

The system of equation (5.58)–(5.62) can be solved with a 4th order Runge-Kutta method or any other numerical method. Here a simplified iterative method was used. First one neglect the term resulting from cooperation up-conversion. The resulting equation has an analytical steady-state solution which can be obtained by simple substitution. The term of up-conversion is then added iteratively by dividing the ions into two groups, namely paired and unpaired ions. The paired ions modified the lifetime of level 2. One starts with a guess value of the population density at level 2 and iteratively adjust it until convergence is achieved.

The parameters used in equations (5.58)–(5.62) are listed in Table 5.5.

To compute the output power and other characteristics of the DFB fiber laser, the laser is divided into M blocks corresponding to the number of sections as illustrated in Figure 5.40. P_p represents the pump power, whereas R and S represent the forward and backward propagating waves respectively. The number of sections chosen can be arbitrary. A small number of sections will result in inaccurate results whereas a large number of sections improves the accuracy. However, the number of sections cannot be made arbitrary large because the computation time is directly proportional to this number. In the current work the value of 100 gives satisfactory accuracy within a reasonable computation time. As already stated, the grating period Λ, its length L and its coupling coefficient k are the most important parameters. The coupled mode equations describing the propagation of optical wave in a DFB fiber laser constitute a two-point boundary value problem.

The boundary conditions are zero input wave amplitude at both ends of the fiber laser, i.e. $R(0) = S(L) = 0$. The BVP can be easily solved by a shooting algorithm. The algorithm works as follows:

If one assumes forward pumping, then at the left end of the laser, pump power, $P_p(0)$ is known. There is no forward wave entering the laser, i.e $R(0) = 0$, and a guess of backward wave amplitude $S(0)$ at an estimated wavelength λ_L is provided. The algorithm then steps through the first section of the DFB fiber laser. The laser power corresponding to this first section is calculated as $P_0^1 = |R^1|^2 + |S^1|^2$. The laser power obtained with the known pump power allows one to solve the rate equations for population densities $N_1, N_2 \ldots N_6$. The values obtained for the population densities allow one to compute the gain and loss of the section, which in turn are used to calculate transfer matrix elements of the section. Finally, forward and backward

TABLE 5.5

Parameters Used in the Simulation of the Er^{3+}-Yb^{3+} Doped DFB Fiber Laser

Symbol	Parameters	Value	Reference
λ_s	Pump wavelength	980 nm	–
λ_p	Signal wavelength	1550 nm	–
σ_{12}	Absorption cross-section of Er^{3+} at λ_s	8.9×10^{-25} m^2	Measured
σ_{13}	Absorption cross-section of Er^{3+} at λ_p	2×10^{-25} m^2	Measured
σ_{21}	Emission cross-section of Er^{3+} at λ_s	8.7×10^{-25} m^2	Measured
σ_{31}	Emission cross-section of Er^{3+} at λ_p	2×10^{-25} m^2	Measured
σ_{56}	Absorption cross-section of Yb^{3+} at λ_p	8.7×10^{-25} m^2	Measured
σ_{65}	Emission cross-section of Yb^{3+} at λ_p	11.6×10^{-25} m^2	Measured
R_{21}	Spontaneous emission rate of Er^{3+}	100 s^{-1}	Measured
R_{32}	Nonradiative decay rate of Er^{3+}	100000 s^{-1}	(Jiang et al. 2005)
R_{31}	Spontaneous emission rate of Er^{3+}	30000 s^{-1}	(Jiang et al. 2005)
R_{41}	Spontaneous emission rate of Er^{3+}	100000 s^{-1}	(Jiang et al. 2005)
R_{43}	Nonradiative decay rate of Er^{3+}	100 s^{-1}	(Jiang et al. 2005)
R_{65}	Spontaneous emission rate of Yb^{3+}	1000 s^{-1}	Measured
C_{22}	Cooperative up-conversion coefficient of Er^{3+}	1.2×10^{-24} m^{3}s^{-1}	(Taccheo et al. 1999)
K_{tr}	Energy transfer coefficient Yb^{3+} to Er^{3+}	5×10^{-22} m^{3}s^{-1}	(Vienne et al. 1998; Poole et al. 1989; Taccheo et al. 1999)
α_s	Background loss at λ_s	0.15 m^{-1}	(Kuthan Yelen et al. 2005)
α_p	Background loss at λ_p	0.20 m^{-1}	(A. Azmi et al. 2009)
N_{Er}	Total Erbium ion population	1.2×10^{25} m^{-3}	–
N_{Yb}	Total Ytterbium ion population	24×10^{25} m^{-3}	–
r	Core Radius	2.3 μm	Measured
Γ_p	Overlap factor at λ_p	0.64	Measured
Γ_s	Overlap factor at λ_s	0.43	Measured

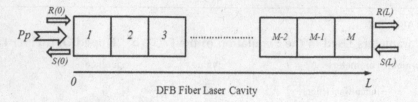

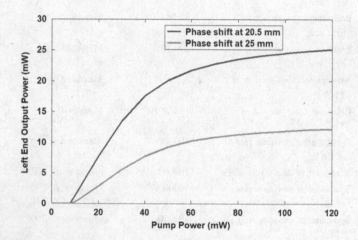

FIGURE 5.40 Schematic representation of a DFB fiber laser divided into the M sections for solving the coupled equations.

FIGURE 5.41 Output power as a function of pump power of a DFB Er3+-Yb3+-doped fiber laser with symmetric and asymmetric phase shift.

optical waves at the input of the next section are determined from the transfer matrix relation:

$$\begin{bmatrix} R^{m+1} \\ S^{m+1} \end{bmatrix} = \begin{bmatrix} T_{11}^m & T_{12}^m \\ T_{21}^m & T_{22}^m \end{bmatrix} \begin{bmatrix} R^m \\ S^m \end{bmatrix} \tag{5.66}$$

At the position of the phase shift, the propagating waves of the next section are obtained by multiplying the accumulated values with the phase-shift matrix provided in equation (5.54). The process is repeated until the last section is reached. At the end of the last section the boundary condition is tested. If the amplitude of the forward wave at the end of the section fulfils the boundary condition, the algorithm is stopped and the guessed value of the amplitude of the backward wave corresponds to the square root of the backward laser power. If the boundary condition is not fulfilled, the difference is used to determine a new guess using a one-dimensional minimisation technique such as the secant method. The process is repeated until convergence is achieved. It is worth mentioning that because the equations are coupled, once the backward amplitude value of the field is obtained, the amplitude of the forward propagating wave is also determined. Also, values like intracavity power, gain, and population density distribution are easily obtained with the same algorithm.

5.3.5.1 Pump Absorption

Figure 5.42 illustrates the power absorption of a 980 nm optical beam inside a 50 mm long DFB fiber laser with a π phase shift in its centre. The grating structure is transparent to the pump power because its wavelength of 980 nm is largely out of the grating stopband. Therefore, free propagation is expected as illustrated in the figure. In the simulation, the pump power was varied from 0 mW to 120 mW. Two values of pump power are represented, namely a typical value of 100 mW and a value of 20 mW just above the threshold.

The strong absorption in a short distance is a result of the high doping concentration. It can be seen that at low powers, the pump is completely depleted. Such a situation leads to reabsorption of laser power and efficiency degradation. On the other hand, pumping at 100 mW results in a residual power of almost 20 mW after the 50 mm which is an advantage when such fiber lasers are cascaded like in the case of serial multiplexing of sensors. However, experience has shown that 1480 nm pumping is best suited for serial multiplexing of sensors. The curve also shows that the rate of change of pump power as a function of position dP_p/dz is maximum at the phase shift position and minimum towards the ends. This situation is a consequence of the particularly large intensity of the propagating wave at this point.

5.3.5.2 Internal Intensity Distribution

The internal intensity distribution of the fundamental transverse mode is illustrated in Figure 5.43. This intensity shows peak maximum value at the position of the phase shift and decreases almost exponentially and equally in both directions as one moves away from the phase shift.

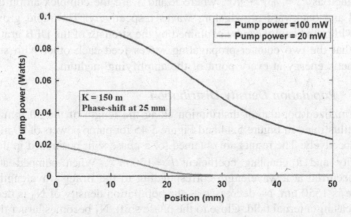

FIGURE 5.42 Pump absorption inside the DFB fiber laser for 20 and 100 mW.

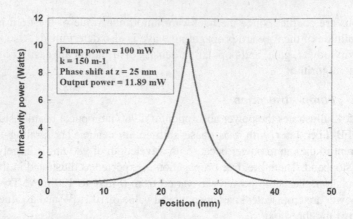

FIGURE 5.43 Intensity distribution inside the cavity of a 50 mm long Er-Yb co-doped fiber laser.

The variation of intensity as a function of position z can be approximated by the following analytical expression:

$$I_{(z)} = I_0 \exp(-2|f(z)|) \tag{5.67}$$

where I_0 is the maximum intensity and $f(z) = k|z|$, where k is the AC coupling coefficient and z the distance from the phase shift. The intensity inside the cavity of a DFB fiber laser can reach excessively high values. Values corresponding to kiloWatts of power are not uncommon for example, a value of 2.9 kW was reported for 10 cm long Erbium-doped fiber lasers (V. C. Lauridsen et al. 1998). The high intensity increases the temperature of the laser, causing a chirp in the grating because of thermal expansion and in terms altering the emitting frequency of the laser. This rather deleterious effect is often reduced by mounting the DFB fiber laser on a heating sink to control its internal temperature. The internal power in the cavity can be obtained as $Ps = |S|^2 + |R|^2$, where R and S are the complex amplitudes of the forward and backward propagating waves respectively. The rapid growth of the waves within the cavity can be explained by the strength of the DFB grating, which means that the two counter-propagating waves feed each other with strong electromagnetic energy at every point of the amplifying medium.

5.3.5.3 *Population Density Distribution*

The normalized population distribution along the length of the 50 mm DFB fiber laser is illustrated in Figure 5.44 and Figure 5.45 for pump powers of 20 mW and 100 mW respectively. The results are obtained for a phase shift positioned in the centre of the cavity and a coupling coefficient $k = 150 \ m^{-1}$. When pumped at 980 nm, lasing occurred at a wavelength corresponding to the Bragg wavelength which in this case is 1550 nm. N_2 decreases as the population density of N_2 is decreased by the increasing internal field. Close to the phase shift, N_1 becomes larger than N_2 and population inversion is suppressed. This situation occurs because the low pump

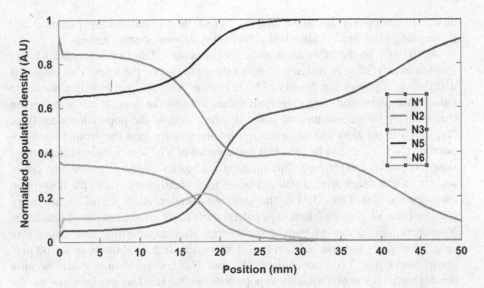

FIGURE 5.44 Relative population density distribution within the cavity of the DFB fiber laser at a pump power of 20 mW. At a distance of around 20 mm, the population density N_1 becomes larger than N_2, which means no population inversion.

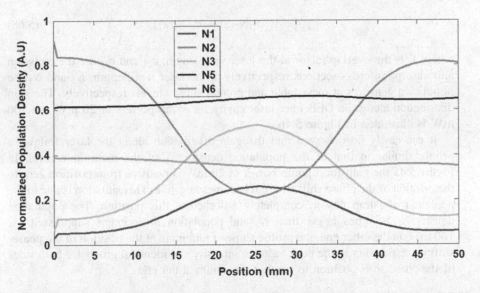

FIGURE 5.45 Relative population density distribution within the cavity of the Erbium-Ytterbium DFB fiber laser at 100 mW pump power. The population density N_2 is larger than N_1 for the majority of positions. At the position of the phase shift, N_2 is reduced and becomes smaller than N_2. This behaviour is explained by the strong intensity at this position which quickly depopulates the metastable level N_2.

power is completely absorbed and N_2 cannot be repopulated by pumping and propagating laser field is absorbed contributing to weak output power.

At 100 mW on the other hand, the effect of pump exhaustion is eliminated, and there is even a 20 mW residual pump at the right end of the laser. Therefore, the distribution of population density of N_2 follows a "bathtub" profile with a minimum value at the phase shift and maximum values towards the ends. It can also be seen that contrary to the assumption made in many models the population density of $^4I_{11/2}$ (N_3) is not zero and can even reach values larger than the ground level population (N_1). It can also be seen that the majority of Yb^{3+}ions remain in the ground state (N_6 quickly fall to zero). This situation can be explained by the emission cross-section of Ytterbium around 980 nm being significantly larger than the absorption cross-section (see Table 5.5) at the same wavelength. Also, the efficient energy transfer from Yb^{3+} to Er^{3+} ions depopulates the Ytterbium excited state because the Ytterbium ions decay in their ground state after transferring their energy to neighbouring ground state erbium ions. Finally, Figure 5.44 shows that at 100 mW pump power, the Yb^{3+} populations (N_5 and N_6) do not change much because the higher pump power quickly repopulates the N_6 level to compensate for the relatively large population decay rate upon energy transfer.

5.3.5.4 *Gain Distribution*

The gain of the DFB fiber laser is given by the following analytical expression,

$$\gamma_{(z)} = \Gamma_s(\sigma_{21}N_2(z) - \sigma_{12}N_1(z)), \tag{5.68}$$

where Γ_s is the overlap factor at the laser wavelength, σ_{21} and σ_{12} are the emission and absorption cross-sections respectively at the laser wavelength, N_2 and N_1 the population densities at metastable and ground energy levels respectively. The gain distribution along the DFB fiber laser cavity for pump powers of 20 mW and 100 mW is illustrated in Figure 5.46.

It can easily be observed that the gain distribution along the laser follows a profile similar to that of the population density N_2 of the metastable level. In Figure 5.45 the gain for a pump power of 20 mW is positive from position zero to the position of the phase shift where it becomes negative. This situation is due to the depletion of pump power, completely absorbed at this position. The population density N_1 becomes larger than N_2 and population inversion is suppressed. At 100 mW on the other end, the profile shows a minimum at the position of the phase-shift corresponding to the peak value in intensity and identical grows for both sides of the phase-shift position to reach a maximum at the ends.

5.3.5.5 *Output Power*

A classic DFB Fiber laser consists of a uniform refractive index grating with constant amplitude and constant period, written in an active medium. This type of DFB fiber laser operates at two fundamental longitudinal wavelengths, corresponding to the edges of the grating bandgap and gives symmetric output powers from both ends equally divided between these two modes. Such a cavity provides a

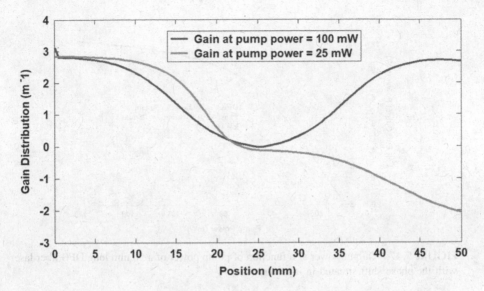

FIGURE 5.46 Gain distribution inside the cavity of a 50 mm long Erbium-Ytterbium doped DFB fiber laser.

dual-wavelength bidirectional operation. In practice, however, a single-wavelength operation is desirable. This is achieved by introducing a π-shift in the spatial phase of the grating. If the phase shift is located in the middle of the grating, due to the symmetry of the cavity, the output powers at both ends are equal. Such a cavity provides single-wavelength operation coinciding with the grating Bragg wavelength and bidirectional operation. In Figure 5.47, the output power as a function of pump power for a 50 mm long DFB fiber laser is illustrated. The grating is uniform, its coupling coefficient is 150 m^{-1} and the phase shift is located in its middle.

The pump power was varied from 0 mW to 120 mW in the simulation. The threshold power was ≈10 mW. After reaching the lasing threshold, the laser output power increased steadily with a slope efficiency of 60%. As pump power reached 40 mW the slope efficiency starts decreasing rapidly. Beyond 60 mW the slope efficiency is only 11%, therefore the output power remains almost constant with respect to increased pump power. This decrease in slope efficiency is caused by the finite energy transfer rate between Yb^{3+} and Er^{3+}, which leads to the reduction of absorbed pump photons relative to launch pump power. The relatively high threshold of the laser can be explained by homogeneous up-conversion. When the pump power is low HUC depopulate the metastable level of the active medium at a rate higher than pump rate and pump photons are lost instead of being converted in lasing photons, thus the cavity loss is higher than gain and lasing is impossible.

5.3.5.6 *DFB Fiber Laser Power Optimization*

In the previous section, we showed that a DFB fiber laser with a π phase shift incorporated in its centre has a single wavelength and bidirectional operation at the grating Bragg wavelength. However, in addition to single-frequency emission, unidirectionality

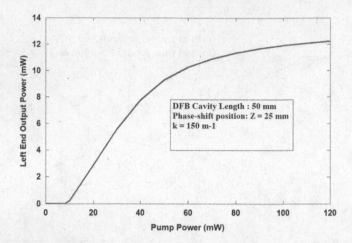

FIGURE 5.47   Output power as a function of pump power of a 50 mm long DFB fiber laser with the phase shift situated in its centre.

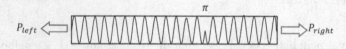

FIGURE 5.48   Schematic diagram of the asymmetric design of the DFB fiber laser. The larger output power is delivered from the end closer to the phase shift position P_{right}.

is a very desirable feature of high-performance lasers. Unidirectional operation of a DFB fiber laser can be achieved by placing the phase shift asymmetrically with respect to the centre of the laser. In this asymmetric design, the larger output power is obtained from the end which is closer to the phase shift location as illustrated in Figure 5.48.

The maximum power from the desired end depends on the coupling coefficient k, and phase shift position z_π. First, we simulated the optimum coupling coefficient that gave the highest output power in the directional configuration by varying the coupling coefficient k over the reasonable range and computing the corresponding output powers. The results are shown in Figure 5.8 where one can see that the output power increases with the coupling coefficient. This behaviour can be explained by a simple working principle of a laser. In fact, as seen from the phase shift position there are two gratings in both directions of equal lengths corresponding to half of the total length of the laser cavity (Figure 5.49).

The two gratings have similar reflectivity, and the DFB fiber laser can be considered as a Fabry-Perot cavity with a mirror which reflectivity are equal to that of the grating, as shown in Figure 5.51.

If the coupling coefficient is weak, the reflectivity of the two gratings is low and the field is weakly confined in the cavity. Thus the feedback is not enough to overcome the cavity losses and trigger the laser action. As the coupling coefficient

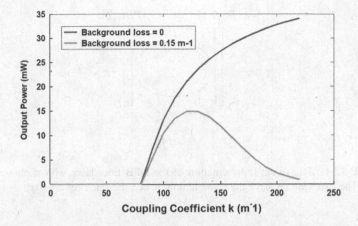

FIGURE 5.49   Output power as a function of the coupling coefficient for 50 mm long DFB fiber laser with a π phase shift in its centre. A) With background loss = 0.15 m^{-1}, B) without background loss.

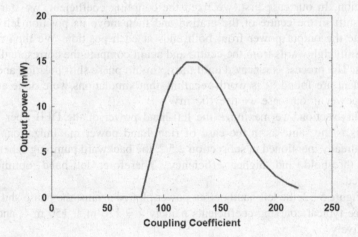

FIGURE 5.50   Output power as a function of the grating coupling coefficient for an Er^{3+}-Yb^{3+} DFB fiber laser.

of the grating is increased, its strength which is given by $k.L$, where k is the coupling coefficient and L the length of the grating, also increases. The increased strength corresponds to higher reflectivity, hence the field is more confined in the cavity and lasing action can take place with higher powers for higher coupling coefficient which is the same as higher reflectivity. However, it can be seen in Figure 5.50, that there exists an optimum value of the coupling coefficient. Beyond this optimum value output power start to decrease with the increment of the coupling coefficient. In this simulation with the parameters reported in Table 5.5, the optimum coupling coefficient was found to be 135 m^{-1}. The background loss, in this case, was equal to 0.15 m^{-1}.

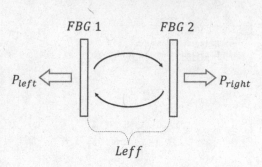

FIGURE 5.51 Equivalent representation of the DFB fiber laser with π phase-shift in its centre. The two gratings have the same length, hence the same reflectivity.

5.3.5.7 *Optimum Phase Shift and Coupling Coefficient*

As discussed in the previous section there exist an optimum π phase shift position and coupling coefficient for maximum unidirectional output power. These optimum values of the parameters z_π and k are found by varying them over a range by simulation. In our case first, we fixed the coupling coefficient, we start with the phase shift in the centre of the grating and then move its position leftwards and compute the output power from both ends at each position. We then moved the phase shift rightwards from the centre and again compute the corresponding output powers. The process is repeated until the optimum phase shift position and coupling coefficient are found. It is worth recalling that simulations were done at a typical pump power in our case we use 100 mW.

In this section, we maximize the left-hand power of the DFB fiber laser. The analysis is the same as in the case of right-hand power maximization. However, as we already mentioned in subsection 2.5.2, the backward pumping scheme has the lower threshold and higher efficiency. Therefore, left-hand optimization is preferred.

In Figure 5.52, the left-hand output power plotted against the phase shift positions for three typical coupling coefficients namely $k = 120$ m^{-1}, 150 m^{-1}, and 180 m^{-1} respectively.

Using equation 5.6 we can establish the correspondence between phase shift position and left-hand side reflectivity. The output power as a function of left-hand side reflectivity is shown in Figure 5.13. The graph is similar to the output power as a function of the coupling ratio of a Fabry-Perot laser. The output power increases with the coupling ratio of the reflector until reaching an optimum value and decreases quickly. The reflectivity optimum value is 0.99 for k = 150 m^{-1} (Figure 5.53).

In Figure 5.14 the reflectivity of the left-hand side grating is plotted against phase shift position. Below 20.5 mm, for example, the reflectivity of left-hand side grating is zero which corresponds to zero output and beyond 40 mm this value is ≈1 also corresponding to zero output because the grating becomes reflective (Figure 5.54).

In Figure 5.15 the output power as a function of pump power is depicted for the optimized and non-optimized DFB fiber laser respectively. The highest left end

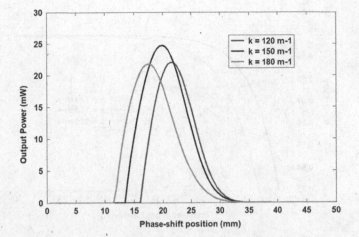

FIGURE 5.52 Left end output power as a function of phase shift position. The highest output power is found for $k = 150$ m^{-1} and phase shift position at 20.5 mm which is 41% of total cavity length of a 50 mm long DFB fiber laser.

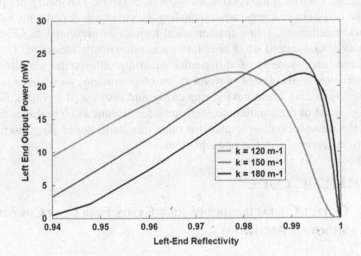

FIGURE 5.53 Output power as a function of the left-hand side reflectivity for k values of 120 m^{-1}, 150 m^{-1}, and 180 m^{-1}.

output power for the 50 mm long DFB fiber laser with $k = 150$ m^{-1} was found by moving the phase shift away from the centre at $z = 20.5$ mm. Comparison with the centrally located phase shows that the output power and slope efficiency considerably increase. However, the laser threshold is not affected by these changes because it depends only on population density distribution inside the DFB fiber laser.

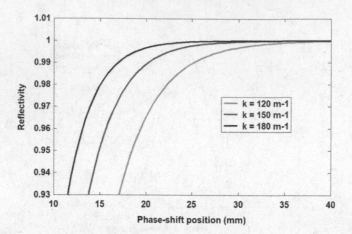

FIGURE 5.54 Reflectivity of the left-hand side grating as a function of phase shift position for $k = 120$ m^{-1}, $k = 150$ m^{-1}, and 180 m^{-1}.

5.4 CONCLUSION

This chapter discussed the continuous wave fiber lasers. The theory of operation of this type of lasers was presented, including its rare-earth doped gain medium and feedback mechanism. Using mathematical formalism presented in Chapter 4, the behaviour of continuous wave fiber lasers was numerically modelled. The solutions to the resulting systems of differential equations allows the simulation of the characteristics of the lasers. Several types of continuous-wave fiber lasers were simulated including linear cavity, ring cavity and distributed feedback fiber lasers. Although most of the simulations were done for erbium and Ytterbium-doped fiber lasers, the same formalism applies for other rare-earth doped fiber lasers, because the only difference is in the rate equations.

5.5 MATLAB® CODE

5.5.1 OUTPUT CHARACTERISTICS OF RING CAVITY FIBER LASER COMPUTED WITH SHOOTING ALGORITHM

```
function [PP,Ps,N2,N1,N,t] = LaserOutput(Pump,leng,-
coupler,Nt)
% This function solves the rate equations and
propagation
% equations for a ring cavity fiber laser using the MATLAB
ODE 45 routine
% and a shooting algorithm
%
%
```

```
% Inputs:
% -Pump: Pump power
% -leng: Length of the doped fiber
% -coupler: Coupling ratio of the output coupler
%
% Outputs:
% -PP : Pump power distribution
% -Ps: Laser power
% -N2: Population density at the metastable level
% -N1: Population density at ground level
% -N : Total population density of the doped fiber
%
% Comments:
% - Computation is done in the ideal case of a two level
energy system
% equations analytically.
% References:
%
%
% Written by Justice Sompo, University of Johannesburg,
South Africa
close all
clc
format long
% Check inputs
if nargin <4
  Nt = 1.26e25;
  if nargin < 3
    coupler = 0.1;
    if nargin < 2
      leng = 5;
      if nargin < 1
                                          error(message
('MATLAB:LaserOutput:NotEnoughInputs'));
      end
    end
  end
end
%=====================================================
========================
%Rate equations parameters
%=========================================
=================================
sigma12S = 2.64e-25;
sigma21S = 2.11e-25;
```

```
sigma12P = 2e-25;
% Lifetimes
t21 = 10e-3;
A21 = 1/t21;
% Concentrations
Ner = Nt;
% Overlap factors and background losses
Gammap = 0.81;
Gammas = 0.6;
lambdap = 980e-9; % Pumb wavelength
lambdas = 1550e-9; %signal wavelength
% Other physical constants
r = 2.3e-6;
h = 6.626e-34;
cel = 3e8;
% Calculated parameters
A = pi*r^2;
F_s = cel/lambdas;
den1 = A*h*F_s;
F_p = cel/lambdap;
den2 = A*h*F_p;
%===================================
=========================================
% Implementation of the shooting secant algorithm to
solve propagation
% equations in steady-state
%==============================
===============================================
epsilon = 1e-6; % tolerance for computation
count = 0; % iniltialise iterations before the while loop
RL = coupler; % reflectivity of mirror at L.
Pp = Pump;
x1 = 10e-3; % first signal power guess
x2 = 100e-3; % second signal power guess
L = leng;
%tspan = [0 6];
tspan = linspace(0,L,41); % to obtain solutions at each
position for 100 positions one can write linspace
(0,L,100)
[t1,y1]=ode45(@propa,tspan,[Pp x1]); %y1 will have the
values of Pp in its first column and the values of Ps in its
second column.
[t2,y2]=ode45(@propa,tspan,[Pp x2]);
i = 1;
Ps11=y1(end,2); %  Ps1 power at z = L
```

```
Psl2=y2(end,2); % Ps2 power at z = L
P01 = Psl1*RL;
P02 = Psl2*RL;
m1 = P01-x1;
m2 = P02-x2;
while (abs(x2-x1) > epsilon)
  tmp=x2;
  count = count+1;
  x2 = x1-(x1-x2)/(m1-m2)*m1;
  x1 = tmp;
  [t1,y1]=ode45(@propa,tspan,[Pp x1]);
  [t2,y2]=ode45(@propa,tspan,[Pp x2]);
  Psl1=y1(end,2); % Ps power at z = L
  Psl2=y2(end,2); % Ps power at z = L
  P01 = Psl1*RL;
  P02 = Psl2*RL;
  m1 = (P01-x1);
  m2 = (P02-x2);
  i = i+1;
  t = t2;
  PP = y2(:,1); % Pump power distribution
  Ps = y2(:,2); % Signal power distribution
  %Population distribution
  W12 = ((sigma12S*Gammas)*Ps)/den1;
  W21 = ((sigma21S*Gammas)*Ps)/den1;
  R12 = ((sigma12P*Gammap)*PP)/den2;
  N1 = Ner*(W21+A21)./(W12+R12+W21+A21);
  N2 = Ner*(W12+R12)./(W21+A21+W12+R12);
  N = N2./Ner;
end
%===========
====================
==================
========================
% Plotting
%==========================================
===================================
figure(1)
plot(t,N2,'m','linewidth',2)
hold on
plot(t,N1,'g','linewidth',2)
figure(2)
plot(t,Ps,'r-','linewidth',2)
hold on
plot(t,PP,'linewidth',2)
```

5.5.2 FUNCTION "PROPA" CALLED IN THE ABOVE FUNCTION

```
function dpdz = propa(~,p)
% PROPA function represents the system of pump and laser
propagation equations
%
% Inputs:
% -The function takes the pump power and initial guesses
of output powers
%
% Outputs:
%   dpdz(1): Represents the pump propagating power
%   dpdz(2): Represents the laser propagating power
%
% Comments:
%   - Computation is done in the ideal case of a two level
energy system
%
% References:
%
% Written by Justice Sompo, University of Johannesburg,
South Africa
close all
clc
%
format longe
% Rate equation Parameters
sigma12S = 2.33e-25;   % Laser absorption cross section
sigma21S = 2.64e-25;   % Laser emission cross section
sigma12P = 2e-25;      % Pump absorption cross section
sigma21P = 0;          % Pump emission cross section
% Lifetimes
t21 = 10e-3;           % Lifetime of the metastable energy
level of Erbium
A21 = 1/t21;
% Concentrations
Ner = 1.2e25;          % Erbium ions concentration
% Overlap factors and background losses
Gammap = 0.64;         % Overlap factor at pump wavelength
Gammas = 0.43;         % Overlap factor at signal wavelength
alphap = 0.005;        % Background loss at pump wavelength
alphas = 0.005;        % Background loss at laser wavelength
lambdap = 980e-9; % Pumb wavelenght
lambdas = 1550e-9; %signal wavelength
```

```
% Other physical constants
r = 2.3e-6;          % Doped fiber core
h = 6.626e-34;       % Planck's constant
cel = 3e8;           % speed of light in free space
% Calculated parameters
A = pi*r^2;
F_s = cel/lambdas;den1 = A*h*F_s;
F_p = cel/lambdap;  den2 = A*h*F_p;
% Fiber Bragg Grating Parameters
W12 = ((sigma12S*Gammas)*(p(2)))/den1;
W21 = ((sigma21S*Gammas)*(p(2)))/den1;
R12 = ((sigma12P*Gammap)*(p(1)))/den2;
%Rate equations solving
N1 = Ner*(W21+A21)./(W12+R12+W21+A21);  % Population
density of the ground energy level
N2 = Ner*(W12+R12)./(W21+A21+W12+R12);  % Population
density of themetastable energy level
dpdz = zeros(2,1);
dpdz(1)= Gammap*(sigma21P*N2-sigma12P*N1)*p(1)-alph-
ap*p(1);
dpdz(2)= Gammas*(sigma21S*N2-sigma12S*N1)*p(2)-alph-
as*p(2);
```

5.5.3 Script to Compute Fabry-Perot Fiber Laser Characteristics in the Case of Forward Pumping

```
function POUT = OutputPowerForward(Pump,leng,Roc)
% This function implements a relaxation algorithm to
solve the propagation
% equations of a Fabry-Perot erbium doped fiber laser, in
the forward pumping
% configuration.
% The cavity of the laser is divided into 100 sections and
power in each
% section is computed using a fourth-order Runge-Kutta
shcheme.
%
%
% Inputs:
%  Pump: Launched pump power
%  leng: Doped fiber length
%  Roc : Reflectivity of the output mirror
```

```
%
% Outputs:
%  POUT: Laser output power
%
% Comments:
%
%
% References:
%
% Written by Justice Sompo, University of Johannesburg,
South-Africa
close all
clc
format longe
tic
% Check inputs
if nargin < 3
  Roc = 0.1;
  if nargin < 2
    leng = 10;
    if nargin < 1
                                                error(message
('MATLAB:OutputPowerBackward:NotEnoughInputs'));
    end
  end
end
%
%------------------------COMPUTATION
PARAMETERS----------------------------
FiberLength = leng;        % meters
n = 100;                % Number of sections
step = (FiberLength-0)/n;  % step size for computation
R1 = 0.98;              % Reflectivity of mirror 1
R2 = Roc;              % Reflectivity of mirror 2
%============================================
==================================
%             TRIP 1 : FORWARD PROPAGARTION
%========================
=====================================================
%----------------------------INITIAL
CONDITIONS----------------------------
ppf = Pump;                % Forward launched power at z = 0
(100 mW)
z_initial = FiberLength;     % origin of longitudinal
distance (z=L for backward displacement)
```

```
pp_initial = 15;        % Initial guess of the pump power at
z = L
psf_initial = 1e-3;        % initial guess of the forward
propagating signal at z = L
psb_initial = 10e-3;        % initial guess of the backward
propagating signal at z = L
z = z_initial;
pp = pp_initial;
psf = psf_initial;
psb = psb_initial;
%=============================
=================================================
%      WHILE LOOP TO COMPUTE LASER PARAMETERS
%==============
=============================
=============================
%
err1 = 10;            % Arbitrarily set value to enter the
while loop
err2 = 10;
err3 = 10;
tol = 1e-4;
IterationNumber = 0;      % Initialise the counter
while (err1 >= tol && err2 >= tol || err3 >= tol) &&
IterationNumber <= 40
        %----------------------CREATE   VECTORS   FOR
PLOTTING---------------------
  Z = z_initial;
  PP = pp_initial;
  PSF = psf_initial;
  PSB = psb_initial;
    %----------------Iterative calculations using RK4
method---------------
  p = n-1 ;          % Integration from z = L to z = 0
  for w = p:-1:0        % Number of loops
,   z = z-step;        % Doing Reverse Integration from "L"
to Zero(0)
    k21 = step*feval('PPF',z,pp,psf,psb);
    f21 = step*feval('PSF',z,pp,psf,psb);
    b21 = step*feval('PSB',z,pp,psf,psb);
    k22 = step*feval('PPF',(z+0.5*step),(pp+0.5*k21),
(psf+0.5*f21),(psb+0.5*b21));
    f22 = step*feval('PSF',z+0.5*step,pp+0.5*k21, psf
+0.5*f21, psb+0.5*b21);
```

```
    b22 = step*feval('PSB',z+0.5*step,pp+0.5*k21, psf
+0.5*f21, psb+0.5*b21);
    k23 = step*feval('PPF',(z+0.5*step),(pp+0.5*k22),
(psf+0.5*f22), (psb+0.5*b22));
    f23 = step*feval('PSF',(z+0.5*step),(pp+0.5*k22),
(psf+0.5*f22), (psb+0.5*b22));
    b23 = step*feval('PSB',(z+0.5*step),(pp+0.5*k22),
(psf+0.5*f22), (psb+0.5*b22));
            k24  = step*feval('PPF',z+step,pp+k23,psf
+f23,psb+b23);
            f24  = step*feval('PSF',z+step,pp+k23,psf
+f23,psb+b23);
            b24  = step*feval('PSB',z+step,pp+k23,psf
+f23,psb+b23);
 .  pp = pp-(1/6)*(k21+2*(k22+k23)+k24);   % Pump power
evolution
    psf = psf -(1/6)*(f21+2*(f22+f23)+f24);  % Forward
laser power evolution
    psb = psb -(1/6)*(b21+2*(b22+b23)+b24);  % Backward
laser power evolution
    Z = [Z,z];
    PP = [PP,pp];
    PSF = [PSF, psf];
    PSB = [PSB, psb];
  end
   %------------STORE THE VALUES OF PUMP, SIGNAL AND
POSITION--------------
  z_FirstPass = z;
  pp_FirstPass = pp;
  psf_FirstPass = psf;
  psb_FirstPass = psb;
      %----------------TEST  CONVERGENCE  AFTER  THE
FIRST PASS-----------------
  err1 = abs(psf-R1*psb);
  err2 = abs(pp-ppf);
  if err1 <=tol && err2 <=tol
    break;
  end
               %========================================
==========================
========
  %            TRIP 2 : BACKWARD PROPAGARTION
               %=================================
=============================================
```

```
        %-------------------APPLY  BC  FOR  NEW  INITIAL
CONDITIONS-----------------
  z = 0;
  pp = ppf;
  psf = psb_FirstPass*R1;
  psb = psb_FirstPass;
            %--------------------CREATE   VECTORS   FOR
PLOTTING-----------------------
  Z = z;
  PP = pp;
  PSF = psf;
  PSB = psb;
  for w = 0:n-1
    z = z+step;      % Integration is from z = 0 to z = L
    k31 = step*feval('PPF',z,pp,psf,psb);
    f31 = step*feval('PSF',z,pp,psf,psb);
    b31 = step*feval('PSB',z,pp,psf,psb);
    k32 = step*feval('PPF',(z+0.5*step),(pp+0.5*k31),
(psf+0.5*31),(psb+0.5*b31));
     f32 = step*feval('PSF',z+0.5*step,pp+0.5*k31, psf
+0.5*f31, psb+0.5*b31);
     b32 = step*feval('PSB',z+0.5*step,pp+0.5*k31, psf
+0.5*f31, psb+0.5*b31);
    k33 = step*feval('PPF',(z+0.5*step),(pp+0.5*k32),
(psf+0.5*f32),(psb+0.5*b32));
    f33 = step*feval('PSF',(z+0.5*step),(pp+0.5*k32),
(psf+0.5*f32),(psb+0.5*b32));
    b33 = step*feval('PSB',(z+0.5*step),(pp+0.5*k32),
(psf+0.5*f32),(psb+0.5*b32));
            k34  =  step*feval('PPF',z+step,pp+k33,psf
+f33,psb+b33);
            f34  =  step*feval('PSF',z+step,pp+k33,psf
+f33,psb+b33);
            b34  =  step*feval('PSB',z+step,pp+k33,psf
+f33,psb+b33);
    pp = pp+(1/6)*(k31+2*(k32+k33)+k34);   % Pump power
evolution
     psf = psf +(1/6)*(f31+2*(f32+f33)+f34);  % Forward
laser power evolution
     psb = psb +(1/6)*(b31+2*(b32+b33)+b34); % Backward
laser power evolution
     Z = [Z,z];
     PP = [PP,pp];
     PSF = [PSF, psf];
     PSB = [PSB, psb];
```

```
  end
  %--------------STORE THE VALUES OF PUMP, SIGNAL AND
POSITION-----------
  z_SecondPass = z;
  pp_SecondPass = pp;
  psf_SecondPass = psf;
  psb_SecondPass = psb;
  err3 = abs(psb-psf*R2); % Test convergence criteria
  z = z_SecondPass;
  pp = pp_SecondPass;
  psb = psf_SecondPass*R2;
  psf = psf_SecondPass;
  IterationNumber = (IterationNumber + 1) ;
end
%--------------------COMPUTE POPULATIONS DENSITY----
---------------------
POUT = psf*(1-R2);
[N1,N2] = PopulationDensity(PP,PSB,PSF);
%====================
=================
================
================
===
%                      PLOTTING
%=================
=================
=================
=================
======
figure(1)
plot(Z,N1,'m','linewidth',2)
hold on
plot(Z,N2,'g','linewidth',2)
xlabel('Fiber length (m)')
ylabel('Population density (ions/m^3)')
figure(2)
plot(Z,PSF*1000,'linewidth',2)
hold on
plot(Z,PSB*1000,'r','linewidth',2)
hold on
plot(Z,PP*1000,'m','linewidth',2)
xlabel('Fiber length (m)')
ylabel('Power (mW))')
toc;
```

5.5.4 Script to Compute the Characteristics of Fabry-Perot Fiber Laser in the Backward Pumping Scheme

```
function POUT = OutputPowerBackward(Pump,leng,Roc)
% This function implements a relaxation algorithm to
solve the propagation
% equations of a Fabry-Perot erbium doped fiber laser, in
the backward pumping
% configuration.
% The cavity of the laser is divided into 100 sections and
power in each
% section is computed using a fourth-order Runge-Kutta
shcheme.
%
%
% Inputs:
%  Pump: Launched pump power
%  leng: Doped fiber length
%  Roc : Reflectivity of the output mirror
%
% Outputs:
%  POUT: Laser output power
%
% Comments:
%
%
% References:
%
% Written by Justice Sompo, University of Johannesburg,
South-Africa
close all
clc
format longe
tic
% Check inputs
if nargin < 3
  Roc = 0.1;
  if nargin < 2
    leng = 10;
    if nargin < 1
                                        error(message
('MATLAB:OutputPowerBackward:NotEnoughInputs'));
    end
  end
```

```
end
%
%----------------------------COMPUTATION
PARAMETERS---------------------------
FiberLength = leng;        % meters
n = 100;                % Number of sections
step = (FiberLength-0)/n;  % step size for computation
R1 = 0.98;              % Reflectivity of mirror 1
R2 = Roc;               % Reflectivity of mirror 2
%==================
================
================
================
======
%             TRIP 1 : FORWARD PROPAGARTION
%==================
================
================
================
======
%----------------------------INITIAL
CONDITIONS---------------------------
ppb = Pump;             % Backward launched power at z = L
(100 mW)
z_initial = 0;          % origin of longitudinal distance
(z=0 for forward displacement)
pp_initial = 15e-3;     % Initial guess of the pump power
at z = 0
psf_initial = 1e-3;     % initial guess of the forward
propagating signal at z = 0
psb_initial = 1e-3;     % initial guess of the backward
propagating signal at z = 0
z = z_initial;
pp = pp_initial;
psf = psf_initial;
psb = psb_initial;
%==================
================
================
================
======
%      WHILE LOOP TO COMPUTE LASER PARAMETERS
%==================
================
================
```

```
==================
======
%
err1 = 10;            % Arbitrarily set value to enter the
while loop
err2 = 10;
err3 = 10;
tol = 1e-4;
IterationNumber = 0;    % Initialise the counter
while (err1 >= tol && err2 >= tol || err3 >= tol) &&
IterationNumber <= 40
        %----------------------CREATE    VECTORS    FOR
PLOTTING----------------------
  Z = z_initial;
  PP = pp_initial;
  PSF = psf_initial;
  PSB = psb_initial;
   %----------------Iterative calculations using RK4
method----------------
  for w = 0:n-1        % Number of loops
    z = z+step;       % The integration is from zero(0) to L
    k31 = step*feval('PPB',z,pp,psf,psb);
    f31 = step*feval('PSF',z,pp,psf,psb);
    b31 - step*feval('PSB',z,pp,psf,psb);
    k32 = step*feval('PPB',(z+0.5*step),(pp+0.5*k31),
(psf+0.5*31),(psb+0.5*b31));
    f32 = step*feval('PSF',z+0.5*step,pp+0.5*k31, psf
+0.5*f31, psb+0.5*b31);
    b32 = step*feval('PSB',z+0.5*step,pp+0.5*k31, psf
+0.5*f31, psb+0.5*b31);
    k33 = step*feval('PPB',(z+0.5*step),(pp+0.5*k32),
(psf+0.5*f32), (psb+0.5*b32));
    f33 = step*feval('PSF',(z+0.5*step),(pp+0.5*k32),
(psf+0.5*f32), (psb+0.5*b32));
    b33 = step*feval('PSB',(z+0.5*step),(pp+0.5*k32),
(psf+0.5*f32), (psb+0.5*b32));
         k34 = step*feval('PPB',z+step,pp+k33,psf
+f33,psb+b33);
         f34 = step*feval('PSF',z+step,pp+k33,psf
+f33,psb+b33);
         b34 = step*feval('PSB',z+step,pp+k33,psf
+f33,psb+b33);
    pp = pp+(1/6)*(k31+2*(k32+k33)+k34);    % Pump power
evolution
```

```
    psf = psf +(1/6)*(f31+2*(f32+f33)+f34);   % Forward
laser power evolution
    psb = psb +(1/6)*(b31+2*(b32+b33)+b34);   % Backward
laser power evolution
    Z = [Z,z];
    PP = [PP,pp];
    PSF = [PSF, psf];
    PSB = [PSB, psb];
  end
  %------------STORE THE VALUES OF PUMP, SIGNAL AND
POSITION--------------
  z_FirstPass = z;
  pp_FirstPass = pp;
  psf_FirstPass = psf;
  psb_FirstPass = psb;
      %----------------TEST CONVERGENCE AFTER THE
FIRST PASS-----------------
  err1 = abs(psb-R2*psf);
  err2 = abs(pp-ppb);
  if err1 <=tol && err2 <=tol
    break;
  end
  %====================================================
====================
  %              TRIP 2 : BACKWARD PROPAGARTION
    %====================================================
======================
      %-------------------APPLY BC FOR NEW INITIAL
CONDITIONS----------------
  z = z_FirstPass;
  pp = ppb;
  psf = psf_FirstPass;
  psb = psf_FirstPass*R2;
        %--------------------CREATE   VECTORS   FOR
PLOTTING-----------------------
  LL = 0;
  Z = z;
  PP = pp;
  PSF = psf;
  PSB = psb;
  L = 1;
  p = n-1 ;
  for w = p:-1:0
    LL = [LL,L];
    L = L+1;
```

```
   z = z-step;        % Doing Reverse Integration from z = L
to z = 0
   k21 = step*feval('PPB',z,pp,psf,psb);
   f21 = step*feval('PSF',z,pp,psf,psb);
   b21 = step*feval('PSB',z,pp,psf,psb);
   k22 = step*feval('PPB',(z+0.5*step),(pp+0.5*k21),
(psf+0.5*f21),(psb+0.5*b21));
    f22 = step*feval('PSF',z+0.5*step,pp+0.5*k21, psf
+0.5*f21, psb+0.5*b21);
    b22 = step*feval('PSB',z+0.5*step,pp+0.5*k21, psf
+0.5*f21, psb+0.5*b21);
   k23 = step*feval('PPB',(z+0.5*step),(pp+0.5*k22),
(psf+0.5*f22), (psb+0.5*b22));
   f23 = step*feval('PSF',(z+0.5*step),(pp+0.5*k22),
(psf+0.5*f22), (psb+0.5*b22));
   b23 = step*feval('PSB',(z+0.5*step),(pp+0.5*k22),
(psf+0.5*f22), (psb+0.5*b22));
           k24 = step*feval('PPB',z+step,pp+k23,psf
+f23,psb+b23);
           f24 = step*feval('PSF',z+step,pp+k23,psf
+f23,psb+b23);
           b24 = step*feval('PSB',z+step,pp+k23,psf
+f23,psb+b23);
   pp = pp-(1/6)*(k21+2*(k22+k23)+k24);    % Pump power
evolution
    psf = psf -(1/6)*(f21+2*(f22+f23)+f24);  % Forward
laser power evolution
    psb = psb -(1/6)*(b21+2*(b22+b23)+b24);  % Backward
laser power evolution
    Z = [Z,z];
    PP = [PP,pp];
    PSF = [PSF, psf];
    PSB = [PSB, psb];
 end
  %--------------STORE THE VALUES OF PUMP, SIGNAL AND
POSITION-----------
 z_SecondPass = z;
 pp_SecondPass = pp;
 psf_SecondPass = psf;
 psb_SecondPass = psb;
 err3 = abs(psf-R1*psb);
 z = z_SecondPass;
 pp = pp_SecondPass;
 psf = psb_SecondPass*R1;
 psb = psb_SecondPass;
```

```
   IterationNumber = (IterationNumber + 1) ;
end
POUT = psf*(1-R2);
[N1,N2] = PopulationDensity(PP,PSB,PSF);
figure(1)
plot(Z,N1,'m','linewidth',2)
hold on
plot(Z,N2,'g','linewidth',2)
xlabel('Fiber length (m)')
ylabel('Population density (ions/m^3)')
figure(2)
plot(Z,PSF*1000,'linewidth',2)
hold on
plot(Z,PSB*1000,'r','linewidth',2)
hold on
plot(Z,PP*1000,'m','linewidth',2)
xlabel('Fiber length (m)')
ylabel('Power (mW))')
toc;
```

5.5.5 PUMP POWER FORWARD

```
function PP = PPF(z,pp,psf,psb)
% PPF function represents forward pump field propagation
equation
%
% Inputs:
% - z : Position along the longitudinal distance of the
cavity
% - pp: pump power
% - psf: Forward laser power
% - psb: Backward laser power
%
% Outputs:
%   PP: Forward pump power as a function of distance z
%
% Comments:
%   - Computation is done in the ideal case of a two level
energy system
%
% References:
```

```
%
% Written by Justice Sompo, University of Johannesburg,
South Africa
close all
clc
%
format longe
% Check inputs
if nargin <4
  error(message('MATLAB:PPF:NotEnoughInputs'));
end
% Rate equation Parameters
sigma12S = 2.33e-25;  % Laser absorption cross section
sigma21S = 2.64e-25;  % Laser emission cross section
sigma12P = 2e-25;     % Pump absorption cross section
sigma21P = 0;         % Pump emission cross section
% Lifetimes
t21 = 10e-3;          % Lifetime of the metastable energy
level of Erbium
A21 = 1/t21;
% Concentrations
Ner = 1.2e25;         % Erbium ions concentration
% Overlap factors and background losses
Gammap = 0.64;        % Overlap factor at pump wavelength
Gammas = 0.43;        % Overlap factor at signal wavelength
alphap = 0.005;       % Background loss at pump wavelength
alphas = 0.005;       % Background loss at laser wavelength
lambdap = 980e-9; % Pumb wavelenght
lambdas = 1550e-9; %signal wavelength
% Other physical constants
r = 2.3e-6;           % Doped fiber core
h = 6.626e-34;        % Planck's constant
cel = 3e8;            % speed of light in free space
% Calculated parameters
A = pi*r^2;
F_s = cel/lambdas;den1 = A*h*F_s;
F_p = cel/lambdap;  den2 = A*h*F_p;
% Fiber Bragg Grating Parameters
W12 = ((sigma12S*Gammas)*(psf+psb))/den1;
W21 = ((sigma21S*Gammas)*(psf+psb))/den1;
R12 = ((sigma12P*Gammap)*(pp))/den2;
%Rate equations solving
N1 = Ner*(W21+A21)./(W12+R12+W21+A21);
N2 = Ner*(W12+R12)./(W21+A21+W12+R12);
```

```
PP = (pp.*Gammap).*(sigma21P*N2-sigma12P*N1)+(alph-
ap.*pp);
```

5.5.6 PUMP POWER BACKWARD

```
function PP = PPB(z,pp,psf,psb)
% PPB function represents backward pump field propagation
equation
%
% Inputs:
% - z : Position along with the longitudinal distance of
the cavity
% - pp: pump power
% - psf: Forward laser power
% - psb: Backward laser power
%
% Outputs:
%   PP: Backward pump power as a function of distance z
%
% Comments:
%   - Computation is done in the ideal case of a two-level
energy system
%
% References:
%
% Written by Justice Sompo, University of Johannesburg,
South Africa
close all
clc
%
format longe
% Check inputs
if nargin <4
  error(message('MATLAB:PPB:NotEnoughInputs'));
end
% Rate equation Parameters
sigma12S = 2.33e-25;  % Laser absorption cross section
sigma21S = 2.64e-25;  % Laser emission cross section
sigma12P = 2e-25;     % Pump absorption cross section
sigma21P = 0;         % Pump emission cross section
% Lifetimes
```

```
t21 = 10e-3;          % Lifetime of the metastable energy
level of Erbium
A21 = 1/t21;
% Concentrations
Ner = 1.2e25;         % Erbium ions concentration
% Overlap factors and background losses
Gammap = 0.64;        % Overlap factor at pump wavelength
Gammas = 0.43;        % Overlap factor at signal wavelength
alphap = 0.005;       % Background loss at pump wavelength
alphas = 0.005;       % Background loss at laser wavelength
lambdap = 980e-9;  % Pumb wavelenght
lambdas = 1550e-9; %signal wavelength
% Other physical constants
r = 2.3e-6;           % Doped fiber core
h = 6.626e-34;        % Planck's constant
cel = 3e8;            % speed of light in free space
% Calculated parameters
A = pi*r^2;
F_s = cel/lambdas;den1 = A*h*F_s;
F_p = cel/lambdap;  den2 = A*h*F_p;
% Fiber Bragg Grating Parameters
W12 = ((sigma12S*Gammas)*(psf+psb))/den1;
W21 = ((sigma21S*Gammas)*(psf+psb))/den1;
R12 = ((sigma12P*Gammap)*(pp))/den2;
%Rate equations solving
N1 = Ner*(W21+A21)./(W12+R12+W21+A21);
N2 = Ner*(W12+R12)./(W21+A21+W12+R12);
PP=     (-1)*(pp.*Gammap).*(sigma21P*N2-sigma12P*N1)-
(alphap.*pp);
```

5.5.7 Laser Power Forward

```
function PF = PSF(z,pp,psf,psb)
% PSF function represents forward laser field propagation
equation
%
% Inputs:
% - z : Position along the longitudinal distance of the
cavity
% - pp: pump power
% - psf: Forward laser power
```

```
% - psb: Backward laser power
%
% Outputs:
%   PF : Forward signal power as a function of distance z
%
% Comments:
%   - Computation is done in the ideal case of a two level
energy system
%
% References:
%
% Written by Justice Sompo, University of Johannesburg,
South Africa
close all
clc
%
format longe
% Check inputs
if nargin <4
  error(message('MATLAB:PSF:NotEnoughInputs'));
end
% Rate equation Parameters
sigma12S = 2.33e-25;  % Laser absorption cross section
sigma21S = 2.64e-25;  % Laser emission cross section
sigma12P = 2e-25;     % Pump absorption cross section
sigma21P = 0;         % Pump emission cross section
% Lifetimes
t21 = 10e-3;          % Lifetime of the metastable energy
level of Erbium
A21 = 1/t21;
% Concentrations
Ner = 1.2e25;         % Erbium ions concentration
% Overlap factors and background losses
Gammap = 0.64;        % Overlap factor at pump wavelength
Gammas = 0.43;        % Overlap factor at signal wavelength
alphap = 0.005;       % Background loss at pump wavelength
alphas = 0.005;       % Background loss at laser wavelength
lambdap = 980e-9; % Pumb wavelenght
lambdas = 1550e-9; %signal wavelength
% Other physical constants
r = 2.3e-6;           % Doped fiber core
h = 6.626e-34;        % Planck's constant
cel = 3e8;            % speed of light in free space
% Calculated parameters
A = pi*r^2;
```

```
F_s = cel/lambdas;den1 = A*h*F_s;
F_p = cel/lambdap;  den2 = A*h*F_p;
% Fiber Bragg Grating Parameters
W12 = ((sigma12S*Gammas)*(psf+psb))/den1;
W21 = ((sigma21S*Gammas)*(psf+psb))/den1;
R12 = ((sigma12P*Gammap)*(pp))/den2;
%Rate equations solving
N1 = Ner*(W21+A21)./(W12+R12+W21+A21);
N2 = Ner*(W12+R12)./(W21+A21+W12+R12);
PF=  (psf.*Gammas).*(sigma21S*N2-sigma12S*N1)-(alph-
as.*psf);
```

5.5.8 Laser Power Backward

```
function PB = PSB(x,y,uf,ub)
% PSF function represents backward laser field propaga-
tion equation
%
% Inputs:
% - z : Position along with the longitudinal distance of
the cavity
% - pp: pump power
% - psf: Forward laser power
% - psb: Backward laser power
%
% Outputs:
%  PB : Backward signal power as a function of distance z
%
% Comments:
%  - Computation is done in the ideal case of a two-level
energy system
%
% References:
%
% Written by Justice Sompo, University of Johannesburg,
South Africa
close all
clc
%
format longe
% Check inputs
```

```
if margin <4
  error(message('MATLAB:PSB:NotEnoughInputs'));
end
% Rate equation Parameters
sigma12S = 2.33e-25;  % Laser absorption cross section
sigma21S = 2.64e-25;  % Laser emission cross section
sigma12P = 2e-25;     % Pump absorption cross section
sigma21P = 0;         % Pump emission cross section
% Lifetimes
t21 = 10e-3;          % Lifetime of the metastable energy
level of Erbium
A21 = 1/t21;
% Concentrations
Ner = 1.2e25;         % Erbium ions concentration
% Overlap factors and background losses
Gammap = 0.64;        % Overlap factor at pump wavelength
Gammas = 0.43;        % Overlap factor at signal wavelength
alphap = 0.005;       % Background loss at pump wavelength
alphas = 0.005;       % Background loss at laser wavelength
lambdap = 980e-9;  % Pumb wavelenght
lambdas = 1550e-9; %signal wavelength
% Other physical constants
r = 2.3e-6;           % Doped fiber core
h = 6.626e-34;        % Planck's constant
cel = 3e8;            % speed of light in free space
% Calculated parameters
A = pi*r^2;
F_s = cel/lambdas;den1 = A*h*F_s;
F_p = cel/lambdap;  den2 = A*h*F_p;
% Fiber Bragg Grating Parameters
W12 = ((sigma12S*Gammas)*(uf+ub))/den1;
W21 = ((sigma21S*Gammas)*(uf+ub))/den1;
R12 = ((sigma12P*Gammap)*(y))/den2;
%Rate equations solving
N1 = Ner*(W21+A21)./(W12+R12+W21+A21);
N2 = Ner*(W12+R12)./(W21+A21+W12+R12);
PB=     (-1)*(ub.*Gammas).*(sigma21S*N2-sigma12S*N1)+
(alphas.*ub);
```

5.5.9 POPULATION DENSITY FUNCTION

```matlab
function [N1,N2] = PopulationDensity(Pp,Psb,Psf)
% PopulationDensity function represents the system of
linear equations for
% population density calculation.
%
% Inputs:
% -The function takes the pump power and initial guesses
of output powers
%
% Outputs:
% N1: Population density of ground level as a function of
distance z
%    N2: Population density of metastable level as a
function of
%   distance z
%
% Comments:
%   - Computation is done in the ideal case of a two-level
energy system
%
% References:
%
% Written by Justice Sompo, University of Johannesburg,
South Africa
close all
clc
%
format longe
% Check inputs
if nargin <3
        error(message('MATLAB:PopulationDensity:Not-
EnoughInputs'));
end
% Rate equation Parameters
sigma12S = 2.33e-25;  % Laser absorption cross section
sigma21S = 2.64e-25;  % Laser emission cross section
sigma12P = 2e-25;     % Pump absorption cross section
sigma21P = 0;         % Pump emission cross section
% Lifetimes
t21 = 10e-3;          % Lifetime of the metastable energy
level of Erbium
A21 = 1/t21;
```

```
% Concentrations
Ner = 1.2e25;          % Erbium ions concentration
% Overlap factors and background losses
Gammap = 0.81;         % Overlap factor at pump wavelength
Gammas = 0.6;          % Overlap factor at signal wavelength
alphap = 0.005;        % Background loss at pump wavelength
alphas = 0.005;        % Background loss at laser wavelength
lambdap = 980e-9;  % Pumb wavelenght
lambdas = 1550e-9; %signal wavelength
% Other physical constants
r = 2.3e-6;            % Doped fiber core
h = 6.626e-34;         % Planck's constant
cel = 3e8;             % speed of light in free space
% Calculated parameters
A = pi*r^2;
F_s = cel/lambdas;
den1 = A*h*F_s;
F_p = cel/lambdap;
den2 = A*h*F_p;
% Fiber Bragg Grating Parameters
W12 = ((sigma12S*Gammas).*(Psb+Psf))/den1;
W21 = ((sigma21S*Gammas).*(Psb+Psf))/den1;
R12 = ((sigma12P*Gammap).*(Pp))/den2;
%Rate equations solving
N1 = Ner*(W21+A21)./(W12+R12+W21+A21);
N2 = Ner*(W12+R12)./(W21+A21+W12+R12);
```

5.5.10 UNIFORM GRATING

```
% Simulation of a uniform fiber Bragg grating. The
characteristics
% of the grating can be modified by the user to obtain
various characteristics
%==================
==================
==================
==================  .
======
clear all
close all
clc
```

```
tic
format longe
%=================
================
===============
===============
======
% Fiber simulation parameters
walD = 1.55e-6;
wal1 = 0.999*walD;
wal2 = 1.001*walD;
step = 500;
wal = wal1:(wal2-wal1)/step:wal2;
%=================
================
===============
===============
======
% For a grating of maximum reflectance R = 0.2
Rmax = 0.9;
rmax = sqrt(Rmax);
kacl = atanh(rmax);
c = 3e8;
h = 25e-9;
v = 1;
%=================
================
===============
===============
======
% Implementation of the transfer matrix method for so-
lution of coupled mode
% equations
nef = 1.47;
L = 5000e-6;
M = 100;
dz = L/M;
dzo = -L+10.69e-3;
kac = kacl/L;
kdc = 2*kac/v;
for r=1:step+1
  w = wal(r);
  F = [1 0;0 1];
  for s = 1:M
    det = 2*pi*nef*(1/w-1/walD);
```

```
     gdc = det+kdc;
     p1 = sqrt(kac^2-gdc^2);
     p2 = gdc^2/kac^2;
     f11 = cosh(p1*dz)-1i*(gdc/p1)*sinh(p1*dz);
     f12 =-1i*(kac/p1)*sinh(p1*dz);
     f21 = 1i*(kac/p1)*sinh(p1*dz);
     f22 = cosh(p1*dz) + 1i*(gdc/p1)*sinh(p1*dz);
     ff = [f11 f12;f21 f22];
     F = ff*F;
  end
r3(r) = F(2,1)/F(1,1);
R3(r) = (abs(r3(r)))^2;
r4(r) = 1/F(1,1);
R4(r) =(abs(r4(r)))^2;
end
%==================
=================
=================
=================
======
% Plots for reflection and transmission spectra for Bragg
reflector
figure (1)
subplot(2,1,1)
plot(wal*1e9,R3,'r','Linewidth',2)
axis([1549 1551.5 0 1])
title('Reflection Spectrum of Bragg Reflector')
xlabel('Wavelength (nm)')
ylabel('Power (p.u)')
subplot(2,1,2)
plot(wal*1e9,R4,'b','Linewidth',2)
axis([1549 1551 0 1])
title ('Transmission spectrum of Bragg Reflector')
xlabel('wavelength(nm)')
ylabel('reflectivity')
```

5.5.11 π-PHASE SHIFTED FIBER BRAGG GRATING

```
% Simulation of a phase-shifted fiber Bragg grating with
phase shift in it
```

```
% centre. The characteristics of the grating can be
modified by the user to
% obtain various characteristics
%==================
================
================
================
======
clear all
close all
clc
%==================
================
================
================
======
% Fiber simulation parameters
lambdaB = 1.55e-6;          % design wavelength
wal1 = 0.9976*lambdaB;
wal2 = 1.00032*lambdaB;
step = 1000;
wal = wal1:(wal2-wal1)/step:wal2;
L = 0.015; % grating lengh in meters
dn =1.0e-4; % induced index change
neff= 1.447; % effective index change for the mode
M = 100;
dz = L/M;
kac = pi*dn/lambdaB;
%==================
================
================
================
======
% implementation of the transfer matrix method for so-
lution of coupled
% mode equations
%==================
================
================
================
======
for tx = 1:step+1
  w = wal(tx);
  F = [1 0;0 1];
  phaseshift =[exp(-1i*pi/2) 0;0 exp(1i*pi/2)];
```

```
   for s = 1:M
      dets = 2*pi*neff*(1/w-1/lambdaB);
      q = -sqrt((dets)^2-kac^2);
      r = (q-dets)/kac;
      ra = abs(r);
             f11 = (1/(1-r^2))*(exp(1i*q*dz)-(r^2)*exp
(-1i*q*dz));
      f21 = r/(1-r^2)*(exp(1i*q*dz)-exp(-1i*q*dz)) ;
      f12 = -f21;
      f22 = conj(f11);
      ff = [f11 f12;f21 f22];
      if s == 50
         F = phaseshift*F;
      else
         F = ff*F;
      end
   end
  r3(tx) = F(2,1)/F(1,1);
  R3(tx) = (abs(r3(tx)))^2;
  r4(tx) = 1/F(1,1);
  R4(tx) =(abs(r4(tx)))^2;
end
%==================
=================
=================
=================
======
% Plot for reflection and transmission spectra for bragg
reflector
figure(1)
subplot(2,1,1)
plot(wal*1e9,R3,'r','Linewidth',2)
axis([1549 1551 0 1.2])
title ('Reflection spectrum of Bragg Reflector')
xlabel('wavelength(nm)')
ylabel('Reflectivity')
subplot(2,1,2)
plot(wal*1e9,R4,'b','Linewidth',2)
axis([1549 1551 0 1])
title ('Transmission spectrum of Bragg Reflector')
xlabel('wavelength(nm)')
ylabel('reflectivity')
```

5.5.12 Shooting Algorithm for Simple DFB Fiber Laser Without CUP

```
function [S,Ps] = shootingDfbLaser(Pp)
% ShootingDfbLaser function computes the output power of
a DFB fiber laser
% using a shooting algorithm with two initial guesses and
a secant method
% for guess refinement
%
% Inputs:
% - Pp: Pump power
%
%
% Outputs:
%  S: Forward internal electric field
%  Ps: Forward output power
% Comments:
%  - The function calls the function dfbSimulation
%  - The function "FB simulation" is called multiple time
to compute
%  different values like:
%   - N1: Population density distribution at the ground
level of Erbium
%   - N2: Population density distribution at the me-
tastable level of Erbium
%   - N3: Population density distribution at the upper
level of Erbium
%   - N5: Population density distribution at the ground
level of Ytterbium
%   - N6: Population density distribution at the upper
level of Erbium
%  - Gain distribution inside the cavity
%  - PP: Pump power distribution
%  - PS: Laser power distribution inside the cavity
%
% References:
%
% Written by Justice Sompo, University of Johannesburg,
South Africa
%
%--------------Initial guess to initiate the Shooting
algorithm------------
P01 = 1e-3;
P02 = 10e-3;
L = 50; % length of the cavity
ii = 1;
```

```
S1 = dfbSimulation(Pp,P01); % final backward propagation
wave for initial guess P01
S2 = dfbSimulation(Pp,P02); % final backward propagation
wave for initial guess P02
while(abs(P01-P02)>0.000001)
  tmp = P02;
  P02 = P01-(P01-P02)/(S1-S2)*S1;
  P01 = tmp;
  S1 = dfbSimulation(Pp,P01); % final backward propaga-
tion wave for initial guess P01
  S2 = dfbSimulation(Pp,P02); % final backward propaga-
tion wave for initial guess P02
  ii = ii+1;
end
ii;
S = S2;
Ps = abs(P02);
%-----------------------------PLOTTING-------------
-------------------
[S,GAMMA,PP,PS,N1,N2,N3,N5,N6] = feval('dfbSimulat-
ion',Pp,Ps);
z = linspace(0,L);
figure(1)
subplot(2,2,1)
plot(z,N2,'r','Linewidth',2)
xlabel('Position (mm)')
ylabel('N2 Population (ions/m^3)')
subplot(2,2,2)
plot(z,GAMMA,'m','Linewidth',2)
xlabel('Position (mm)')
ylabel('Gain (m^-1)')
subplot(2,2,3)
plot(z,PP,'g','Linewidth',2)
xlabel('Position (mm)')
ylabel('Pump Power (Watts)')
subplot(2,2,4)
plot(z,PS,'Linewidth',2)
xlabel('Position (mm)')
ylabel('Laser Power (Watts)')
figure(2)
subplot(2,2,1)
plot(z,N1,'r','Linewidth',2)
xlabel('Position (mm)')
ylabel('N1 Population (ions/m^3)')
subplot(2,2,2)
```

```
plot(z,N3,'m','Linewidth',2)
xlabel('Position (mm)')
ylabel('N3 Population (ions/m^3)')
subplot(2,2,3)
plot(z,N5,'g','Linewidth',2)
xlabel('N5 Population (ions/m^3)')
ylabel('Pump Power (Watts)')
subplot(2,2,4)
plot(z,N6,'Linewidth',2)
xlabel('Position (mm)')
ylabel('N6 Population (ions/m^3)')
```

5.5.13   SHOOTING ALGORITHM FOR DFB FIBER LASER WITH CUP

```
function [S,Ps] = shootingDfbLaser2(Pp)
% ShootingDfbLaser2 function computes the output power
of a DFB fiber laser
% using a shooting algorithm with two initial guesses and
a secant method
% for guess refinement
%
% Inputs:
% - Pp: Pump power
%
%
% Outputs:
%  S: Forward internal electric field
%  Ps: Forward output power
% Comments:
%  - The function calls the function dfbSimulation2
%   - The function " dfbSimulation2" is called multiple
time to compute
%  different values like:
%  - N1: Population density distribution at the ground
level of Erbium
%   - N2: Population density distribution at the me-
tastable level of Erbium
%   - N3: Population density distribution at the upper
level of Erbium
%  - N5: Population density distribution at ground level
of Ytterbium
```

```
%   - N6: Population density distribution at upper level
of Erbium
%   - Gain distribution inside the cavity
%   - PP: Pump power distribution
%   - PS: Laser power distribution inside the cavity
%
%
% References:
%
% Written by Justice Sompo, University of Johannesburg,
South Africa
%
% Initial condition to initiate the shooting algorithm
P01 = 1e-3;
P02 = 10e-3;
%
%
L = 50; % length of the cavity
ii = 1;
S1 = dfbSimulation2(Pp,P01); % final backward propaga-
tion wave for initial guess P01
S2 = dfbSimulation2(Pp,P02); % final backward propaga-
tion wave for initial guess P02
while(abs(P01-P02) > 0.000001)
  tmp = P02;
  P02 = P01-(P01-P02)/(S1-S2)*S1;
  P01 = tmp;
   S1 = dfbSimulation2(Pp,P01); % final backward propa-
gation wave for initial guess P01
   S2 = dfbSimulation2(Pp,P02); % final backward propa-
gation wave for initial guess P02
   ii = ii+1;
end
ii;
S = S2;
Ps = abs(P02);
[S,GAMMA,PP,PS,N1,N2,N3,N5,N6]              =          feval
('dfbSimulation2',Pp,Ps);
z = linspace(0,L);
figure(1)
subplot(2,2,1)
plot(z,N2,'r','Linewidth',2)
xlabel('Position (mm)')
ylabel('N2 Population (ions/m^3)')
subplot(2,2,2)
plot(z,GAMMA,'m','Linewidth',2)
xlabel('Position (mm)')
ylabel('Gain (m^-1)')
```

```
subplot(2,2,3)
plot(z,PP,'g','Linewidth',2)
xlabel('Position (mm)')
ylabel('Pump Power (Watts)')
subplot(2,2,4)
plot(z,PS,'Linewidth',2)
xlabel('Position (mm)')
ylabel('Laser Power (Watts)')
figure(2)
subplot(2,2,1)
plot(z,N1,'r','Linewidth',2)
xlabel('Position (mm)')
ylabel('N1 Population (ions/m^3)')
subplot(2,2,2)
plot(z,N3,'m','Linewidth',2)
xlabel('Position (mm)')
ylabel('N3 Population (ions/m^3)')
subplot(2,2,3)
plot(z,N5,'g','Linewidth',2)
xlabel('N5 Population (ions/m^3)')
ylabel('Pump Power (Watts)')
subplot(2,2,4)
plot(z,N6,'Linewidth',2)
xlabel('Position (mm)')
ylabel('N6 Population (ions/m^3)')
```

5.5.14 DFB Simulation

```
function      [S,GAMMA,PP,PS,N1,N2,N3,N5,N6]      =
dfbSimulation(Pp,P0)
% Function dfbsimulation compute the DFB fiber laser in-
ternal gain, pump
% power and laser power distribution given pump power and
laser power
%
%
% Inputs:
% - Pp:Pump power
% - P0:Laser power
%
% Outputs:
% - N1: Population density distribution at ground level
of Erbium
```

```
%   - N2: Population density distribution at metastable
level of Erbium
%   - N3: Population density distribution at upper level
of Erbium
%   - N5: Population density distribution at ground level
of Ytterbium
%   - N6: Population density distribution at upper level
of Erbium
%   - Gain distribution inside the cavity
%   - PP: Pump power distribution
%   - PS: Laser power distribution inside the cavity
%   - GAMMA: Gain distribution
% Comments:
%   - Erbium-Ytterbium co-doped gain medium, rate equa-
tion solved
%   analytically neglecting cooperative upconversion
%
% References:
%
% Written by Justice Sompo, University of Johannesburg,
South Africa
format longe
% Fiber Bragg Grating Parameters
L = 0.050;M = 100;
dn =7.38e-5;
neff= 1.47;
lambdaB = 1547.3661e-9;
period = lambdaB/(2*neff);
dz = L/M;
phaseshift =[exp(-1i*(pi/2)) 0;0 exp(1i*(pi/2))];
kac = (pi*dn)/lambdaB;
phase =0;
% Initial guess
w =1547.3661e-9;
%Pp = 100e-3;
%P0 =18.13e-3;
S0 = sqrt(P0); %Estimated backward laser amplitude
R0 = 0;     % No forwards propagating wave
F = [R0;S0];
Ps = (abs(S0))^2+(abs(R0))^2;
N1 = zeros(1,100);
for ii=1:M
                [n1,n2,n3,n5,n6,Gamma,alpha]      =
rateEquationSolver(Pp,Ps);
  % finish solving the rate equation here
```

```
%Pump power distribution
Pp = Pp*exp(alpha*dz);
% Feedback mechanism
beta = pi/period;
dets = 2*pi*neff*(1/w-1/lambdaB);
detsprime = (dets+1i*Gamma);
p1 = sqrt(kac^2 - detsprime^2);
% Transfer matrix
    f11  = (cosh(p1*dz)-1i*((dets+1i*Gamma)/p1)*sinh
(p1*dz))*exp(-1i*beta*dz);
 f12 = (kac/p1)*sinh(p1*dz)*exp(-1i*(beta*dz+phase));
 f21 = (kac/p1)*sinh(p1*dz)*exp(1i*(beta*dz+phase));
    f22  = (cosh(p1*dz)+1i*((dets+1i*Gamma)/p1)*sinh
(p1*dz))*exp(1i*beta*dz);
 ff = [f11 f12;f21 f22];
 F = ff*F;
 if ii == 50
    F = phaseshift*F;
 end
 phase = phase+(2*pi*dz)/period;
 R=F(1,1);
 S=F(2,1);
 Ps = (abs(S))^2+(abs(R))^2;
 PLb = (abs(S))^2;
 PLf = (abs(R))^2;
 N1(ii)=n1;
 N2(ii)=n2;
 N3(ii)=n3;
 N5(ii)=n5;
 N6(ii)=n6;
 GAMMA(ii)=Gamma;
 PP(ii)=Pp;
 PS(ii)=Ps;
end
```

5.5.15   DFB Simulation2

```
function      [S,GAMMA,PP,PS,N1,N2,N3,N5,N6]      =
dfbSimulation2(Pp,P0)
% Function dfbSimulation2 compute the DFB fiber laser
internal gain, pump
```

```
% power and laser power distribution for given pump and
laser powers
%
%
% Inputs:
% - Pp: Pump power
% - P0: Laser power
%
% Outputs:
%   - N1: Population density distribution at the ground
level of Erbium
%   - N2: Population density distribution at the me-
tastable level of Erbium
%   - N3: Population density distribution at the upper
level of Erbium
%   - N5: Population density distribution at the ground
level of Ytterbium
%   - N6: Population density distribution at the upper
level of Erbium
% - Gain distribution inside the cavity
% - PP: Pump power distribution
% - PS: Laser power distribution inside the cavity
% - GAMMA: Gain distribution
%
% Comments:
%   - Erbium-Ytterbium co-doped gain medium, rate equa-
tion solved
% by a numerical iterative method including cooperative
upconversion
%
% References:
%
% Written by Justice Sompo, University of Johannesburg,
South Africa
format longe
% Fiber Bragg Grating Parameters
L = 0.050;
M = 100;
dn =7.38e-5;
neff= 1.47;
lambdaB = 1547.3661e-9;
period = lambdaB/(2*neff);
dz = L/M;
phaseshift =[exp(-1i*(pi/2)) 0;0 exp(1i*(pi/2))];
kac = (pi*dn)/lambdaB;
```

```
phase = 0;
% Initial guess
w =1547.3661e-9;
%Pp = 100e-3;
%P0 =18.13e-3;
S0 = sqrt(P0); %Estimated backward laser amplitude
R0 = 0;      % No forwards propagating wave
F = [R0;S0];
Ps = (abs(S0))^2+(abs(R0))^2;
N1 = zeros(1,100);
for ii=1:M
                  [n1,n2,n3,n5,n6,Gamma,alpha]        =
rateEquationPairedSolver(Pp,Ps);
  % finish solving the rate equation here
  %Pump power distribution
  Pp = Pp*exp(alpha*dz);
  % Feedback mechanism
  beta = pi/period;
  dets = 2*pi*neff*(1/w-1/lambdaB);
  detsprime = (dets+1i*Gamma);
  p1 = sqrt(kac^2 - detsprime^2);
  % Transfer matrix
    f11 = (cosh(p1*dz)-1i*((dets+1i*Gamma)/p1)*sinh
(p1*dz))*exp(-1i*beta*dz);
 f12 = (kac/p1)*sinh(p1*dz)*exp(-1i*(beta*dz+phase));
  f21 = (kac/p1)*sinh(p1*dz)*exp(1i*(beta*dz+phase));
    f22 = (cosh(p1*dz)+1i*((dets+1i*Gamma)/p1)*sinh
(p1*dz))*exp(1i*beta*dz);
  ff = [f11 f12;f21 f22];
  F = ff*F;
  if ii == 50
    F = phaseshift*F;
  end
  phase = phase+(2*pi*dz)/period;
  R=F(1,1);
  S=F(2,1);
  Ps = (abs(S))^2+(abs(R))^2;
  PLb = (abs(S))^2;
  PLf = (abs(R))^2;
  N1(ii)=n1;
  N2(ii)=n2;
  N3(ii)=n3;
  N5(ii)=n5;
  N6(ii)=n6;
  GAMMA(ii)=Gamma;
```

```
  PP(ii)=Pp;
  PS(ii)=Ps;
end
```

5.5.16 RATE EQUATIONS SINGLE SOLVER

```
function        [N1,N2,N3,N5,N6,Gamma,alpha]        =
rateEquationSolver(Pp,Ps)
% rateEquationsSover function solves the Erbium
Ytterbium rate equations in
% Steady-state without taking into account the influence
of cooperative
% upconversion
%
% Inputs:
% - Pp: Pump power
% - Ps: Laser power
%
% Outputs:
%   N1: Population density of the ground state level of
Erbium
%   N2: Population density of the metastable level of
Erbium
%  N3: Population density of the energy level 3 of Erbium
%   N5: Population density of the ground state of
Ytterbium
%  N6: Population density of the upper level of Ytterbium
%  Gamma: Gain coefficient of the laser
%  alpha: Absorption coefficient of the propagating field
inside the
%  cavity
%
% Comments:
%  - Computation is done in the ideal case of a two-level
energy system
%
% References:
%
% Written by Justice Sompo, University of Johannesburg,
South Africa
%
% Rate equation Parameters
```

```
sigma12 = 0.28e-24;    % Absoption cross section of Erbium
ions at 1550 nm (m^2)
sigma21 = 0.42e-24;    % Emission cross section of Erbium
ions at 1550 nm (m^2)
sigma56 = 0.58e-24;    % Absorption cross section of
Ytterbium ions at 980 nm (m^2)
sigma65 = 2.2e-24;      % Emission cross section of
Ytterbium ions at 980 nm (m^2)
sigma13 = 0.20e-24;     % Absorption cross section of
Erbium at 980 nm (m^2)
sigma31 = 0.20e-24;    % Emission cross section of Erbium
at 980 nm (m^2)
% Lifetimes
t65 = 0.8e-3;          % Lifetime of upper energy level of
Ytterbium
t21 = 10e-3;        % Lifetime of metastable energy level of
Erbium
t31 = 10e-6;           % Lifetime of upper energy level of
Erbium
A6 = 1/t65;
A3 = 1/t31;
A2 = 1/t21;
% ion-ions interactions
Cup = 2.5e-21;     % cooperative upconversion coefficient
Ctr = 5e-21;       % Er-Yb energy transfer coefficient
% Concentrations
Ner = 1.2e25;      % Population density of Erbium ions
Nyb = 24e25;       % Population density of Ytterbium ions
% Overlap factors and background losses
Gammap = 0.82;         % Erbium overlap factor at pump
wavelength
Gammas = 0.73;         % Erbium overlap factor at laser
wavelength
alphap = 0.2;      % Background loss of the Er-Yb fiber at
980 nm
alphas = 0.15;     % Background loss of the Er-Yb fiber at
the laser wavelength of 1550 nm
% pump wavelength
lambdap = 980e-9;
w =1547.3661e-9; % Output wavelength
% Other physical constants
r = 2.3e-6;
hpk = 6.626e-34;
cel = 3e8;
% Calculated parameters
```

```
A = pi*r^2;
F_p = cel/lambdap;
den2 = A*hpk*F_p;
F_s = cel/w;den1 = A*hpk*F_s;
W12 = (sigma12*Ps*Gammas)/den1;
W21 = (sigma21*Ps*Gammas)/den1;
R56 = (sigma56*Pp*Gammap)/den2;
R65 = (sigma65*Pp*Gammap)/den2;
R13 = (sigma13*Pp*Gammap)/den2;
a = W21+A2+A3;
b = R13+A3;
c = W21+A2;
d = W12+W21+A2;
e = R56+R65+A6;
f = Ctr*(a*R13+b*d);
g = Ctr*(a*R56*Nyb-b*c*Ner)+e*a*R13+e*b*d;
h = -e*b*c*Ner;
N1 = -(g/(2*f))+((sqrt(g^2-4*f*h))/(2*f));
N3 = (c*Ner-d*N1)/a;
N2 = Ner-N1-N3;
N6 = (R56*Nyb+R13*N1-b*N3)/e;
N5 = Nyb-N6;
% Gain and loss computation
Gamma = Gammas*(sigma21*N2-sigma12*N1)-alphas;
alpha  =  Gammap*(sigma65*N6-sigma56*N5+sigma31*N3-
sigma13*N1)-alphap;
```

5.5.17 Rate Equation Pair Solver

```
function       [N1,N2,N3,N5,N6,Gamma,alpha]       =
rateEquationPairedSolver(Pp,Ps)
% rateEquationsParedSover function solves the Erbium
Ytterbium rate equations in
% Steady-state taking into account the influence of
cooperative
% upconversion the population density of Erbium ions at
the metastable level
% are divided into two groups (single and paired ions).
The algorithm starts with
% single ions, and paired ions are added by an iterative
process
%
```

```
% Inputs:
% - Pp: Pump power
% - Ps: Laser power
%
% Outputs:
%   N1: Population density of the ground state level of
Erbium
%   N2: Population density of the metastable level of
Erbium
%   N3: Population density of the energy level 3 of Erbium
%     N5: Population density of the ground state of
Ytterbium
%   N6: Population density of the upper level of Ytterbium
%   Gamma: Gain coefficient of the laser
%   alpha: Absorption coefficient of the propagating field
inside the
%   cavity
%
% Comments:
%   - Computation is done in the ideal case of a two level
energy system
%  .
% References:
%       -Distributed Feedback Fiber Laser Strain and
Temperature Sensor,
%   Olivier Hadeler, University of Southampton
%
% Written by Justice Sompo, University of Johannesburg,
South Africa
%
% Rate equation Parameters
sigma12 = 0.28e-24;   % Absoption cross section of Erbium
ions at 1550 nm (m^2)
sigma21 = 0.42e-24;   % Emission cross section of Erbium
ions at 1550 nm (m^2)
sigma56 = 0.58e-24;    % Absorption cross section of
Ytterbium ions at 980 nm (m^2)
sigma65 = 2.2e-24;       % Emission cross section of
Ytterbium ions at 980 nm (m^2)
sigma13 = 0.20e-24;     % Absorption cross section of
Erbium at 980 nm (m^2)
sigma31 = 0.20e-24;   % Emission cross section of Erbium
at 980 nm (m^2)
% Lifetimes
t65 = 0.8e-3;        % Lifetime of upper energy level of
Ytterbium
```

```
t21 = 10e-3;       % Lifetime of metastable energy level of
Erbium
t31 = 10e-6;         % Lifetime of upper energy level of
Erbium
A6 = 1/t65;
A3 = 1/t31;
A2 = 1/t21;
% ion-ions interactions
Cup = 2.5e-21;     % cooperative upconversion coefficient
Ctr = 5e-21;       % Er-Yb energy transfer coefficient
% Concentrations
Ner = 1.2e25;      % Population density of Erbium ions
Nyb = 24e25;       % Population density of Ytterbium ions
Ners = Ner*0.53;     % Population density of Erbium
single ions
Nybs = Nyb*0.53;     % Population density of ytterbium
single ions
Nerp = Ner*0.47;     % Population density of Erbium
paired ions
Nybp = Nyb*0.47;     % Population density of Ytterbium
paired ions
% Overlap factors and background losses
Gammap = 0.82;       % Erbium overlap factor at pump
wavelength
Gammas = 0.73;       % Erbium overlap factor at laser
wavelength
alphap = 0.2;      % Background loss of the Er-Yb fiber at
980 nm
alphas = 0.15;       % Background loss of the Er-Yb fiber at
laser wavelength of 1550 nm
% pump wavelength
lambdap = 980e-9;
w =1547.3661e-9; % Output wavelength
% Other physical constants
r = 2.3e-6;
hpk = 6.626e-34;
cel = 3e8;
% Calculated parameters
A = pi*r^2;
F_p = cel/lambdap;
den2 = A*hpk*F_p;
F_s = cel/w;
den1 = A*hpk*F_s;
% Solve rate equations for single ions
```

```
W12 = (sigma12*Ps*Gammas)/den1;
W21 = (sigma21*Ps*Gammas)/den1;
R56 = (sigma56*Pp*Gammap)/den2;
R65 = (sigma65*Pp*Gammap)/den2;
R13 = (sigma13*Pp*Gammap)/den2;
a = W21+A2+A3;
b = R13+A3;
c = W21+A2;
d = W12+W21+A2;
e = R56+R65+A6;
f = Ctr*(a*R13+b*d);
g = Ctr*(a*R56*Nybs-b*c*Ners)+e*a*R13+e*b*d;
h = -e*b*c*Ners;
N1s = -(g/(2*f))+((sqrt(g^2-4*f*h))/(2*f));
N3s = (c*Ners-d*N1s)/a;
N2s = Ners-N1s-N3s;
N6s = (R56*Nybs+R13*N1s-b*N3s)/e;
N5s = Nybs-N6s;
Gammasing = Gammas*(sigma21*N2s-sigma12*N1s);
% Gain and loss computation
%----------------solve      rate      equations      for
paired ions------------------
icount = 0;
maxit = 1000;
err = 10;
toler = 1e-5;
while (err > toler && icount <= maxit)
  icount = icount+1;
  % Solve rate equations
  W12 = (sigma12*Ps*Gammas)/den1;
  W21 = (sigma21*Ps*Gammas)/den1;
  R56 = (sigma56*Pp*Gammap)/den2;
  R65 = (sigma65*Pp*Gammap)/den2;
  R13 = (sigma13*Pp*Gammap)/den2;
  a = W21+A2+A3;
  b = R13+A3;
  c = W21+A2;
  d = W12+W21+A2;
  e = R56+R65+A6;
  f = Ctr*(a*R13+b*d);
  g = Ctr*(a*R56*Nybp-b*c*Nerp)+e*a*R13+e*b*d;
  h = -e*b*c*Nerp;
  N1p = -(g/(2*f))+((sqrt(g^2-4*f*h))/(2*f));
  N3p = (c*Nerp-d*N1p)/a;
```

```
N2p = Nerp-N1p-N3p;
N6p = (R56*Nybp+R13*N1p-b*N3p)/e;
N5p = Nybp-N6p;
A2new = 100 + Cup*N2p;
err = abs((A2new-A2)/A2new);                              \
A2 = A2new;
end
Gammapair = Gammas*(sigma21*N2p-sigma12*N1p);
%------------------Total                      population
densities-------------------------
N1 = N1s+N1p;
N2 = N2s+N2p;
N3 = N3s+N3p;
N5 = N5s+N5p;
N6 = N6s+N6p;
Gamma = Gammas*(sigma21*N2-sigma12*N1)-alphas;
alpha  =  Gammap*(sigma65*N6-sigma56*N5+sigma31*N3-
sigma13*N1)-alphap;
```

REFERENCES

Achtenhagen, Martin, Robert James Beeson, Feng Pan, Bruce Nyman, and Amos Hardy. 2001. "Gain and Noise in Ytterbium-Sensitized Erbium-Doped Fiber Amplifiers: Measurements and Simulations." *Journal of Lightwave Technology* 19 (10): 1521–1526. 10.1109/50.956139.

Azmi, A.I. and G.D. Peng. 2008. "Performance Analysis of Apodized DFB Fiber Laser." *2008 IEEE PhotonicsGlobal at Singapore, IPGC 2008*, no. mm. 10.1109/IPGC.2008.4781457.

Azmi, A., Deep Sen, and Gang-Ding Peng. 2009. "Output Power and Threshold Gain of Apodized DFB Fiber Laser." *Proc SPIE* 7386 (June). 10.1117/12.836890.

Bai, Xiaolei, Quan Sheng, Haiwei Zhang, Shijie Fu, Wei Shi, and Jianquan Yao. 2015. "High-Power All-Fiber Single-Frequency Erbium-Ytterbium Co-Doped Fiber Master Oscillator Power Amplifier." *IEEE Photonics Journal* 7 (6). 10.1109/JPHOT.2015. 2490484.

Ball, G. a., W.W. Morey, and W.H. Glen. 1991. "StandingWave Monornode Erbium Fiber Laser." *IEEE Photonics Technology Letters* 3 (7): 613–615. 10.1109/68.87930.

Beregovski, Yuri, Oliver Hennig, Mahmoud Fallahi, Francisco Guzman, Rosalie Clemens, Sergio Mendes, and Nasser Peyghambarian. 1998. "Design and Characteristics of DBR-Laser-Based Environmental Sensors." *Sensors and Actuators, B: Chemical* 53 (1–2): 116–124. 10.1016/S0925-4005(98)00301-3.

Canning, J., N. Groothoff, E. Buckley, T. Ryan, K. Lyytikainen, and J. Digweed. 2003. "All-Fiber Photonic Crystal Distributed Bragg Reflector (PC-DBR) Fiber Laser." *Optics Express* 11 (17): 1995. 10.1364/oe.11.001995.

Denker, B., B. Galagan, V. Osiko, S. Sverchkov, A.M. Balbashov, J.E. Hellström, V. Pasiskevicius, and F. Laurell. 2007. "Yb3+,Er3+:YAG at High Temperatures: Energy

Transfer and Spectroscopic Properties." *Optics Communications* 271 (1): 142–147. 10.1016/j.optcom.2006.09.046.

Erdogan, Turan. 1997. "Fiber Grating Spectra - Lightwave Technology, Journal Of." *Lightwave* 15 (8): 1277–1294.

Federighi, M., F. Di Pasquale, and F. Di Pasquale. 1995. "The Effect of Pair-Induced Energy Transfer on the Performance of Silica Waveguide Amplifiers with High Er3+ /Yb3+ Concentrations." *IEEE Photonics Technology Letters* 7 (3): 303–305. 10.1109/68.372753.

Flannery, Brian P, William H Press, Saul A Teukolsky, and William Vetterling. 1992. "Numerical Recipes in C." *Press Syndicate of the University of Cambridge, New York* 24: 78.

Foster, Scott B., Geoffrey A. Cranch, Joanne Harrison, Alexei E. Tikhomirov, and Gary A. Miller. 2017. "Distributed Feedback Fiber Laser Strain Sensor Technology." *Journal of Lightwave Technology* 35 (16): 3514–3530. 10.1109/JLT.2017.2689821.

Kenneth, B . 1985. *Basic Algebra*By Nathan Jacobson. Volume I (1974) and II (1980). W. H. Freeman, San Francisco. Algebra (second edition). By Saunders Mac Lane and Garrett Birkhoff. The American Mathematical Monthly.

Geng, Jihong, Jianfeng Wu, Shibin Jiang, and Jirong Yu. 2007. "Efficient Single-Frequency Thulium Doped Fiber Laser near 2-Mm." *Optics InfoBase Conference Papers*, 2–4.

Giles, C Randy and Emmanuel Desurvire. 1991. "Modeling Erbium-Doped Fiber Amplifiers." *Journal of Lightwave Technology* 9 (2): 271–283.

Ha, Sung N. 2001. "A Nonlinear Shooting Method for Two-Point Boundary Value Problems." *Computers & Mathematics with Applications* 42 (10–11): 1411–1420.

Hanna, D.C., R.M. Percival, R.G. Smart, and A.C. Tropper. 1990. "Efficient and Tunable Operation of a Tm-Doped Fiber Laser." *Optics Communications* 75 (3–4): 283–286. 10.1016/0030-4018(90)90533-Y.

Hanna, D.C. 1988. "CW Oscillation of a Monomode Tm-Doped Fiber Laser." *Electronics Letters* 24 (19): 1222–1223.

Hemming, Alexander, Shayne Bennetts, Nikita Simakov, John Haub, and Adrian Carter. 2012. "Development of Resonantly Cladding-Pumped Holmium-Doped Fiber Lasers." *Fiber Lasers IX: Technology, Systems, and Applications* 8237: 82371J. 10.1117/12.909458.

Hemming, Alexander, Nikita Simakov, Alan Davidson, Michael Oermann, Len Corena, Neil Carmody, John Haub, Robert Swain, and Adrian Carter. 2014. "Development of High Power Holmium-Doped Fiber Amplifiers." *SPIE Photonics West 2014-LASE: Lasers and Sources* 8961: 1–6. 10.1117/12.2042963.

"High Performance Single Frequency FBG Based Er3+Yb3+ Fiber Laser.Pdf." n.d.

Hill, Kenneth O and Gerald Meltz. 1997. "Fiber Bragg Grating Technology Fundementals and Overview." *IEEE Journal of Lightwave Technology* 15 (8): 1263–1276. 10.1109/50.618320.

Horak, P., N.Y. Voo, M. Ibsen, and W.H. Loh. 2005. "Dominant Causes of Linewidth in DFB Fiber Lasers." *(CLEO). Conference on Lasers and Electro-Optics*, 2005. 3: 5–7. 10.1109/CLEO.2005.202200.

Ibsen, M., E. Ronnekieiv, G.J. Cowle, M.O. Berendt, Oliver Hadeler, Michalis Zervas, and Richard Laming. 1999. *Robust High Power (>20 MW) All-Fiber DFB Lasers with Unidirectional and Truly Single Polarisation Outputs*. in Conference on Lasers and Electro-Optics, C. Chang-Hasnain, W. Knox, J. Kafka, and K. Vahala, eds., OSA Technical Digest (Optical Society of America, 1999), paper CWE4.

Jeong, Yoonchan, Seongwoo Yoo, Christophe A. Codemard, Johan Nilsson, Jayanta K. Sahu, David N. Payne, and R. Horley, et al. 2007. "Erbium:Ytterbium Codoped Large-Core Fiber Laser with 297-W Continuous-Wave Output Power." *IEEE Journal on Selected Topics in Quantum Electronics* 13 (3): 573–578. 10.1109/JSTQE.2007.897178.

Jiang, Chun, Weisheng Hu, and Qingji Zeng. 2005. "Improved Gain Performance of High Concentration Er3+ - Yb3+-Codoped Phosphate Fiber Amplifier." *IEEE Journal of Quantum Electronics* 41 (5): 704–708. 10.1109/JQE.2005.845355.

Kim, J.W., A. Boyland, J.K. Sahu, and W.A. Clarkson. 2009. "Ho-Doped Silica Fiber Laser in-Band Pumped by a Tm-Doped Fiber Laser." *CLEO/Europe - EQEC 2009 -*

European Conference on Lasers and Electro-Optics and the European Quantum Electronics Conference 6084 (2006): 6873. 10.1109/CLEOE-EQEC.2009.5194642.

Kogelnik, H. 1988. "Theory of Optical Waveguides " Guided-Wave Optoelectronics. In, edited by Theodor Tamir, 7–88. Berlin, Heidelberg: Springer Berlin Heidelberg. 10. 1007/978-3-642-97074-0_2.

Kogelnik, H. and C.V. Shank. 1972. "Coupled-Wave Theory of Distributed Feedback Lasers." *Journal of Applied Physics* 43 (5): 2327–2335. 10.1063/1.1661499.

Kringlebotn, Jon Thomas, David Neil Payne, Laurence Reekie, and Jean Luc Archambault. 1998. "Optical Fiber Distributed Feedback Laser."

Laming, R.I. 1989. "Efficient Pump Wavelenghts of Erbium Doped Fiber Optical Amplifiers." *IEEE* 25 (1): 12–14. 10.1049/el:19890009.

Lauridsen, V.C., J.H. Povlsen, and P. Varming. 1998. "Design of DFB Fiber Lasers." *Electronics Letters* 34 (21): 2028–2030. 10.1049/el:19981446.

Lauridsen, V.C., J.H. Povlsen, and P. Varming. 1999. "Optimising Erbium-Doped DFB Fiber Laser Length with Respect to Maximum Output Power." *Electronics Letters* 35 (4): 300–302. 10.1049/el:19990192.

Lauridsen, Vibeke Claudia, Thomas Sondergaard, Poul Varming, and Jørn Hedegaard Povlsen. 1997. "Design of Distributed Feedback Fiber Lasers." In *Integrated Optics and Optical Fiber Communications, 11th International Conference on, and 23rd European Conference on Optical Communications (Conf. Publ. No.: 448)*, 3: 39–42. IET.

Li, Nanxi, Jonathan D Bradley, Gurpreet Singh, E Salih Magden, Jie Sun, and Michael R Watts. 2015. "Self-Pulsing in Erbium-Doped Fiber Laser". 1–25.

Liau, Jiun-Jie, Nai-Hsiang Sun, Chih-Cheng Chou, Shih-Chiang Lin, Ru-yen Ro, Po-Jui Chiang, Jung-Sheng Chiang, Hung-Wen Chang, and 1. 2015. "Numerical Approaches for Solving Coupled Mode Theory-Part I: Uniform Fiber Bragg Gratings." *Statewide Agricultural Land Use Baseline 2015* 1 (July). 10.1017/CBO9781107415324.004.

Lin, Qian, Mackenzie A. Van Camp, Hao Zhang, Branislav Jelenković, and Vladan Vuletić. 2012. "Long-External-Cavity Distributed Bragg Reflector Laser with Subkilohertz Intrinsic Linewidth." *Optics Letters* 37 (11): 1989–1991.

Lyu, Chengang, Chuang Wu, Hwa Tam, Chao Lu, and Jianguo Ma. 2013. "Polarimetric Heterodyning Fiber Laser Sensor for Directional Acoustic Signal Measurement." *Optics Express* 21 (July): 18273–18280. 10.1364/OE.21.018273.

McCall, M. 2000. "On the Application of Coupled Mode Theory for Modeling Fiber Bragg \ngratings." *Journal of Lightwave Technology* 18 (2): 236–242. 10.1109/50.822798.

Mears, R.J., L. Reekie, S.B. Poole, and D.N. Payne. 1985. "Neodymium-Doped Silica Single-Mode Fiber Lasers." *Electronics Letters* 21 (17): 738–740. 10.1049/el:19850521.

Miniscalco, W.J. 1991. "Erbium-Doped Glasses for Fiber Amplifiers at 1500-Nm." *IEEE Journal of Lightwave Technology* 9 (2): 234–250. 10.1109/50.65882.

Moghaddam, M.R.A., Sulaiman Wadi Harun, Roghaieh Parvizi, Z.S. Salleh, Hamzah Arof, Asiah Lokman, and Harith Ahmad. 2011. "Experimental and Theoretical Studies on Ytterbium Sensitized Erbium-Doped Fiber Amplifier." *Optik-International Journal for Light and Electron Optics* 122 (20): 1783–1786.

Morrison, David D., James D. Riley, and John F. Zancanaro. 1962. "Multiple Shooting Method for Two-Point Boundary Value Problems." *Communications of the ACM* 5 (12): 613–614.

Ndergaard, Thomas. 2000. "Photonic Crystal Distributed Feedback Fiber Lasers with Bragg Gratings." *Journal of Lightwave Technology* 18 (4): 589.

Ng, Sebastian W.S., David G. Lancaster, Tanya M. Monro, Peter C. Henry, and David J. Ottaway. 2016. "Air-Clad Holmium-Doped Silica Fiber Laser." *IEEE Journal of Quantum Electronics* 52 (2). 10.1109/JQE.2015.2507518.

Nie, Qiuhua, Xunsi Wang, Tiefeng Xu, Xiang Shen, and Liren Liu. 2006. "Spectroscopic Properties from Er / Yb Co-Doped Tellurite Glass and Fiber," In 4th IEEE International Conference on Industrial Informatics. 1129–1134.

Parekhan, M.A., and O.R. Banaz. 2008. "Design Optimization for Efficient Erbium-Doped Fiber Amplifiers." *World Academy of Science, Engineering and Technology* 46 (2008): 40–43.

Pátek, K. 1970. *Glass Lasers.* Cleveland, Ohio: CRC Press.

Percival, R.M., Daryl Szebesta, and S.T. Davey. 1992. "Highly Efficient and Tunable Operation of Two Colour Tm-Doped Fluoride Fiber Laser." *Electronics Letters* 28 (7): 671–673. 10.1049/el:19920424.

Peterka, Pavel, Ivan Kasik, Anirban Dhar, Bernard Dussardier, and Wilfried Blanc. 2011. "Theoretical Modeling of Fiber Laser at 810 Nm Based on Thulium-Doped Silica Fibers with Enhanced 3H4 Level Lifetime." *Optics Express* 19 (3): 2773–2781. 10.1364/OE.19.002773.

Poole, S.B., J.E. Townsend, D.N. Payne, M.E. Fermann, G.J. Cowle, R.I. Laming, and P.R. Morkel. 1989. "Characterization of Special Fibers and Fiber Devices." *Journal of Lightwave Technology* 7 (8): 1242–1255. 10.1109/50.32389.

Pradhan, Shilpa, Graham Town, and Ken Grant. 2006. "Dual-Wavelength DBR Fiber Laser." *Photonics Technology Letters, IEEE* 18 (September): 1741–1743. 10.1109/LPT.2006.880799.

Prajzler, V., V. Jeřábek, O. Lyutakov, I. Hüttel, J. Špirková, V. Machovič, J. Oswald, D. Chvostová, and J. Zavadil. 2008. "Optical Properties of Erbium and Erbium/Ytterbium Doped Polymethylmethacrylate." *Acta Polytechnica* 48 (5). 10.14311/1047.

Quintela, M.A. and I. Laarossi. 2014. "Polarimetric DBR Fiber Laser Sensor for Strain-Temperature Discrimination' I".Proceedings of SPIE - The International Society for Optical Engineering 9157: 1–4. 10.1117/12.2059551.

Reekie, L., et al. 1987. "Diode-laser-pumped Nd3+-doped fiber laser operating at 938 nm." *Electronics Letters* 23(17):884–885.

Rodríguez-Cobo, Luis, María Ángeles Quintela, and José Miguel López-Higuera. 2014. "DBR Fiber Laser Sensor with Polarization Mode Suppression." *IEEE Journal on Selected Topics in Quantum Electronics* 20 (5). 10.1109/JSTQE.2014.2300050.

Semmalar, S. and S. Malarkkan. 2013. "Output Signal Power Analysis in Erbium-Doped Fiber Amplifier with Pump Power and Length Variation Using Various Pumping Techniques." *ISRN Electronics* 2013 (Article ID 312707): 1–6. 10.1155/2013/312707.

Shi, Wei2015. "Single-Frequency Fiber Lasers Using Rare-Earth-Doped Silica,".SPIE Newsroom.9–10. 10.1117/12.809482.

Shimizu, M., H. Suda, and M. Horiguchi. 1987. "High-Efficiency Nd-Doped Fiber Lasers Using Direct-Coated Dielectric Mirrors." *Electronics Letters* 23 (15): 1–2. 10.1049/el:19870545.

Simakov, N., Z. Li, S. Alam, P. Shardlow, D. Jain, J.K. Sahu, A. Clarkson, D.J. Richardson, N. Simakov, and A. Hemming. 2015. "Holmium-Doped Fiber Amplifier for Optical Communications at 2.05-2.132."*OFC*, Conference Paper session Tu2C.6., 6–8.

Simakov, Nikita, Alexander Hemming, W. Andrew Clarkson, John Haub, and Adrian Carter. 2013. "A Cladding-Pumped, Tunable Holmium Doped Fiber Laser." *Optics Express* 21 (23): 28415. 10.1364/oe.21.028415.

Sorin, Miclos, Savastru Dan, and Lancranjan Ion. 2010. "Numerical Analysis of an Active FBG Sensor." *International Conference on Applied Computer Science - Proceedings*, 490–497.

Spiegelberg, Christine, Jihong Geng, Yongdan Hu, Yushi Kaneda, Shibin Jiang, and N. Peyghambarian. 2004. "Low-Noise Narrow-Linewidth Fiber Laser at 1550 Nm (June 2003)." *Journal of Lightwave Technology* 22 (1): 57–62. 10.1109/JLT.2003.822208.

Strohhöfer, Christof, and Albert Polman. 2001a. "Relationship between Gain and Yb 3+ Concentration in Er 3+–Yb 3+ Doped Waveguide Amplifiers." *Journal of Applied Physics* 90 (9): 4314–4320.

Strohhöfer, Christof, and Albert Polman. 2001b. "Relationship between Gain and Yb3+ Concentration in Er3+-Yb3+ Doped Waveguide Amplifiers." *Journal of Applied Physics* 90 (9): 4314–4320. 10.1063/1.1406550.

Strohhöfer, Christof, and Albert Polman. 2003. "Absorption and Emission Spectroscopy in Er3+ - Yb3+ Doped Aluminum Oxide Waveguides." *Optical Materials* 21 (4): 705–712. 10.1016/S0925-3467(02)00056-3.

Taccheo, S., G. Sorbello, S. Longhi, and P. Laporta. 1999. "Measurement of the Energy Transfer and Upconversion Constants in Er ± Yb-Doped Phosphate Glass" *Applied Physics B* 74(3):233–236.

Thyagarajan, K. 2006. *Erbium-Doped Fiber Amplifiers. Guided Wave Optical Components and Devices.* India: Elsevvier. 10.1016/B978-012088481-0/50009-7.

Urquhart, Paul. n.d. "Review of Rare-Earth Doped Fiber Lasers and Ampl If Iers."

Varming, Poul, Jörg Hübner, and Martin Kristensen. 1997. "DFB Fiber Laser as Source for Optical Communication Systems." In *Conference on Optical Fiber Communications*, WL7. 1997 OSA Technical Digest Series. Dallas, Texas: Optical Society of America.

Vienne, G.G., J.E. Caplen, Liang Dong, J.D. Minelly, J. Nilsson, and D.N. Payne. 1998. "Fabrication and Characterization of Yb/Sup 3+/:Er/Sup 3+/ Phosphosilicate Fibers for Lasers." *Journal of Lightwave Technology* 16 (11): 1990–2001. 10.1109/50.730360.

Wang, Shyh. 1974. "Principles of Distributed Feedback and Distributed Bragg Reflector Lasers." *IEEE J.ournal of Quantum Electronics* 10 (4): 413–427.

Wong, Allan C L, W.H. Chung, Chao Lu, and Hwa Yaw Tam. 2010a. "Composite Structure Distributed Bragg Reflector Fiber Laser for Simultaneous Two-Parameter Sensing." *IEEE Photonics Technology Letters* 22 (19): 1464–1466. 10.1109/LPT.2010.2062500.

Wong, Allan C L, W.H. Chung, Chao Lu, and Hwa-yaw Tam. 2010b. "Composite Structure Distributed Bragg Reflector Fiber Laser for Simultaneous Two-Parameters Sensing_PTL10.Pdf". *IEEE Photonics Technology Letters* 22 (19): 1464–1466. 10.1109/LPT.2010.2062500

Yamada, Makoto, and Kyohei Sakuda. 1987. "Analysis of Almost-Periodic Distributed Feedback Slab Waveguides via a Fundamental Matrix Approach." *Applied Optics* 26 (16): 3474. 10.1364/ao.26.003474.

Yamamoto, T., Y. Miyajima, and T. Komukai. 1994. "1.9 Micrometer Tm-Doped Silica Fiber Laser Pumped at 1.57 Micrometer." *Electronics Letters* 30 (3): 220–221.

Yelen, K., M.N. Zervas, and L.M.B. Hickey. 2005. "Fiber DFB Lasers with Ultimate Efficiency." *Journal of Lightwave Technology* 23 (1): 32–43. 10.1109/JLT.2004.840037.

Yelen, Kuthan, Louise M B Hickey, and Mikhail N. Zervas. 2004. "A New Design Approach to Fiber DFB Lasers with Improved Efficiency." *IEEE Journal of Quantum Electronics* 40 (6): 711–720. 10.1109/JQE.2004.828257.

Yelen, Kuthan, Louise M B Hickey, and Mikhail N Zervas. 2005. "Experimentally Verified Modeling of Erbium-Ytterbium Co-Doped DFB Fiber Lasers." *Journal of Lightwave Technology* 23 (3): 1380.

Zhang, Yuning, Can Li, Shanhui Xu, Huaqiu Deng, Zhouming Feng, Changsheng Yang, Xiang Huang, Yuanfei Zhang, Jiulin Gan, and Zhongmin Yang. 2016. "A Broad Continuous Temperature Tunable DBR Single-Frequency Fiber Laser at 1064 Nm." *IEEE Photonics Journal* 8 (2). 10.1109/JPHOT.2016.2539826.

6 Q-switched Fiber Laser

6.1 INTRODUCTION: WORKING PRINCIPLE

Several applications require a pulsed laser. One of the most effective ways of producing these high-energy pulses is using a Q-switched fiber laser. Q-switching is done by temporarily switching the quality factor of the laser cavity. The quality factor is defined as the ratio between the energy stored in the resonator and the energy lost in a return cycle. In practice, this is done by periodically modulating the losses inside the cavity. High loss corresponds to the low Q-factor of the cavity and low loss corresponds to the high Q-factor of the cavity. When the pump source is switched on, the loss in the cavity is maintained at a high value. Population inversion in the gain medium increases to reach values far beyond threshold values obtainable in the continuous wave regime. After a specific duration, the loss in the cavity is dropped to a minimum value and all the accumulated photons due to population inversion are released in a short time corresponding to several round trips, creating a large pulse at the output of the laser. After releasing the pulse, the loss in the cavity is switched back to a high value and the process is repeated. This is done periodically, releasing a train of pulses. Switching the loss values can be accomplished by using an external switch in the cavity that periodically obstructs the propagation of the photons in the cavity as illustrated in Figure 6.1. This type of Q-switched fiber laser is known as an active Q-switched fiber laser. On the other hand, a saturable absorber can be used to obtain Q-switching. In this case, the transmission property of the saturable absorber is modulated by the propagating field intensity.

As illustrated in Figure 6.2 the process leading to pulse generation in Q-switching lasers can be divided into two intervals namely, the pumping interval and the pulse output interval. In the figure, the x-axis represents time and the gain, loss and output power are represented on the y-axis. During the pumping interval, energy is constantly brought to the system by an optical pumping mechanism, while the loss in the cavity is kept to a high level. Gain grows as a result of the high value of population inversion. At the end of this pumping interval, cavity losses are abruptly brought to a minimum value. The photons accumulated in the cavity began to be released, their quantity increasing exponentially. At the same time, the gain drops as result of stimulated emission. When the gain reaches a value below the threshold, there remains an important quantity of photons in the cavity, which must be released. Because of the absence of the gain, these photons will be in an interval of time, which is a function of the cavity time factor. After this time, a pulse is completely released; the cavity returns to its initial state and the process can be restarted.

DOI: 10.1201/9781003256380-6

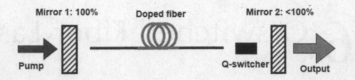

FIGURE 6.1   Schematic representation of a Q-switched fiber laser.

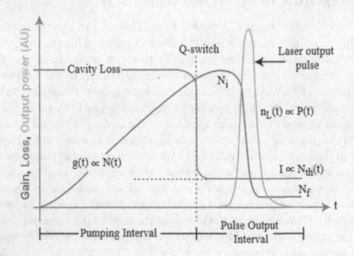

FIGURE 6.2   Illustration of the Q-switched process.

6.2   FUNDAMENTAL MATHEMATICAL DESCRIPTION

In the traditional point model, the laser rate equations used to describe the resonator photon density and gain medium inversion density for rare-earth-doped lasers are given by:

$$\frac{d\phi(t)}{dt} = v[N_2(t)\sigma_e(\lambda_s) - N_1(t)\sigma_a(\lambda_s)]\phi(t) - \frac{\phi(t)}{t_c} \tag{6.1}$$

$$\frac{dN_2(t)}{dt} = R_p - v[N_2(t)\sigma_e(\lambda_s) - N_1(t)\sigma_a(\lambda_s)]\phi(t) - \frac{N_2(t)}{\tau} \tag{6.2}$$

$$N_t = N_1 + N_2 \tag{6.3}$$

where N_t is the rare-earth ions total population density, ϕ the photon density, v the light velocity in the core of the fiber, τ the lifetime of the excited state level, and t_c the time constant of the cavity, which depends on the cavity length and loss as:

$$t_c = \frac{2l_{cav}}{v} \frac{1}{\left[ln\frac{1}{R_{oc}} + \delta \right]}$$ (6.4)

where l_{cav} is the optical length of the cavity, R_{oc} the output mirror reflectivity, and δ the round trip background loss that include scattering, the insertion loss of the components, aberration due to thermal lensing, a possible acousto-optic diffraction loss etc. Equation (6.4) shows that the time constant of the cavity is proportional to the photon round trip time in the cavity given by $t_r = 2l_{cav}/v$, which in turn is a function of the cavity length. Therefore, a shorter cavity will have a shorter time constant, in other words the cavity will be quickly emptied after the switch is closed.

The density of population inversion is defined as:

$$N = N_2 - \frac{\sigma_a(\lambda_s)}{\sigma_e(\lambda_s)} N_1$$ (6.5)

And the parameter γ defined as:

$$\gamma = 1 + \sigma_a(\lambda_s)/\sigma_e(\lambda_s)$$ (6.6)

Substituting equations (6.5) and (6.6) in (6.1) and (6.2) yields:

$$\frac{d\phi(t)}{dt} = v\sigma_e(\lambda_s)N(t)\phi(t) - \frac{\phi(t)}{t_c}$$ (6.7)

$$\frac{dN(t)}{dt} = R_p - \gamma\sigma_e(\lambda_s)N(t)\phi(t) - \frac{N(t)}{t}$$ (6.8)

The parameter γ is equalled to 2 and 1 for an ideal 3 level and 4 level system respectively. The parameter γ is equalled to 2 and 1 for an ideal 3 level and 4 level system respectively. Considering the most common 3 level system and replacing $v\sigma_e(\lambda_s)$ by a constant K, equations (6.7) and (6.8) become:

$$\frac{d\phi(t)}{dt} = KN(t)\phi(t) - \frac{\phi(t)}{t_c}$$ (6.9)

$$\frac{dN(t)}{dt} = R_p - 2KN(t)\phi(t) - \frac{N(t)}{t}$$ (6.10)

Equations (6.9) and (6.10) are identical to the elementary rate equations for the cavity photon number and the inverted population difference of a laser defined by Siegman (Anthony E. Siegman 1986). This set of coupled equations exactly describes the dynamics of the population inversion $N(t)$ and the photon population $\phi(t)$.

It is worth mentioning that Q-switching is obtained by periodically modulating the loss in the cavity. To explicitly describe the dependence of photon density on cavity loss one can substitute equation (6.4) into (6.7), resulting in equation (6.11) below:

$$\frac{d\phi(t)}{dt} = \phi\left(v\sigma_e(\lambda_s)N(t) - \frac{\varepsilon(\phi, t)}{t_r}\right) \qquad (6.11)$$

where all parameters are defined as before and the total loss in the cavity resonator is expressed as:

$$\varepsilon = -lnR + \delta + \zeta(t, \phi) \qquad (6.12)$$

where $-lnR$ represents the output coupling loss from the mirror reflectivity R, the term δ has been defined and represents the fiber background losses such as absorption and scattering and the last term $\zeta(t, \phi)$ represents the time-dependent loss of the Q-switcher. The function $\zeta(t, \phi)$ is obtained by modulating the loss as a function of time using typical switches including a rotating mirror, electro-optic or acousto-optic modulator in the case of an active Q-switched fiber laser or modulating the photon density using a saturable absorber in the case of a passively Q-switched fiber laser.

Q-switching is a two intervals process, the first interval is the "high-loss interval" and the second interval is the "low-loss interval". The behaviour of gain and Q-switched laser output power during the two intervals can be obtained by solving equations (6.9) and (6.10) with appropriate boundary conditions. During the high-loss interval, the system is pumped while oscillation is prevented by introducing loss into the cavity. This high loss interval corresponds also to the low Q-factor of the resonator. This interval often corresponds to a duration of one to few upper-level lifetimes, thus, population inversion builds to values several orders of magnitude higher than the corresponding threshold value and a large amount of energy is stored in the cavity. Since laser oscillation is not taking place, the photon population density becomes negligible and equation (6.9) vanishes and equation (6.10) reduces to

$$\frac{dN(t)}{dt} = R_p - \frac{N(t)}{t} \qquad (6.13)$$

which has the analytical solution (Anthony E. Siegman 1986)

$$N(t) = N_\infty[1 - exp(-t/\tau)] \qquad (6.14)$$

where $N_\infty = R_p\tau_L$ is the maximum asymptotic value of the inverted population. Equation (6.14) shows that during the high-loss interval, the population inversion grows exponentially towards the asymptotic maximum of N_∞ as illustrated in Figure 6.2.

The second switching interval, known as the low-loss switching interval, starts when cavity loss is suddenly reduced to a minimum value. The drop of cavity losses is almost immediately followed by a rapid build-up of photon density to a very large value, due to the large population inversion available at this moment, releasing a large pulse, while population inversion gets depleted. This interval is so short that the pump rate and spontaneous emission don't have much impact. Therefore, these two terms can be neglected from the coupled equations (6.9) and (6.10) which can thus be written as:

$$\frac{d\phi(t)}{dt} = KN(t)\phi(t) - \frac{\phi(t)}{t_c} \qquad (6.15)$$

$$\frac{dN(t)}{dt} = -2KN(t)\phi(t) \qquad (6.16)$$

Equations (6.15) and (6.16) are used to analyse the behaviour of the laser pulse. The pulse build-up time is difficult to define in part because the Q-switcher itself can have a finite opening time and the pulse itself has a finite rise time and duration. However, the build-up time can be approximately evaluated by assuming that at time $t = 0$, when the cavity loss are switched from high to a low value, the laser has an initial inversion N_i which is x times larger than the CW threshold value N_{th}. On the other hand, the photon density at this instant is extremely small and the rate of inversion depletion is also small. The pulse build-up time is also significantly short compared to the upper-level lifetime, therefore, the population inversion $N(t)$ remains almost equal to its initial value throughout the pulse build-up time. The photon rate equation (6.15) can thus be approximated to (Anthony E. Siegman 1986):

$$\frac{d\phi(t)}{dt} \approx K[N_i - N_{th}]\phi(t) \qquad (6.17)$$

which has an analytical solution given by:

$$\phi(t) = \phi_i exp\left[\left(\frac{N_i}{N_{th}} - 1\right)/t_c\right] \qquad (6.18)$$

which shows that the photons density in the cavity grows exponentially from an initial value to a maximum value which is a function of the ratio between the initial population density and the threshold population density. Using the same theoretical analysis, the photon rate equation during the pulse releasing interval can be approximated as:

$$\frac{d\phi(t)}{dt} = K[N(t) - N_{th}]\phi(t) \qquad (6.19)$$

And the inversion rate equation as:

$$\frac{dN(t)}{dt} = -2KN(t)\phi(t) \tag{6.20}$$

Dividing equation (6.19) by (6.20) yield the relation:

$$\frac{d\phi}{dN} = \frac{N_{th} - N}{2N} \tag{6.21}$$

Equation (6.21) can be integrated from the cavity switch on time $t = t_i$ to any arbitrary time t as:

$$2\int_{\phi i}^{\phi(t)} d\phi = \int_{Ni}^{N(t)} \left(\frac{N_{th}}{N} - 1\right)dN \tag{6.22}$$

To easily solve the integral (6.22) one takes into account the fact that the initial photon density number n_i is negligible compared to any photon density number $n(t)$ at any given time t, during the laser pulse release. Therefore, one can set the lower limit of the left-hand side of integral (6.22) to zero. With these new limits the result of the integral can be obtained as the following implicit relation:

$$2\phi(t) \approx N_i - N(t) - \ln\left(\frac{N_i}{N(t)}\right)N_{th} \tag{6.23}$$

Equation (6.23) can be used to obtain the main characteristics of the Q-switched pulse, namely peak power, pulse width, and pulse energy.

One of the most important parameters towards obtaining these characteristics is the final inversion N_f remaining after the Q-switch pulse has been completely released. This value can be obtained by setting, the photon flux density $\phi(t) \approx 0$. Under these conditions, equation (6.23) becomes:

$$\left(\frac{N_f}{N_i}\right) = exp\left(\frac{N_f - N_i}{N_{th}}\right) \tag{6.24}$$

Equation (6.24) is a transcendental equation linking the initial and final population inversion densities N_i and N_f. It is worth recalling that N_{th} is the threshold population inversion. The instantaneous output power is related to the photon density number by:

$$P_{out}(t) = N(t)h\nu\left(\frac{cT}{2Ln}\right), \tag{6.25}$$

where $h\nu$ is the photon energy at the wavelength ν, h the Planck's constant, c the speed of light in vacuum, T the transmittance of the output coupler, L the length of the doped fiber and n its refractive index.

The peak output power is obtained by setting $dP_{out}/dn = 0$ in equation (6.25) which result in the equality,

$$N(t) = N_{th} \qquad (6.26)$$

Substituting (6.23) and (6.26) in (6.25) yields:

$$P_{peak} = h\nu \left(\frac{cT}{2Ln}\right)\left[N_{th}\,ln\frac{N_{th}}{N_i} - (N_{th} - N_i)\right] \qquad (6.27)$$

The total output pulse energy, on the other hand, is obtained by integrating the instantaneous power P_{out} in the interval between the initial time t_i to a final time t_f as follows:

$$E_{out} = \int_{ti}^{tf} P_{out}(t)\,dt \qquad (6.28)$$

Changing the variables from time to population density yields:

$$E_{out} = \int_{N_i}^{N_f} \frac{P_{out}}{dN/dt}dN \qquad (6.29)$$

Substituting P_{out} in (6.29), P_{out} by its value obtained from (6.25) the integral yields:

$$E_{out} = \frac{T}{\delta}h\nu(n_i - n_f) \qquad (6.30)$$

Approximating the output pulse with a triangular shape, the full-with-at-half-maximum (FWHM) can be obtained as $\Delta t = E_{out}/P_{peak}$. Substituting E_{out} and P_{peak} by their respective values obtained in equations (6.30) and (6.27), one can obtain:

$$\Delta t = \tau_c \frac{N_i - N_f}{N_{th}\,ln\left(\frac{N_{th}}{N_i}\right) - (N_{th} - N_i)} \qquad (6.31)$$

Finally, the initial population density N_i is related to the pump power by the relation:

$$N_i = \left(\frac{\tau}{h\nu_p}\right)P_p \qquad (6.32)$$

where τ is the lifetime of the upper energy level of a two energy levels system, ν_p the frequency of pump radiation and P_p the pump power. The pump power threshold, on the other hand, can be obtained as:

$$P_{th} = \frac{h\nu_p}{\sigma\tau} \frac{\delta_1}{2} \frac{A_f}{F} \qquad (6.33)$$

The point model described above is good for understanding laser dynamics but does not accurately model the rare-earth-doped Q-switched fiber laser.

6.3   SWITCHING METHODS

As stated before, there are two main methods of obtaining Q-switched laser pulses, namely, by active and passive Q-switching. Both two techniques have advantages and disadvantages. Active Q-switched lasers allow more control on the pulse generated, as one can determine precisely when the pulse must be released. However, they require an external driver which in case of a Pockels cells used in electro-optic modulators, may require voltages in the range of kilovolts to work. On the other hand, a simple saturable absorber that does not require any external driver is sufficient to operate a passively Q-switched laser. As a result, passive Q-switched fiber lasers are more compact and less costly than their active counterparts. The choice of one or the other method will be dictated by three main considerations, namely the cost, the size, controllability of triggering and the pulse energy. Because saturable absorbers do not require any sort of electronic driver and are relatively easy to realise, passively switched fiber lasers are less expensive to build than active ones. In addition to their low cost, passive Q-switched fiber lasers constructed using saturable absorbers have a small size in general. For this reason, they are used in microchip lasers for example as it is not uncommon to build optical cavities of the order of a millimetre in length. By contrast, the majority electro-optic or acousto optic-based Q-switched fiber lasers may be up to 10 centimetres in length.

The main disadvantage of a passive Q-switched laser is the lack of control on the switching time. In this type of lasers, switching, hence the laser repetition rate, solely depends on the saturable absorber. This makes these systems prone to detrimental jitter effects. An active Q-switched fiber laser, on the other hand, provides the possibility of fully controlling the time where the pulse is triggered as well as the repetition rate. Although most of the modern passive Q-switch fiber lasers possesses an internal photodiode allowing synchronization to an external device, the pulse control is not as flexible as the one obtained from an actively Q-switched device.

Finally, active Q-switched lasers tend to produce much higher energy pulses than passive Q-switched lasers. This is because, with active Q-switched fiber lasers, switching can be controlled to build a maximum population inversion before releasing the pulse. In a passive Q-switch on the other hand, the pulse will be released once the saturable absorber reaches saturation independently of whether the population inversion has reached its maximum or not.

6.4 ACTIVE Q-SWITCHED FIBER LASERS

6.4.1 MECHANICAL DEVICES

The simplest way of building a Q-switched fiber laser is to use a mechanical device that periodically obstructs the propagation of the optical field in the cavity and prevent the occurrence of lasing. This mechanical device can be a shutter, a chopper wheel, or a spinning mirror installed inside the cavity. The biggest advantage of a mechanical shutter over other switching techniques is its high extinction ratio which ensures that undesirable lasing does not take place during the pumping interval when the cavity loss is high. Also, in this switching technique, insertion loss is zero when cavity loss is low, meaning that the only loss experienced by the pulse is the one at the output mirror, resulting in higher output and low threshold. The disadvantage of a mechanically switched Q-switched laser is the relatively low switching speed compared to other switching techniques. Mechanical noise and vibration are more prevalent in mechanically switched lasers.

6.4.2 ELECTRO-OPTIC MODULATOR

Because of the complexity involved in their constructions and the disadvantages mentioned in the previous paragraph, Q-switched fiber lasers using mechanical devices are not very common. The other alternative to building an active Q-switched laser is to use an electro-optic modulator (EOM). As the name implies, an electro-optic modulator is a type of switch controlled by applying a voltage to it. Electro-optical modulators are mostly based on Pockels cells. The Pockels effect consists of inducing a birefringence into a non-centrosymmetric crystal as a result of the applied electric field. The induced birefringence rotate the polarisation of the light passing through it. The Pockels effect used in conjunction with a polariser can be an excellent switch for active Q-switched lasers. An applied voltage is applied to the Pockels cell in such a way that a quarter-wave plate is formed within the crystal. Light propagating in such a crystal will undergo a 90° rotation at each round-trip rotation resulting in a mismatch with the polarisation imposed by the polariser which in turn stops its propagation. Switching the cavity to a lower loss is achieved by suppressing the applied electric field. The process can be reversed by inserting a quarter-wave plate in the cavity and lowering the losses by applying an electric field to it. The second arrangement is advantageous because the voltage is only applied to the cell for a shorter period when the pulse builds up. Compared to other switching techniques, electro-optic Q-switching possess the advantage of faster Q-switching, and high extinction ratios (95–99%). Its main drawback is the high switching voltage, (swinging from 0 kV to 5 kV in some cases) which must be carefully controlled to prevent induced interference to nearby equipment in addition to its high cost.

6.4.3 ACOUSTO-OPTIC MODULATOR

Active Q-switching can also be achieved by using an Acousto-Optic Modulator (AOM). Acousto-optic modulators use the acousto-optic effect which is the modification of the

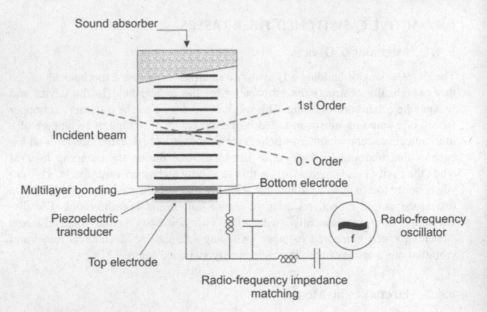

FIGURE 6.3 Schematic diagram of an acousto-optic modulator.

refractive index of an optical medium by the oscillating mechanical pressure of a sound wave. An acousto-optic modulator is constructed using an optical material such as a glass or crystal with a piezoelectric transducer attached to its end. If an RF signal is applied to the transducer, the transducer converts electric energy into a sound wave that travels through the optical components of the AOM. The sound wave creates a phase grating into the crystal with a period equal to the wavelength of the propagating acoustic wave as seen in Figure 6.3.

The incident beam is diffracted by the grating into un-diffracted 0th order and un-diffracted 1st order components. The Bragg condition for diffraction is given by:

$$sin\theta_B = \frac{\lambda f}{2v} \tag{6.34}$$

where f is the acoustic frequency, v the velocity of the acoustic wave, λ its wavelength and θ_B is the Bragg wavelength achieved when the angles of incidence and of diffraction are equals.

Equation (6.34) shows that the Bragg condition is a function of the type of acousto-optic modulator. For example, in a Tellurium Dioxide (TeO_2) acousto-optic modulator, the acoustic wave propagates with a velocity of 4190 m/s (Yariv 1991), therefore, if operated at 1550 nm, with a drive frequency of 110 MHz, the Bragg angle will be 1.2°.

It is worth mentioning that all the angles are described with respect to the acoustic wavefronts as illustrated in Figure 6.4.

One of the most important characteristics of the acousto-optic modulator is diffraction efficiency.

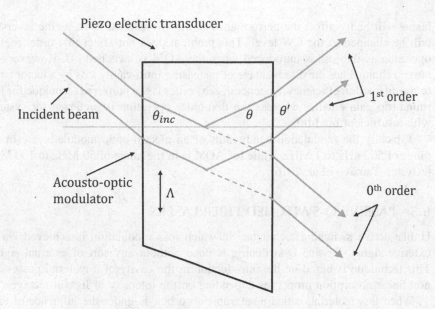

Piezo electric transducer

Incident beam

θ_{inc}

θ θ'

1st order

Acousto-optic
modulator

Λ

0th order

FIGURE 6.4 Schematic representation of the AOM with the different angles represented.

An acousto-optic modulator can be used in two ways depending on the order of the exploited beam. The acousto-optic modulator can be positioned in such a way when no RF signal is applied, the cavity loss is high and becomes low when a signal is applied or the reverse, depending on which order of diffraction beam is fed back into the cavity.

Acousto-optic Q-switched lasers offer many advantages such as fast switching and high repetition rates (10 to 100 ns). However, when operating at 1550 nm, they suffer from low diffraction efficiencies, especially in the region of 60–80%. The switching time of an AOM is a function of the time required by the acoustic beam to cross the optical beam. If one considers the previous case of a TeO$_2$ medium, in the ideal case where a Gaussian beam with a beam waist of 50 µm is coupled into the AOM, the resulting crossing time will be approximately 12 ns. On the other hand, the diffraction efficiency of the AOM depends on a couple of parameters namely, the interaction length and the overlap between optical and acoustic beam as well as the RF power. However, for a well-focused beam it will result in improved rise time, the diffraction efficiency, on the other hand, is reduced, due to the acoustic beam failure to entirely intercept the optical beam. Therefore, a trade-off must be reached between the modulator rise time and diffraction efficiency.

An active Q-switch laser using an acousto-optic modulator can be operated in two regimes, namely, a zero-order regime, where the un-diffracted beam and the first-order regime provide the cavity feedback where the cavity feedback is provided by the diffracted beam. The choice between the two regimes depends on the available round-trip cavity gain and the diffraction efficiency of the acousto-optic modulator. In the zero-order regime, the increase of the loss in the cavity is triggered by switching on the AOM. In this case, care must be taken to ensure that no CW lasing takes place during the opening time of the AOM. Failing to prevent CW lasing will badly affect the performance of the Q-switched laser, as the inversion will be clamped to the CW level. This problem does not affect first-order regime operation as the loss is introduced when the AOM is switched off. However, the latter technique has the disadvantage of increasing intra-cavity loss by a factor equal to the diffraction efficiency. In general, zero-order is the preferred technique for low round-trip gain cavities whereas the first-order operation is preferred for systems with insufficient hold-off.

Typically the modulation bandwidths of an electro-optic modulator are in the range of 500 kHz to 1 MHz, while for AOM is in the range of 50 MHz to 100 MHz (Álvarez-Tamayo et al. 2016).

6.5 PASSIVE Q-SWITCHED FIBER LASERS

Unlike active switching techniques in which loss modulation is achieved via an external signal, Passive Q-switching is done without any sort of external signal. This technique is based on the introduction in the cavity of a material possessing non-linear absorption properties depending on the intensity of light it receives.

When this material, called a saturable absorber, is under the influence of low-intensity light, its transmission coefficient is minimum and prevent the optical wave from propagating. This process introduces a huge loss in the cavity while population inversion grows (Spühler et al. 1999). At a specific value of absorbed optical power, the transmission of the saturable absorber drops, and it becomes transparent to optical radiation when the transmission reaches its maximum (Kurkov 2011). This situation leads to a sudden drop of the cavity loss, the energy accumulated during the pumping interval is then released in a short interval of time creating a huge pulse at the output. The saturable absorber is characterized by the modulation depth, which can be defined as the difference of transmission between the initial state of absorption and the saturation state.

The saturable absorber can be bulk (J. Y. Huang et al. 2007; Filippov et al. 2004) or an optical fiber doped with absorber ions such as chrome or vanadium (Adel et al. 2003; Kurkov et al. 2010). Energy absorption occurs when the saturable absorber is submitted to the influence of a light beam. The first criteria of the selection of a saturable absorber are that loss modulation must occur at the laser wavelength. Also, the transmission varies as a function of the intensity of the laser wave incident on it. Semiconductors can also be used as bulk saturable absorbers (Keller et al. 1992; Lan et al. 2010), or as quantum dots (Lecourt et al. 2006; J. Y. Huang et al. 2009), or as semi-conductors saturable absorber mirror commonly known as SESAM.

SESAM mirrors (Fluck, Häring, et al. 1998) are made of a Bragg grating obtained by stacking layers of semiconductor material and a layer of quantum dot saturable absorber (Häring et al. 2001; Spühler et al. 1999).

Passive Q-switching is simple, more effective and done without the contribution of external electronics. Figure 6.5 illustrates the evolution of the saturable absorber absorption coefficient as a function of photon flux density in the cavity. The saturable absorber introduces important intra-cavity losses α_{off} when the photon flux density is weak. When the gain medium is pumped, the cavity accumulates energy until the gain becomes equal to the loss. At this point, the laser oscillating condition

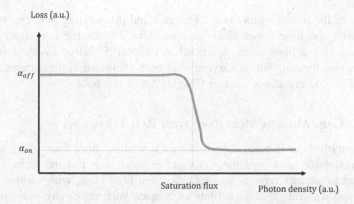

FIGURE 6.5 Schematic representation of the loss of a saturable absorber as a function of photon flux density flowing through it.

is fulfilled, and lasing can start. This induces a fast increase of the intracavity photon flux density that saturate the saturable absorber resulting in a less stable state of the laser because the gain is much higher than the losses α_{ON}. Stimulated emission generates a huge pulse that saturates the gain. The latter reduces and becomes less than the loss α_{ON}, which in terms reduces laser oscillation and the saturable absorber, becomes opaque again. This process is repeated periodically as long as the pump keeps delivering energy to the gain medium.

In passive Q-switched fiber lasers, the repetition rate depends on the pump power and the concentration of the saturable absorber. At a given pump power, the peak power increases with the concentration of the saturable absorber because more energy can be absorbed in higher concentrations. This phenomenon is only limited by the gain saturation of the amplifying medium.

Passive Q-switched fiber lasers are built in such a way that intracavity losses are initially high. Q-switch lasing is based on keeping the loss at a high level during the pumping interval to store energy within the cavity and to reduce this loss to release the stored energy in the form of a pulse of light. This pulse reaches its maximum value when the saturable absorber becomes saturated and the gain becomes unable to compensate for the loss. At this point, the saturable absorber is said to be transparent.

6.6 THEORETICAL ANALYSIS OF ACTIVE Q-SWITCHED FIBER LASER

The output pulse of an active Q-switch fiber laser or any type of Q-switched laser has important characteristics such as peak power, time duration and pulse energy, pulse shape and repetition rate. These characteristics depend on several parameters. To fully characterise the pulse, the influence of these parameters on the pulse must be carefully studied. Theoretical studies of the fiber laser are often done by using numerical modelling. For fiber laser, the numerical modelling includes solving rate equations together with the propagation equation. The simple point model is often used to solve a numerical model of a bulk laser but is not appropriate for fiber because of the nature of its cavity. The gain and propagating field can vary hugely along with the long-doped fiber gain medium. Unlike the numerical model for continuous-wave fiber lasers described in Chapter 5, pulsed lasers have their population density distribution varying with time, therefore in these cases, the steady solution of rate equations used in Chapter 5 will not hold.

6.6.1 GAIN MEDIUM MODELLING WITH RATE EQUATIONS

For illustration, an Erbium-doped active Q-switched fiber laser using an acousto-optic modulator as a switching element is modelled, but the technique can be extended to another type of active Q-switched fiber laser, with various rare-earth-doped species and other switching techniques such as electro-optic modulators. Also, for the sake of simplicity, the Erbium gain medium is treated as a two-level energy system. Under this condition, the rate equations describing the interaction between the optical field and Erbium ions for a linear cavity active fiber laser illustrated in Figure 6.6, can be described by the following system of partial differential equations in which doped fiber are assumed to be a single mode for the sake of simplicity:

$$\frac{\partial N_2(z, t)}{\partial t} + \frac{N_2(z, t)}{\tau} = \frac{\Gamma_p \lambda_p}{hcA_c} [\sigma_{ap} N_1(z, t) - \sigma_{ep} N_2(z, t)] P_p^-(z, t)$$

$$+ \sum_k \frac{\Gamma_k \lambda_k}{hcA_c} [\sigma_{ak} N_1(z, t) - \sigma_{ek} N_2(z, t)] (P_k^+(z, t) + P_k^-(z, t))$$

(6.35)

$$N_t = N_1(t) + N_2(t)$$

(6.36)

where N_t is the total population density of Erbium ions, N_1 and N_2 the ground and upper-level population density respectively, τ the fluorescence lifetime of the upper energy level, A_c the core area of the erbium-doped fiber Γ_p and Γ_k represent the overlap factor at pump and laser wavelength respectively, $P_p^\pm$ and $P_k^\pm$ represent the forward and backward pump and laser power respectively, and σ_{ap}, σ_{ep}, σ_{ak} and σ_{ek} are the absorption and emission cross-sections of the Erbium ion at the pump wavelength and emission wavelength respectively.

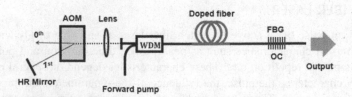

FIGURE 6.6 Schematic representation of the linear cavity Q-switched fiber laser with an acousto-optic modulator as the switching element, the output mirror her is a Fiber Bragg grating, the first-order diffracted beam is coupled into the fiber.

Equations (6.35) and (6.36) are completed by the following equations describing the propagation of pump and laser waves inside the doped fiber as:

$$\pm \frac{\partial P_p^\pm(z, t)}{\partial z} + \frac{1}{v_p}\frac{\partial P_p^\pm(z, t)}{\partial t} = \Gamma_p[\sigma_{ep}N_2(z, t) - \sigma_{ap}N_1(z, t)]P_p^\pm(z, t) - \alpha_p P_p^\pm(z, t)$$

$$(6.37)$$

$$\pm \frac{\partial P_k^\pm(z, t)}{\partial z} + \frac{1}{v_k}\frac{\partial P_k^\pm(z, t)}{\partial t} = \Gamma_k[\sigma_{ek}N_2(z, t) - \sigma_{ak}N_1(z, t)]P_k^\pm(z, t) - \alpha_k P_k^\pm(z, t)$$

$$+ 2\sigma_{ek}\frac{hc^2}{\lambda_k^3}\Delta\lambda_k N_2(z, t) \qquad (6.38)$$

$$k = 1, ..., K,$$

where v_k is the group velocity of the wave of wavelength k, the last term of equation (6.38) represents the contribution from spontaneous emission and K the number of channels taken into account in the laser gain spectrum.

It is also assumed that the doped fiber is much longer than the undoped fiber that completes the cavity. Because standard fiber has very low background loss (0.02 dB/Km) and does not significantly increase the length of the cavity, its influence can be neglected.

Equations (6.35)–(6.36) constitute the travelling wave model and are completed by the boundary conditions of the linear active Q-switched fiber laser as illustrated in Figure 6.6:

$$P_p^+(0) = \eta_p P_p$$
$$P_k^+(0, t) = P_k^-(0, t)[R_{HR}(\lambda_k)T(t)\alpha_k] \qquad (6.39)$$
$$P_k^-(L, t) = P_k^+(L, t)[R_{OC}(\lambda_k)\alpha_k]$$

where R_{HR} is the reflectivity of the high reflectivity mirror, R_{OC} the reflectivity of the output coupler η_p the loss related to the pump power coupling into the core of the fiber, $T(t)$ the transmission coefficient of the Acousto-optic modulator and α_k the sum of all loss including background loss of the fiber, splicing loss and insertion loss of various components.

For a unidirectional ring cavity fiber laser the propagation equations (6.35)–(6.38) become:

$$\frac{\partial N_2(z, t)}{\partial t} + \frac{N_2(z, t)}{\tau} = \frac{\Gamma_p\lambda_p}{hcA_c}[\sigma_{ap}N_1(z, t) - \sigma_{ep}N_2(z, t)]P_p(z, t)$$

$$+ \sum_k \frac{\Gamma_k\lambda_k}{hcA_c}[\sigma_{ek}N_1(z, t) - \sigma_{ak}N_2(z, t)]P_k(z, t) \quad (6.40)$$

$$\frac{\partial P_p(z, t)}{\partial z} + \frac{1}{v_p}\frac{\partial P_p(z, t)}{\partial t} = \Gamma_p[\sigma_{ep}N_2(z, t) - \sigma_{ap}N_1(z, t)]P_p(z, t) - \alpha_p P_p(z, t)$$

$$(6.41)$$

$$\frac{\partial P_k(z, t)}{\partial z} + \frac{1}{v_k}\frac{\partial P_k(z, t)}{\partial t} = \Gamma_k[\sigma_{ek}N_2(z, t) - \sigma_{ak}N_1(z, t)]P_k(z, t) - \alpha_k P_k(z, t)$$

$$+ 2\sigma_{ek}\frac{hc^2}{\lambda_k^3}\Delta\lambda_k N_2(z, t) \qquad (6.42)$$

Equations (6.40)–(6.45) are subject to the following boundary conditions:

$$P_p^+(0) = \eta_p P_p$$
$$P_k(0, t) = P_k(L, t)[R_{OC}(\lambda_k)T(t)\alpha_k]$$

$$(6.43)$$

Because, unlike linear cavities, a ring cavity might be significantly long (Figure 6.7), it is common to include the propagation equations in the undoped part of the fiber that complete the cavity.

The pump and laser propagation equation inside the undoped fiber can be written as:

$$\frac{\partial P_p(z, t)}{\partial z} + \frac{1}{v_p}\frac{\partial P_p(z, t)}{\partial t} = -\alpha_p P_p(z, t) \qquad (6.44)$$

$$\frac{\partial P_k(z, t)}{\partial z} + \frac{1}{v_k}\frac{\partial P_k(z, t)}{\partial t} = -\alpha_k P_k(z, t) \qquad (6.45)$$

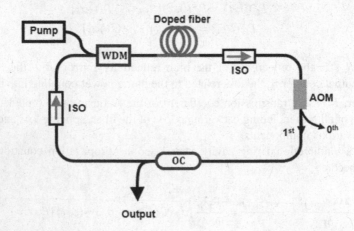

FIGURE 6.7 Unidirectional ring cavity fiber laser used in the theoretical study.

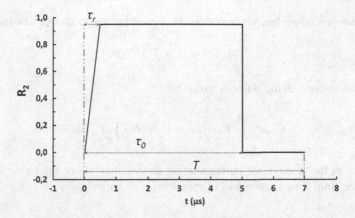

FIGURE 6.8 Simplified transmission function of the acousto-optic modulator in one period.

The transmission function of the acousto-optic modulator is obtained from the analysis of its working principle. The behaviour of this acousto-optic modulator in one period is represented in Figure 6.8.

The equation representing the transmission function of Figure 6.8 can be written as:

$$T(t) = \begin{cases} \frac{T_{max}}{\tau_r} \times t & 0 \leq t \leq \tau_r \\ T_{max} & \tau_r \leq t \leq \tau_0 \\ 0 & t \geq \tau_0 \end{cases} \qquad (6.46)$$

where T_{max} is the maximum transmission of the modulator, this modulation of the loss triggers the process of Q-switching. The initial time when the modulator is switched on is considered time 0. At a time $t < 0$, the pump power is applied to the laser cavity, but the acousto-optic modulator is off. At $t \geq 0$ the modulator is switched on and its transmission coefficient grows linearly to a maximum value in the interval of time τ_r known as the "rise time" of the modulator. During the interval comprised between τ_r and τ_0 (the opening time of the modulator), the transmission coefficient keeps a constant value corresponding to the maximum value of the transmission. Finally, for time $t \geq \tau_0$, the modulator is off, and its transmission falls to zero until the end of the period. In practice, however, during the rising time, the transmission function of the modulator is not linear but has a sinusoidal shape. In this condition, equation (6.46) becomes:

$$T(t) = \begin{cases} 0.5 + 0.5sin\left[\pi(t - 0.5\tau_r)/\tau_r\right] & 0 \leq t \leq \tau_r \\ T_{max} & \tau_r \leq t \leq \tau_0 \\ 0 & t \geq \tau_0 \end{cases} \qquad (6.47)$$

From the preceding boundary condition, the output pulse can be obtained as:

$$P_{out}(t) = (1 - R_1)P_s^-(0, t) \tag{6.48}$$

The total energy of the pulse is given by:

$$E_{out} = \int_{t_i}^t P_{out}(t)dt = (1 - R_1) \int_{t_i}^t P_s^-(0, t)dt \tag{6.49}$$

where t_i in the initial time corresponding to the time where the pulse starts rising and t the final time required to release the pulse entirely.

6.6.2 SOLUTION ALGORITHM

For the sake of simplicity, we are going to illustrate the solution of an active Q-switched Erbium-doped fiber laser in a ring cavity configuration. The algorithm for the linear cavity is similar to the one for the ring cavity except for the boundary conditions that differ from the two configurations. The active Q-switched ring cavity fiber laser is numerically modelled by the partial differential equation (6.41)–(6.45). The solutions to this system of equations are functions of position and time. The simplest way to solve systems of partial differential equations is the finite difference method. The essence of the finite difference method is to approach the derivatives to infinitesimal finite differences. The gain medium is divided into several sections and the distributions of pump and laser powers as well as the gain are computed for each section and for each time interval. It is also worth mentioning that to implement the method, initial values of pump power, laser power, and the population density of the ground and upper level must be provided for each section. The initial conditions are obtained by finding the steady-state solution of equations (6.35), (6.41), and (6.42) with appropriate boundary conditions for steady-state operation as defined in Chapter 5. It worth recalling that the steady-state solution of rate equations is obtained by setting the time derivative in equation (6.35) to zero. Equation (6.35) then becomes:

$$0 = \frac{\Gamma_p \lambda_p}{hcA_c}[\sigma_{ap}N_1(z, t) - \sigma_{ep}N_2(z, t)]P_k(z, t)$$

$$+ \sum_k \frac{\Gamma_k \lambda_k}{hcA_c}[\sigma_{ek}N_1(z, t) - \sigma_{ak}N_2(z, t)]P_k(z, t) - \frac{N_2(z, t)}{\tau} \tag{6.50}$$

Knowing (6.36) the values of N_1 and N_2 can be found for any given pump and laser power. By using Runge-Kutta methods combined with algorithms like shooting method or relaxation methods employed in Chapter 5, the steady-state values of pump power, laser power and the population density distributions at the ground and upper energy level can be easily obtained and constitute initial values for the finite difference scheme.

The spatial steps and time slots are labelled Δz and Δt respectively. To have a stable simulation the equation, it is recommended to have:

$$\Delta t = \Delta z / v \qquad (6.51)$$

where $v = c/n$, is the speed of light in the core of the fiber, c is the speed of light in free space at n the refractive index of the core. Several schemes exist to discretise the system of differential equations. In this work, however, we use the forward-space scheme (FTTS) for its simplicity. Equations (6.41)–(6.45) are discretised as:

$$N_2(z, t + \Delta t) = N_2(z, t)\left(1 - \frac{\Delta t}{\tau}\right) + \frac{\Gamma_p \lambda_p}{hcA_c}[\sigma_{ap}N_1(z, t) - \sigma_{ep}N_2(z, t)]P_p(z, t)\Delta t$$

$$+ \frac{\Gamma_k \lambda_k}{hcA_c}[\sigma_{ak}N_1(z, t) - \sigma_{ek}N_2(z, t)]P_k(z, t)\Delta t + P_k(z, t)\Delta t \quad (6.52)$$

$$\frac{P_p(z + \Delta z, t + \Delta t) - P_p(z + \Delta z, t)}{\Delta z} + \frac{1}{v_p}\frac{P_p(z + \Delta z, t + \Delta t) - P_p(z + \Delta z, t)}{\Delta t}$$

$$= \Gamma_p[[\sigma_{ep}N_2(z, t) - \sigma_{ap}N_1(z, t)]P_p(z, t) - \alpha_p P_p(z, t)] \qquad (6.53)$$

$$\frac{P_k(z + \Delta z, t + \Delta t) - P_k(z + \Delta z, t)}{\Delta z} + \frac{1}{v_k}\frac{P_k(z + \Delta z, t + \Delta t) - P_k(z + \Delta z, t)}{\Delta t}$$

$$= \Gamma_k[[\sigma_{ek}N_2(z, t) - \sigma_{ak}N_1(z, t)]P_k(z, t) - \alpha_k P_k(z, t)] + 2\sigma_{ek}\frac{hc^2}{\lambda_k^3}\Delta\lambda_k N_2(z, t)$$

$$(6.54)$$

The number of sections is related to the computation speed and the accuracy of the results. A small number of sections corresponding to a fast computation speed, but the result can be inaccurate. On the other hand, a large number of sections will result in slow computation but increased accuracy. In this work, the simulations were done with MATLAB®. Using MATLAB, 200 sections result in computation times of around six minutes which is very long. Reducing the number of sections to 50 reduced the computation time to acceptable values of tens of seconds while not affecting too much the accuracy of the results. Another factor affecting computation time is the duration of the simulation. The simulation was run for a typical time of 300 microseconds for all the results. However, this duration can be insufficient when one wants to observe, for example, multiple pulses at low repetition rates. For a repetition rate of 1 kHz, the intervals between two consecutive switchings on of the acousto-optic modulator is 1 millisecond which is far above the "typical" value of 300 microseconds, hence this time has to be increased to some value above 1 ms, to be able to track just two pulses, which results in a simulation time of up to 30 minutes. Therefore, there is a limit of repetition rates that can be simulated within reasonable computation time.

TABLE 6.1

Parameters Used in the Active Q-switched Fiber Laser Simulation

Symbol	Parameter	Value	Reference
λ_s	Emission wavelength	1550 nm	Assumed
λ_p	Pump wavelength	980 nm	Assumed
α_{ak}	Absorption cross section at laser wavelength	3.15×10^{-25} m^2	manufacturer
σ_{ek}	Emission cross section at laser wavelength	4.44×10^{-25} m^2	manufacturer
α_{ap}	Absorption cross section at pump wavelength	2.44×10^{-24} m^2	manufacturer
α_{ep}	Emission cross section at pump wavelength	1.8×10^{-25} m^2	manufacturer
α_k	Background loss at laser wavelength	0.005 m^{-1}	manufacturer
α_p	Background loss at pump wavelength	0.005 m^{-1}	manufacturer
τ	Excited state lifetime	10 ms	manufacturer
N_t	Erbium fiber concentration	4×10^{24} ions/m^3	Assumed
n_{fiber}	Refractive index of the fiber	1.45	manufacturer
d_c	Core diameter	3 μm	Assumed
NA	Numerical aperture	0.2	manufacturer
Γ_k	Overlap factor at laser wavelength	0.7	Assumed
Γ_p	Overlap factor at laser wavelength	0.83	Assumed
L	Fiber length	3 m	Assumed
R	Output coupling ratio	0.8	Assumed
c	Speed of light in vacuum	3×10^8 m/s	–
t_r	Rise time of the AOM	5 ns	Assumed
t_o	Open time of the AOM	40 μs	Assumed

6.6.3 PARAMETERS USED IN THE SIMULATION

The parameters used in the simulation are represented in Table 6.1.

6.7 CHARACTERISTICS OF THE ACTIVE Q-SWITCHED FIBER LASER

As already mentioned in the previous sections, the characteristics of the active Q-switched Erbium-doped fiber laser will be studied completely with the results obtained from a simulation done in MATLAB.

6.7.1 INFLUENCE OF THE LENGTH OF THE DOPED FIBER

The influence of the doped fiber length can be studied by considering two different lengths, for example, 3 metres and 12 metres lengths were considered in the simulations performed in this book. The results presented in Figure 6.9 were obtained for the pump power of 150 mW. In the case of the short fiber (3 m) the pulse shape is almost Gaussian, whereas, in the case of the long fiber (12 m), the pulse shows a

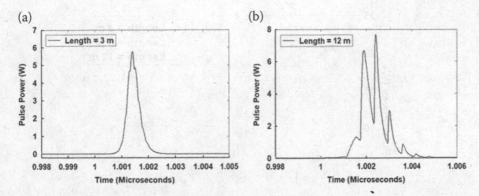

FIGURE 6.9 Output pulse at 150 mW pump power and a) doped fiber length of 3 m, b) doped fiber length of 12 m.

multipeak behaviour. The pulse is made of several, almost distinct, pulses with different full-wave-at-half-maximum (FWHM) values. The time interval between the sub-pulses is equal to almost the photon round trip time in the cavity approximated to 145 ns for a 12 m fiber in a ring cavity if the undoped fiber influence is neglected. This multipeak behaviour has been reported in several works (Adachi and Koyamada 2002; Lees and Newson 1996; Myslinski et al. 1992; Huo et al. 2004). Adachi et al. reported that the origin of this multipeak behaviour is related to the contribution of the amplified spontaneous emission in the formation of the output pulses (Adachi and Koyamada 2002).

The phenomena of multi-peaks have already been reported (Wang, Martinez-Rios, and Po 2003; Adachi and Koyamada 2002; Huo et al. 2004). The separation of the neighbouring peaks is equal to the round-trip time, 47.5 ns for our 5-m fiber. When the AOM is open, the pulse formation begins. However, the reflection on the mirrors leads to the apparition of successive peaks (after each round trip) as long as there is some gain into the amplifying medium. The global pulse is the sum of the different components as indicated in Figure 6.10.

6.7.2 INFLUENCE OF THE PUMP POWER

Another important parameter that influences the shape of the output pulse of an active Q-switched fiber laser is the pump power. This time the simulation is conducted with a fiber length of 8 metres and a concentration 4×10^{24} ions/m^3, the other parameters remain as default parameters described in Table 6.1.

Pulse shape for pump powers of 60 mW (a) and 150 mW (b) are illustrated in figure Figure 6.11. It can be seen that the multipeak behaviour is much pronounced for the pump power of 150 mW compared to 60 mW. The pulse duration (FWHM) was also measured for the two pulses. Values of 110 ns and 35 ns were found for pump powers of 60 mW and 150 mW respectively, showing that the pulse duration of the output pulse is a decreasing function of the pump power. To obtain a good pulse a certain value of pump power must be chosen for a given length of the doped fiber.

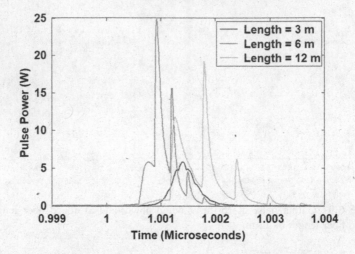

FIGURE 6.10 Comparison of pulse shapes from different lengths of doped fibers namely 3 meters, 6 meters, and 12 meters.

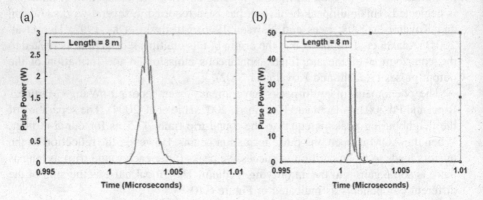

FIGURE 6.11 Output pulse shape for two pulses for two pump powers a) 60 watts and b) 150 watts.

6.7.3 INFLUENCE OF CONCENTRATION

Rare-earth ions concentration is another parameter that has a significant impact on the output pulse shape. Figure 6.12 illustrate the influence of rare-earth ions concentration on the shape of the pulse for 4 different concentrations. The fiber length is 4 metre, the pump power 80 mW, and the core diameter of the doped fiber 3 microns. The rising time of the AOM is 50 ns and the open time is 1.5 µs.

As can be seen, the energy of the pulse increased with rare-earth ion concentration but at the same time, the multi-peak phenomena also appear as the concentration increases. Therefore, the doping concentration of the fiber must be carefully chosen to avoid the occurrence of this detrimental phenomena. In this simulation, a doping concentration of 4×10^{24} ions/m^3 was an excellent value that

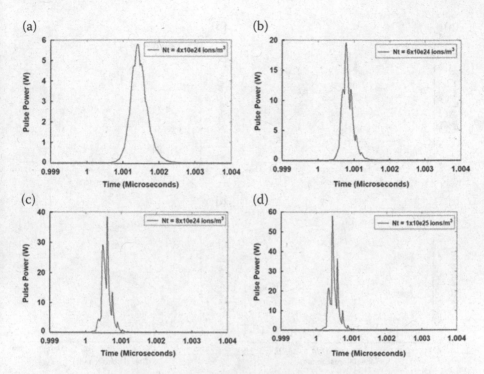

FIGURE 6.12 Influence of rare-earth ions concentration on the shape of the output pulse.

shows less multi-peak phenomena for a large range of pump powers. In this case, also the onset of the multi-peak in the output pulse can be attributed to the influence of amplified spontaneous emission. As can be seen in equation (6.42), amplified spontaneous emission is a function of the population of the excited state N_2. Therefore, for a high value of the population density of N_2 ASE will reach higher values and the multi-peak phenomenon will be more pronounced. In short, if at time $t = 0$, (before opening of the AOM), if the inversion is high, which is more likely to occur at high concentrations, one will observe more multi-peaks in the output pulse. In conclusion, one can say that the multi-peak phenomenon is strongly related to the amount of ASE in the fiber, with a moderate concentration, it is possible to obtain a "clean" pulse with less multiple peaks than with high concentrations.

6.7.4 INFLUENCE OF AOM RISE TIME

Another important parameter of the Active Q-switched fiber laser is the rise time of the AOM. This time has a notable influence on the pulse duration, the pulse shape as well as the pulse energy. The simulation was conducted with the same 4-metre long fiber with an Erbium doping concentration of 4×10^{24} ions/m^3 concentration, a core diameter of 3 μm and a pump power of 100 mW.

As it can be seen in Figure 6.13 that a short rising time corresponds to a pulse with multiple peaks and higher peak powers, whereas a "cleaner" pulse with lower

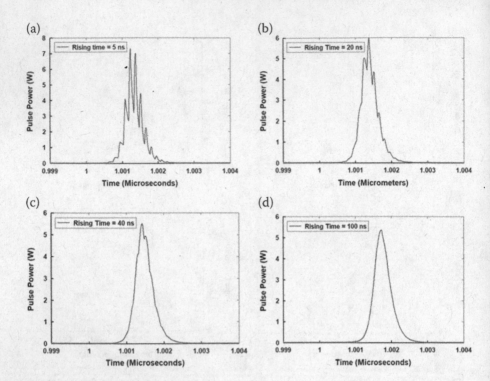

FIGURE 6.13 Output pulse as a function of various acousto-optic modulator rise times.

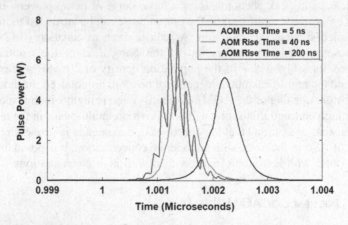

FIGURE 6.14 Superimposed pulses for 3 acousto-optic modulators rise time namely 5 ns, 40 ns, and 200 ns.

peak power is achieved at a longer rising time of the modulator. At 100 ns rise time, the pulse has an almost perfect Gaussian shape. The rise time of the AOM also affects the pulse duration and its energy and the time when it starts to rise as illustrated in Figure 6.14.

It can be easily seen that pulse energy decrease when the rise time of the AOM increases, whereas the pulse duration (at FWHM) decreases with it. In short, one can retain that a short rise time of the AOM triggers a multipeak behaviour, whereas a longer than average rising time of the AOM, like the case of 200 ns shown in Figure 6.14 delays the formation of the pulse and can reduce its energy.

6.7.5 INFLUENCE OF THE CORE DIAMETER

The influence of the doped fiber core diameter is illustrated in Figure 6.15. The simulation was conducted with a 4-metre long Erbium-doped fiber with a concentration of 4×10^{24} ions/m^3 pumped with 100 mW and an AOM rise time of 30 ns. The intensity of the optical wave is inversely proportional to the area of the fiber core. Therefore, larger areas tend to have low intensity and lower transition rates. At constant power, a larger core area will have lower peak energy and longer time duration of the pulse. Also, the pulse is delayed more and more for larger core areas. However, there is a "puzzle" at core diameter of 3 microns. The peak power and energy was supposed to be higher than the one at 4 microns, but simulation showed smaller values.

The smaller value of peak power at a core diameter of 3 microns might be attributed by the saturation of the gain because of the excessive value of intensity that quickly depopulates the excited state of the doped fiber. To avoid this effect, one may decide to reduce the length of the doped fiber to prevent the total depopulation of the excited energy level. Figure 6.16 illustrates the distribution of various quantities inside the cavity.

The population density distribution for level 1 and 2, the gain distribution, pump power and laser power are illustrated for an 8-metre long Erbium-doped fiber with a doping concentration of 4×10^{24} ions/m^3 and a pump power of 120 mW. It can be seen that towards the end of the fiber (starting at 7 m), N_1 the ground state population density becomes larger than that of the excited state, therefore the gain

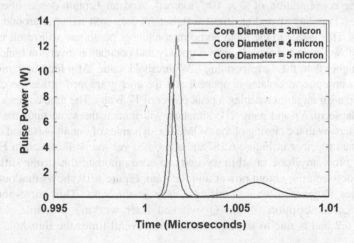

FIGURE 6.15 Output pulse as a function of doped fiber core diameter.

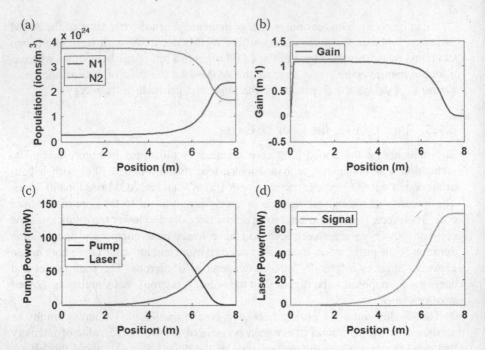

FIGURE 6.16 Population density distribution (a), Gain distribution (b), pump power and laser power distribution (c) inside the laser cavity.

becomes negative and laser power stops growing as seen in Figure 6.16(c). It can also be seen that this point corresponds to the absorption of the majority of the pump energy.

The dynamics of the population densities of the ground and excited-state levels, laser output power and gain as a function of time is illustrated in Figure 6.17.

The working conditions of the laser are the same, namely 100 mW pump power, a doping concentration of 4×10^{24} ions/m^3. And an Erbium-doped fiber length of 8 metres. Pumping started at time $t = 0$, but N_2 only start rising at around 0.5 microseconds. This is due to the finite pump rate, a higher pump rate will result in a quicker rising of N_2. From the time N_2 grows quickly and population inversion builds to values much higher than the corresponding CW threshold value. At a time of 1 microsecond, the acousto-optic modulator is opened, and the energy trapped inside the cavity is released in a strong pulse reaching a peak power of 17 Watts. The time corresponds also to the collapse of N_2 and gain. The situation will remain the same until the next cycle which starts with the closing of the AOM. The dynamics of output pulse and optical gain are plotted together in Figure 6.18. and a zoomed version is illustrated in Figure 6.19.

The plots are done in arbitrary units to accommodate the strong differences in values between the output power and the gain. Figure 6.19 shows that output power continues to grow after a significant drop in the gain. This corresponds to the theoretical description of the Q-switched laser working principle described in Figure 6.2 and is due to the high inversion (several times the threshold inversion)

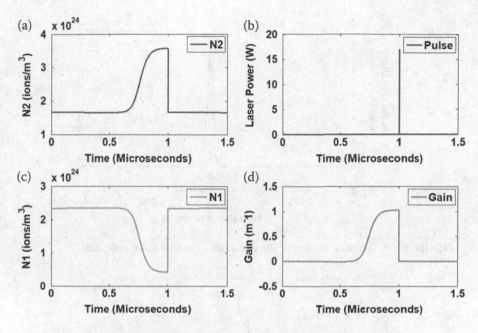

FIGURE 6.17 Dynamics of N2 population density (a), Laser output power (b), N1 population density (c) and Gain (d) of the active Q-switched fiber laser.

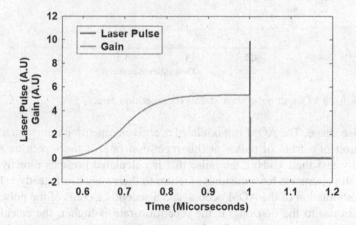

FIGURE 6.18 Dynamic of the output pulse and cavity gain.

achieved. Therefore, even if the gain is low, there is still a large number of photons in the cavity that contribute to the growth of optical intensity.

6.7.6 Influence of the AOM Repetition Rate

The switching frequency of the AOM also plays an important factor because it influences the output pulse characteristics like peak power, energy, time duration

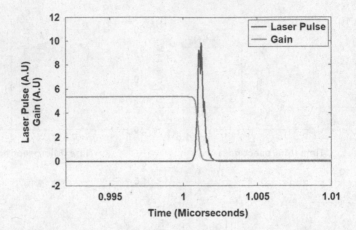

FIGURE 6.19 Zoomed version of the dynamics of output pulse and gain.

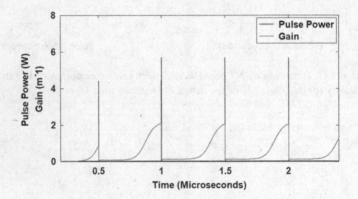

FIGURE 6.20 Output pulse train at 25 kHz repetition rate.

and pulse shape. The AOM is modulated at a given repetition rate, which allows the generation of a train of pulses at this repetition rate. If the repetition rate is low (typically less than 1 kHz), the pulse that is calculated presents directly its steady shape: all parameters have the time to return to their switch-off steady values before a new modulation of the AOM occurs, and a second calculus of the pulse leads to a shape similar to the first one. If the repetition rate is higher, the calculated pulse reaches the steady-state only after 3, 4 or 5 periods of modulation (see discussion in reference Huo et al. 2004). In this case, the calculus of the pulse-shape is much longer as it requires many periods of modulation of the AOM until stabilization of the shape is observed.

Figure 6.20, Figure 6.21, and Figure 6.22 illustrate a pulse train and the corresponding gains for repetition rates of 25 kHz, 30 kHz, and 40 kHz respectively. For 25 and 30 kHz, the peak power does not change for the same pump power. The first pulse in the two cases has a low peak power. This situation is due to the relatively quick opening of the acousto-optic modulator at a time 50 μs when the population is not fully inverted as it can be seen with the gain evolution (color-in-print curve).

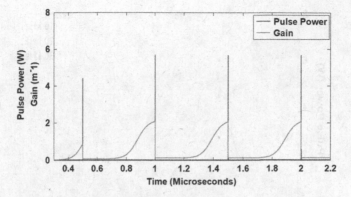

FIGURE 6.21   Output pulse train at AOM repetition rate of 30 kHz.

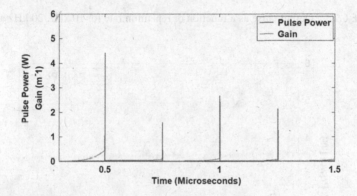

FIGURE 6.22   Output pulse at AOM repetition rate of 40 kHz.

Opening the AOM after a long time will result in a higher peak power for the first pulse. However, showing multiple pulses in this case will mean conducting the simulation for a longer time, which will result in much longer computation time. Several minutes can then be necessary for the computation. In any case, the first pulse is never considered because the laser is not stable. Therefore, opening the AOM at 50 μs is perfectly valid when one observes the behaviour of pulses beyond the first emitted. Figure 6.23 illustrates the influence of the repetition rate on the output pulse.

Unlike pump power and rise time of the AOM, the repetition rate does not trigger multipeak behaviour. However, it has a significant influence on peak power and pulse duration as shown in Figure 6.23. The pulses at 10 and 20 kHz are very similar, whereas at 30 kHz the pulse is strongly delayed and its peak power decreases while the pulse duration increases at the same time. The lower peak power and pulse energy observed at higher repetition rate is due to the lower gain achieved because population inversion "does not have much time" to rebuild after its depletion due to pulse release. In other words, before pumping replenishes the excited energy level, the OAM is opened, and another pulse released. It is worth mentioning

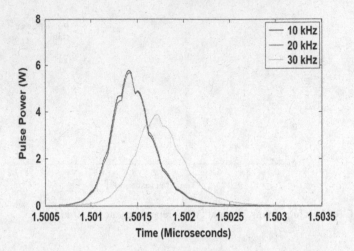

FIGURE 6.23 Output pulse as a function of repetition rate for 10 kHz, 20 kHz and 30 kHz.

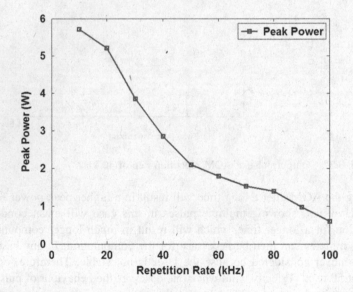

FIGURE 6.24 Peak Power as a function of repetition rate.

that these results were obtained at a pump power of 60 mW and a higher or lower pump power will result in different observations. At higher pumping, for example, the pulse delay, peak power reduction, and increase of pulse duration will occur at much higher repetition rate.

Figure 6.24 illustrate the variation of peak power as a function of repetition rate for an 8-metre-long ring cavity active Q-switched Erbium-doped fiber laser. The simulation was made for repetition rates ranging from 10 to 100 kHz at intervals of 10 kHz.

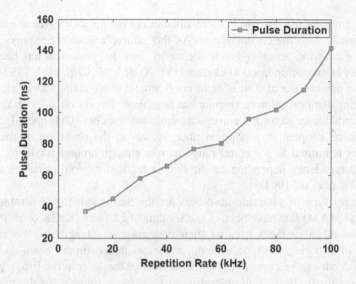

FIGURE 6.25 Pulse duration as a function of Repetition rate

A peak power of 5.78 W was obtained for a repletion rate of 10 kHz. This value then decreased as the repetition rate increases to reach a final value of 0.5 W for the repetition rate of 100 kHz. Beyond 100 kHz, the peak power becomes very low, making the laser strongly inefficient.

Figure 6.25 illustrates the pulse duration as a function of the depletion rate for the active Q-switched fiber laser. Here the 8-metre long Erbium-doped active Q-switched fiber laser was pumped with 60 mW of power with the same rise time for the acousto-optic modulator. The curve shows that the pulse duration constantly increased starting with a value of 35 ns at 10 kHz repetition rate to reach a value of 145 ns for 100 kHz repetition rate.

6.8 MODELLING OF PASSIVE Q-SWITCHED FIBER LASER

6.8.1 TYPE OF SATURABLE ABSORBERS

The first passive Q-switched appeared right after the demonstration of the first Q-switched fiber laser. These lasers used organic dyes. These dyes were largely used because of their simplicity of utilization dissolved in solvents or deposited on glass films then insert them inside the cavity of the laser. However, their use is limited to low power lasers and result in less compact devices.

The most common saturable absorbers are the ones based on crystals doped with Cr^{4+} doped ions. The Yttrium Aluminium Garnet (YAG) Cr^{4+}: YAG (Y.-F. Chen, Huang, and Wang 1997) is the most common. However, other crystals exist such as the Yttrium Scandium Gallium Garnet (YSGG) Cr^{4+}: YSGG (Klimov et al. 2007). This crystal is used as saturable absorbers around the wavelength of 1064 nm along with the Nd^{3+} doped YAG crystal or fibers. The biggest advantage of this absorber is that its technology is now mature and well mastered, however, its installation within a crystal limits its use. Its use is more suitable in configurations where light interacts transversally with the crystal; the use of this saturable absorber requires a complex coupling with the absorber inside the cavity. For this reason, it has been used intensively in microchip lasers (Feldman et al. 2003; Y.-F. Chen et al. 1997) which are compact devices but also in fiber lasers (Laroche et al. 2006; Pan et al. 2007). An interesting demonstration of shaping has been made in printing a Cr^{4+}: YAG using a femtosecond laser to form a waveguide saturable absorber (Okhrimchuk et al. 2009) that can be coupled to an optical fiber. However, the operation of that saturable absorber is limited to a spectral range narrow enough around 1064 nm, which is a little bit broadened depending on the doping ion distribution inside the crystal network (Eilers et al. 1994).

Another type of saturable absorbers are the Semiconductor Saturable Absorber Mirrors (SSAM) that have been in use for almost 23 years (Keller et al. 1996). They have several advantages because their operating wavelength can be adjusted (or tuned) in changing the type of semiconductors that compose them and the modulation depth can be controlled via the number of the laser in the Bragg grating that composes them. These saturable absorbers are more suitable for the realization of chip lasers (Fluck et al. 1998) and (Spuehler et al. 2001) where they are respectively used along with an amplifier medium made of glass doped with Erbium/Ytterbium to emit at a wavelength of 1535 nm or a YAG crystal doped with Ytterbium to emit at 1030 nm. However, their use is limited to external cavity laser, both in chip and fiber form and can necessitate extra effort to couple the output of the laser with other components.

Since from about fifteen years ago, the properties of nonlinear absorption of metallic nanoparticles are studied. In particular, a saturable absorption has been measured on silver nano-spheres at 532 nm deposited on a quartz substrate (Gurudas et al. 2008), on gold nano-sphere in aqueous solution at 532 and 1064 nm (Haitao et al. 2015; Boni et al. 2008) and on platinum nano-spheres at 532 nm in aqueous solution (Gao et al. 2005). This saturable absorber properties have been attributed to the plasmonic resonance of surface and have been verified in the case of gold nanoparticles (Philip et al. 2000). The size or form factor of nanostructures allow the production of a saturable absorber effect on a large range of wavelengths. However, the constancy of the phenomena of the saturable absorption is subject to the control of the nanoparticle dispersion.

Most recently, nanostructured material showed saturable absorption properties. This was first reported in 2002 on carbon nanotubes (Sakakibara et al. 2003; Y.-C. Chen et al. 2002) (single-walled carbon nanotubes SWNT). These nanotubes present the advantage of being capable of being used at different wavelength depending on their diameters. Also, before 2010, graphene showed absorption properties at large band (Dawlaty et al. 2008) and becomes rapidly used as a saturable absorber (Kim et al. 2015; Z. Sun et al. 2010). These absorption properties are independent of frequency and only depend on the fine-structure constant of the graphene. This absorption can be saturated under strong optical power density by a Pauli blocking effect preventing multiplicity of electrons states. Similar properties have been demonstrated on graphene oxide (Liu et al. 2013; Wang et al. 2011) and reduced graphene oxide (Sobon et al. 2012; He et al. 2012) at different wavelengths and in different configurations like end-butting (He et al. 2012), with an insulator clamped between two fibers terminated by connectors (Y Chen et al. 2013), another insulator (Y.-J. Sun et al. 2015), the graphene deposited on a tapered fiber (Liu et al. 2013), insulator put in contact with a side polished fiber (Lee et al. 2015), and finally, graphene deposited by a vapour deposition technique on a planar waveguide of Yb: KYW (Kim et al. 2015).

For long, the absorption properties of several nonlinear bidimensional materials have been studied. Topologic insulators have been used in different configurations to create pulsed lasers. Namely, the Bismuth selenide (Bi_2Se_3) spread out on a mirror placed at the output of a ceramic waveguide operating at 1064 nm (Tan et al. 2015) or on the output surface of a fiber connector operating at 1565 nm (Y Chen et al. 2013); Bismuth Telluride (Bi_2Te_3) spread out on a quartz substrate inserted inside a volumic cavity operating around 1045 nm (Y.-J. Sun et al. 2015) or produced in thin layers placed in evanescent interaction with a side polished fiber (Lee et al. 2015) finally antimony telluride (Sb_2Te_3) in the same last configuration (Sotor et al. 2014).

These properties have been observed in the dichalcogenides of transition metals such as nano-films of Molybdenum Disulphide (MoS_2) (Luo et al. 2014) and Tungsten Disulphide (WS_2) (B. Chen et al. 2015) both of them diluted and dried in a polyvinyl alcohol-based polymer (PVA) inserted between two connectors terminated fiber via sandwiching.

Finally, a phenomenon of saturable absorption has been demonstrated on black phosphorus or phosphorene (an atomic thin film of black phosphorous) since 2015 (Yu Chen et al. 2015) via a sandwiching between two fibers.

To conclude on different configurations analysed here, sandwiching has the advantage of offering a good compacity for the fiber devices, however, adding a fiber connector lead to the onset of additional loss and can create a point of fragility in the device, in particular in the presence of important peak powers. Configurations using an evanescent interaction seems to be a very interesting alternative, namely because the intervals don't undergo the maximum power density taken by the guiding structure while remaining compact solutions. Also, these configurations don't require any reshaping of the beam or external coupling. However, they can require a modification of the guiding structure itself.

6.8.2 Rate Equations of Passive Q-switched Fiber Laser

In the doped fiber, the rate equations can be written as:

$$\frac{\partial N_2(z,t)}{\partial t} + \frac{N_2(z,t)}{\tau} = \frac{\Gamma_p \lambda_p}{hcA_c}[\sigma_{ap}N_1(z,t) - \sigma_{ep}N_2(z,t)]P_p^-(z,t)$$

$$+ \sum_k \frac{\Gamma_k \lambda_k}{hcA_c}[\sigma_{ak}N_1(z,t) - \sigma_{ek}N_2(z,t)](P_k^+(z,t) + P_k^-(z,t))$$

$$(6.55)$$

$$\pm \frac{\partial P_p^{\pm}(z, t)}{\partial z} + \frac{1}{v_p} \frac{\partial P_p^{\pm}}{\partial t} = \Gamma_p [\sigma_{ep} N_2(z, t) - \sigma_{ap} N_1(z, t)] P_p^-(z, t) - \alpha_p P_p^-(z, t)$$

(6.56)

$$\pm \frac{\partial P_k^{\pm}(z, t)}{\partial z} + \frac{1}{v_k} \frac{\partial P_k^{\pm}(z, t)}{\partial t} = \Gamma_k [\sigma_{ek} N_2(z, t) - \sigma_{ak} N_1(z, t)] \quad P_k^{\pm}(z, t)$$

$$- \alpha_k P_k^{\pm} + \xi_k N_2(z, t)$$

(6.57)

$$\xi_k = M \Gamma_k \sigma_{ek} \frac{hc^2}{\lambda_k^3} \Delta \lambda_k$$

(6.58)

$$N_t = N_1(t) + N_2(t)$$

(6.59)

where N_t is the doping concentration of the amplifying medium, N_1 and N_2 the ground and excited state population densities respectively. The overlap factor for pump Γ_p and emission Γ_k are obtained assuming a Gaussian shape for both the pump and signal waves respectively. The factor M in equation (6.58) represents the number of transverse modes in the fiber, ($M = 2$ for single mode fiber). All the other parameters are defined as in equations (6.40)–(6.42). The superscript '±' corresponds to the forward and backward propagations respectively. These equations are completed by the rate and propagation equations in the Cr^{4+}: YAG saturable absorber is given as:

$$\frac{\partial N_2^{sa}(x, t)}{\partial t} + \frac{N_2^{sa}(x, t)}{\tau_{sa}} = \sum_k \frac{\lambda_k}{hcA_{sa}(x)} \sigma_{gsak} N_1^{sa}(x, t) [P_{sak}^+(z, t) + P_{sak}^-(z, t)] \quad (6.60)$$

$$N^{sa} = N_1^{sa}(t) + N_2^{sa}(t)$$

(6.61)

$$\pm \frac{\partial P_{sak}^{\pm}(x, t)}{\partial z} + \frac{1}{v_{sa}} \frac{\partial P_{sak}^{\pm}(x, t)}{\partial t} = [-\sigma_{gsak} N_1^{sa}(x, t) - \sigma_{esak} N_2^{sa}(x, t)]$$

$$P_{sak}^{\pm}(x, t) - \alpha_{sa} P_{sak}^{\pm}(x, t)$$

(6.62)

where x represents the longitudinal position in the saturable absorber, N^{sa} the total concentration of doping ions in the saturable absorber, N_1^{sa} and N_2^{sa} represent the ground and excited state population densities respectively, $P_{sak}^{\pm}$ are the power of forward and backwards propagating laser waves in the saturable absorber, σ_{gsak} and σ_{esak} are the absorption and emission cross state of the saturable absorber ions respectively, τ_{sa} the lifetime of the excited state of the saturable absorber, n_{sa} is the refractive index of the Cr^{4+}: YAG material, $v_{sa} = c/n_{sa}$ is the speed of light in the saturable absorber with c the speed of light in free space and α_{sa} the nonsaturable background loss of the saturable absorber. The Gaussian beam is assumed for the saturable absorber with the beam area at position x in the saturable absorber given by:

$$A_{sak}(x) = \pi \omega_{sak}(x)^2 \qquad (6.63)$$

where $\omega_{sak}(z)$ is the beam radius given by:

$$\omega_{sak}(x) \doteq \omega_{sa0k}\left[1 + \left(\frac{x}{x_0}\right)^2\right]^{1/2} \qquad (6.64)$$

$$x_{0k} = \frac{\pi \omega_{sa0k}^2 n_{sa}}{\lambda_k} \qquad (6.65)$$

where ω_{sa0k} is determined by the mode field diameter inside the doped fiber and the focal ratio in the saturable absorber. Equations (6.55)–(6.65) are solved with the following boundary conditions for a backward pumping:

$$P_p^-(L, t) = P_0 \qquad (6.66)$$

$$P_k^-(L, t) = P_k^+(L, t)R_{oc} \qquad (6.67)$$

$$P_{sak}^-(L_{sa}/2, t) = P_k^-(0, t)(1 - \eta) \qquad (6.68)$$

$$P_{sak}^+(-L_{sa}/2, t) = P_{sak}^-\left(\frac{L_{sa}}{2}, t\right) \qquad (6.69)$$

$$P_k^+(0, t) = P_{sak}^+\left(\frac{L_{sa}}{2}, t\right)(1 - \eta) \qquad (6.70)$$

$$P_{out}(t) = P_k(L, t)(1 - R_{oc}) \qquad (6.71)$$

where P_0 is the launched pump power, L the length of the doped fiber, L_{sa} the thickness of the saturable absorber, R_{oc} the reflectivity of the output coupler, η represent all the loss in the cavity including Fresnel reflection at the fiber faces, lens coupling as well as coating loss, and $P_{out}(t)$ the output power. The parameters used in the equation are listed in table Table 6.2.

Equations (6.55)–(6.65) with boundary conditions (6.66)–(6.71) are solved with a finite difference scheme as described in section 6.6.2. However, in this case the doped fiber and saturable absorber are divided into 100 and 50 sections respectively, with the parameters of Table 6.2. The parameters are obtained from fiber manufacturer specification and additional parameters are obtained from the work of Pan et al. (2010). It is worth mentioning that the parameters are only indicative and are

TABLE 6.2
Default Parameters Used in the Simulation of the Passive Q-switched Fiber Laser

Parameter	Value	Unit	Reference
Doped fiber length	4	m	Assumed
Ytterbium radiative lifetime (τ)	850	µs	Manufacturer
Cr^{4+}: YAG radiative lifetime (τ_{sa})	1	µs	(Pan et al. 2010)
Core diameter	4	µm	Manufacturer
Core numerical aperture (NA)	0.15		Manufacturer
Cladding diameter	125	µm	Manufacturer
Cladding numerical aperture (NA)	0.46		Manufacturer
Fiber refractive index (n)	1.45		Manufacturer
Cr^{4+}: YAG refractive index (n_{sa})	1.82		(Pan et al. 2010)
Normalized frequency (V number)	2.56	–	(Pan et al. 2010)
Mode Field diameter (MFD)	4.22	µm	Manufacturer
Fiber background loss (α_p, α_k)	0.005	m^{-1}	Manufacturer
Cr^{4+}: YAG background loss (α_{sa})	10^{-5}	m^{-1}	(Pan et al. 2010)
Overlap factor at pump wavelength (Γ_p)	0.83	–	(Pan et al. 2010)
Centre lasing wavelength (λ_0)	1064	nm	Assumed
Absorption cross-section at laser wavelength (σ_{ak})	1.75×10^{-27}	m^2	Manufacturer
Emission cross-section at laser wavelength (σ_{ek})	2.4×10^{-25}	m^2	Manufacturer
Absorption and emission cross-section of the Cr^{4+}: YAG (σ_{gsa}, σ_{esa})	3×10^{-22}	m^2	(Pan et al. 2010)
Ytterbium concentration (N_t)	9.8×10^{24}	Ions/m^3	Assumed
Cr^{4+}:YAG concentration (N^{sa})	6.2×10^{23}	Ions/m^3	(Pan et al. 2010)
Cr^{4+}: YAG transmission at the centre lasing wavelength (T_0)	56%		(Pan et al. 2010)
Cr^{4+}: YAG transmission at 1064 nm (T_0)	50%		(Pan et al. 2010)
Cr^{4+}: YAG saturable absorber thickness (L_{sa})	2.5	mm	(Pan et al. 2010)
Focal ratio in the saturable absorber	1:2.4		(Pan et al. 2010)
Focal length of lens 3	6.24	mm	(Pan et al. 2010)
Focal length of lens 4	15	mm	(Pan et al. 2010)
Focal length of lens 5	15	mm	(Pan et al. 2010)
Numerical aperture of lens 3	0.4		(Pan et al. 2010)
Speed of light in vacuum (c)	3×10^8	m/s	
Planck constant (h)	6.626×10^{-34}	m^2kg/s	

generally varied within a large range to simulate the characteristics of the device under investigation. Also, the computation is valid for any two or three-level systems. Therefore, Er^{3+}, Nd^{3+} and another type of rare-earth-doped fiber laser models can be solved with the same algorithm. The laser spectrum was divided into 40 intervals of 2 nm from 1020 nm to 1100 nm. When the laser is pumped and reaches the threshold, the pulse starts building from a seed originated from spontaneous emission. The seed value can be obtained from equation (6.58), if $\Delta\lambda_k$ and central emission wavelength λ_k are known. Also, for a single-mode fiber $M = 2$ to represent the only two possible polarizations of the propagating LP_{01} transverse mode.

Figure 6.26 shows the output train of the Ytterbium-doped passive Q-switched fiber laser simulated using the algorithm described in the previous section. Unlike active Q-switched fiber lasers, the repetition rate of passive Q-switched fiber laser depends on pump power. Simulations were conducted for three different pump powers, namely 2 W, 6 W, and 10 W. At low pump powers, the laser takes a longer interval of time to reach the threshold as it can be seen in Figure 6.26 where 50 microseconds were required for a pump power of 2 W. This time falls to close to 18 microseconds for a pump power of 6 W and only 10 microseconds for a pump power of 10 W. This can be easily justified because the rate of pumping is directly proportional to the value of pump power. For the same reason, the number of photons in the cavity increases slowly at low pump powers, resulting in a low-intensity increment, thus the saturable absorber is slowly bleached, and the

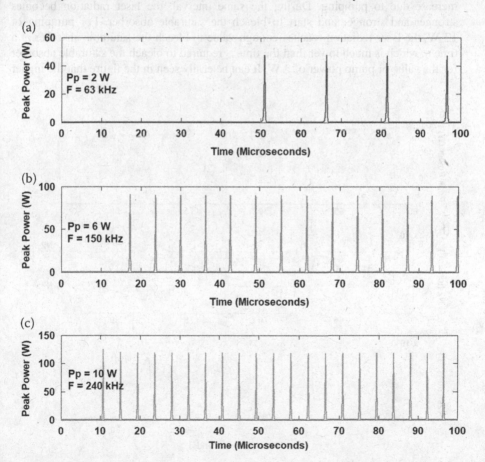

FIGURE 6.26 Pulse train for three pump powers, a) 2 W, b) 6 W, c) 10 W.

repetition rate is slow. The recorded repletion rate was around 63 kHz for a pump power of 2 W, 150 kHz for 6 W, and 240 kHz for 10W. It can also be observed that the peak power increases monotonically at the same time with values of 50 W, 90 W, and 125 W at pump powers of 2 W, 6 W, and 10 W respectively.

Figure 6.27 shows the variation of the excited state population density of the doped fiber $N_2(t)$ and the saturable absorber. To conduct the study, a comparison is made between two pumping powers, namely 3 W (color-in-print curve) and 12 W (Red curve). The 3 W pumping corresponds to a slow pumping rate and population density of the excited state of doped fiber grows slowly. At 12 W pumping of the other hand, the pump rate is high and the population density of the excited state of the doped fiber grows quickly.

The same trend is observed inside the saturable absorber, where longer time is required to bleach the saturable absorber at low pump power and a significantly short time is required to bleach the saturable absorber for a higher pump power. At a time t_0 the two pump powers of 3 W and 12 W achieve equal population inversion densities. As can be seen in Figure 6.27. As time increases, both the value of $N_2(t)$ continue to increase due to pumping. During the same interval, the laser radiation becomes stronger and stronger and start to bleach the saturable absorber. For pumping of 12 W, the laser radiation reaches enough value to bleach the saturable absorber at a time t_1, which is much lower than the time t_2 required to bleach the saturable absorber for the value of pump power of 3 W. It can be easily seen in the figure that the higher

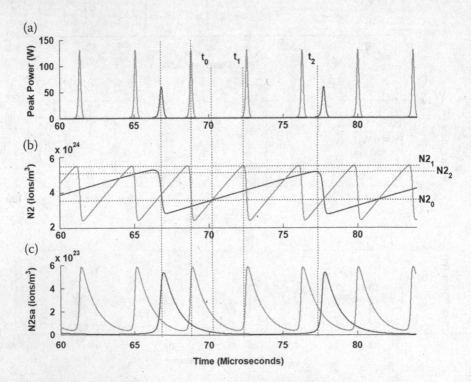

FIGURE 6.27 Dynamics of populations for doped fiber and saturable absorber.

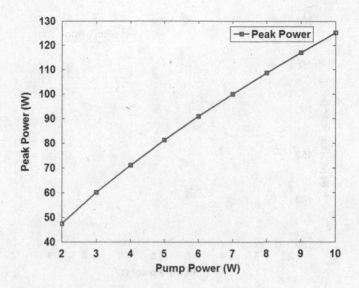

FIGURE 6.28   Peak power as a function of pump power.

pump power of 12 W increases the population inversion to a much higher value than the lower pumping power of 3 W, before triggering the switching process. Therefore, the higher pump will generate a pulse with high peak power and shorter time duration.

Figure 6.28 illustrates the variation of pulse peak power as a function of pump power. As can be seen, the output peak power increases almost linearly with pump power. The relatively high value of pump threshold around 2 W is explained by the large core diameters of the doped fiber used in the simulation. This happens because the rate of absorption of the fiber is inversely proportional to the core diameter. Simulations were conducted until the pump power of 10 W corresponding to a peak power of 125 W.

As it was already illustrated in Figure 6.26, the repetition rate of the passive Q-switched fiber laser grows with pump power. The growing trend is illustrated in Figure 6.29. At a pump power of 2 W, a repetition rate of 65 kHz is recorded. This value grows almost steadily to reach the value of close to 280 kHz at a pump power of 10 W. Pulse duration, on the other hand, decreases almost monotonically within the same interval starting with a value of 288 ns and reaching the value 120.7 ns at the pump power of 10 W is illustrated in Figure 6.30.

Finally, Figure 6.31 illustrates the pulse energy as a function of pump power where pulse energy increases linearly with pump power. As it has already been explained, the higher pump power allows one to reach very high inversion values, which potentially releases high-energy pulses. In the simulation, the pulse energy grows from 150 μJ at 2 W to a value of 289 μJ obtained at 10 W. It is worth mentioning that all the simulated values were obtained with a 4 metres long Ytterbium-doped fiber.

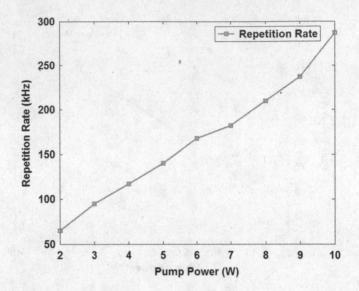

FIGURE 6.29   Repetition rate as a function of pump power.

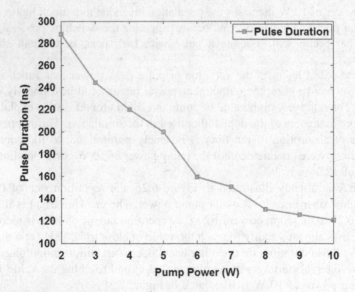

FIGURE 6.30   Pulse duration as a function of pump power.

6.9   Q-SWITCHED FIBER LASER: STATE OF THE ART

Active Q-switched fiber lasers:

The early solid-state active Q-switched laser used electro-optic switches. In 1966, P. Wurtz proposed a theoretical and experimental study of a Neodymium doped glass laser with a Potassium Dihydrogen Phosphate (KH_2PO_4) crystal used as a Pockels

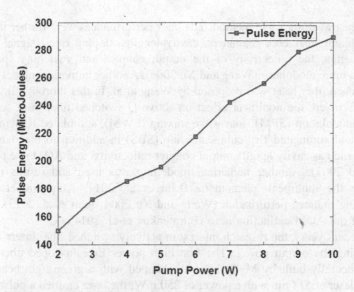

FIGURE 6.31 Pulse energy as a function of pump power.

Cell for switching. The electro-optic modulator was driven by a voltage of 18 kV and the laser emitted at a wavelength of 1060 nm (Wurtz 1966). Q-switched fiber lasers were first developed in 1986 (Mears et al. 1986). Since then, more reliable and efficient Q-switched fiber laser have been built, due to improvements pump sources technology (Seguin and Oleskevich 1993), better design of Acousto-optic modulators as well as the use of doped fiber with better designs. In 1992, Morkel et al. reported an electro-optic Q-switched Nd^{3+} doped fiber laser emitting at 1053 nm when pumped in the 810 nm region. At 22 mW power, they reported a peak power of more than 1 kW and 2 ns time duration for a repetition rate of 1 kHz (Jedrzejewski et al. 1992). In 2003 Ashraf F. et al. reported the first Tm^{3+} doped active Q-switched fiber laser, using an EOM. They use a large core fiber, with a core diameter of 17 μm. The laser emitted at 2 μm with an output pulse of about 3.3 kW peak power at a repetition rate of 70 Hz and corresponding pulse energy of 2.3 mJ (El-Sherif and King 2003). Although EOM Q-switched fiber laser has a fast switching and high extinction ratio, they are less popular than, their AOM counterpart because of their relatively low repetition rate, high cost and the high voltage sources required to operate them.

In 1993, Sejka et al. demonstrated the first ring cavity AOM Q-switched Erbium-doped fiber laser. Peak power of 340 W with a pulse width of 12 ns was achieved with this laser (M Sejka et al. 1993), later they investigated the characteristics of the laser as a function of parameters such as the length of the doped fiber, the AOM modulation frequency and Erbium concentration of the doped fiber (Milan Sejka et al. 1995). Following this first successful demonstration, several improvements were brought to active fiber lasers. The optimum parameters and understanding peculiarities of the lasers were often achieved by the mean of numerical modelling and simulation. Numerical modelling and optimization of double-clad fiber lasers were presented. The dynamic characteristics of pulse energy, pulse width, population inversion, and stored energy at tens of kHz repetition rate were studied by using the travelling wave method. The laser performances were further investigated for different fiber core diameters, cavity lengths, doping rates, signal and pump wavelengths, the reflectivity of the output coupler and switching speed of the acousto-optic modulator (Wang and Xu 2006). Another numerical model of actively Q-switched fiber laser was proposed by Wang et al. In this thorough investigation, they discussed, the nonlinear effect on active Q-switched fiber laser such as self-phase modulation (SPM), four-wave mixing (FWM), stimulated Raman scattering (SRS) and stimulated Brillouin scattering (SBS) in addition to the standard parameters such as cavity length, output coupler reflectivity, and AOM rise time (Wang and Xu 2007). Another numerical modelling was used and results reported to analyse the multipeak phenomena (Lim et al. 2011; Kolpakov et al. 2011), switching induced perturbation (Wang and Xu 2004; Jeon et al. 2013), and influence of the AOM extinction ratio (Barmenkov et al. 2014).

In recent years, the research interest in actively switched fiber lasers focused on increasing the output power. The first high power Erbium-doped fiber laser was experimentally built by Mylinski et al. Pumped with a green light beam from an Argon laser at 514 nm with a power of 250 mW, the laser emitted a pulse with peak power of 230 W and a pulse width of 8 ns (Myslinski et al. 1992). Later on, Ytterbium-doped fiber with large core diameter was found to be the most efficient gain medium for high power fiber lasers. Using such a Ytterbium-doped fiber with 15 μm core diameter in a bidirectional pumping configuration, a laser pulse with the energy of 170 μJ and 2 kW peak power at a low repetition rate of 500 Hz was realised, also, the laser was tunable from 1060 nm to 1100 nm using a diffraction grating (C. C. Renaud et al. 1999). Using a large pitch photonic crystal fiber, sub 60 ns pulse duration with 26 mJ pulse energy and near diffraction-limited beam quality ($M^2 < 1.3$) were obtained from an AOM actively Q-switched fiber laser. Also, the repetition rate was 5 kHz and an average power of 130 W. The core diameter of the fiber was 135 μm with a mode field diameter (MFD) larger than 90 μm (Stutzki et al. 2012). A cladding-pumped Ytterbium-doped AOM Q-switched fiber laser was reported. With a wavelength operation of 1090 nm, the laser was capable of generating 2.3 mJ pulse energy, 5 W average power at a repetition rate of 500 kHz in a high brightness beam ($M^2 = 3$) (Alvarez-Chavez et al. 2000). Several other high power actively Q-switched fiber laser was demonstrated using standard communication erbium-doped fiber with dual pumping (Kolpakov et al. 2014) or Nd^{3+} doped fibers (åWIDERSKI et al. 2005). Ranaud et al. theoretically and experimentally investigated Ytterbium-doped active Q-switched fiber lasers. They compare the extractable energy from two high-energy fiber designs, namely, single of few-mode low-NA. Large Mode Area fibers (LMA) and large-core multimode fiber which may incorporate a fiber taper for brightness enhancement (Cyril C. Renaud et al. 2001). Other areas of active Q-switched fiber laser investigated include single frequency operation (Li et al. 2017; Geng et al. 2009; Kaneda et al. 2004; Álvarez-Tamayo et al. 2016) and tunability (González-García et al. 2013; González García et al. 2015; 2013; Chakravarty et al. 2017; Mears et al. 1986).

Active Q-switched fiber lasers have also been designed for specific applications. Zhang et al. proposed an acousto-optic Q-switched fiber laser for gas detection by using a PAS cell inside the cavity of the fiber laser (Q. Zhang et al. 2017). A Fiber laser with cylindrical vector beam emission was demonstrated, pulse energy of 4.25 μJ, a pulse width of 64 ns at a repetition rate of 20 kHz were obtained (J. Zhang et al. 2018). A laser emitting this kind of beams can find application in material processing, nonlinear optics, lithography, particle acceleration, high-resolution metrology, optical trapping and manipulation, etc. A Q-switched fiber laser suitable for particle image velocimetry (PIV) was proposed. A theoretical design of the laser was provided based on carefully optimizing cavity length and rising time of the acousto-optic modulator to have the original pulse split into two distinct pulse separated by a round trip time (Mgharaz et al. 2009).

Q-switching of fiber laser using bulk element have major drawbacks like reduced mechanical stability and high insertion loss. To overcome these problems and allow the construction of robust, compact and efficient active Q-switched fiber laser, all fiber modulation techniques have been investigated. These techniques require the development of reliable in-fiber cavity losses modulators. Three main approaches have been proposed to this end, namely: direct interaction with the evanescent field of the LP_{01} transverse mode, mechanical tuning of a fiber Bragg grating and in-fiber Acousto-optic interaction (Andrés et al. 2008). Using side polished fiber, Chandonnet et al. developed an Erbium-doped Q-switched fiber laser by controlling the evanescent field from the fiber. When pumped with a source emitting 100 mW at 980 nm, the laser delivered a pulse with 400 W peak power (Chandonnet and Larose 1993). Their intensity modulator was made of a side-polished coupler and a piezoelectric-actuated specially shaped overlay medium to change the losses created by evanescent field coupling as illustrated in Figure 6.32.

Kieu et al. on the other hand used a microsphere resonator coupled through a fiber taper at one end of the cavity and a FBG at the other. Using this technique, they reported a 102 W peak power pulse with a pulse duration of about 160 ns (Kieu and Mansuripur 2006). The previous techniques require accurate positioning of the fiber since the evanescent field extends only a few micrometres away from the core of the fiber. Also, the area where the interaction between the evanescent field and the switcher takes place must be isolated from the influence of the external environment. These challenges make the interaction with the evanescent field difficult to implement.

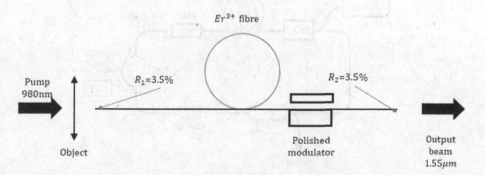

FIGURE 6.32 Schematic diagram of the fiber laser.

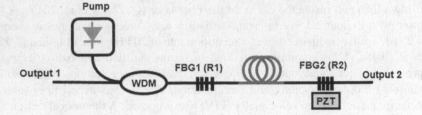

FIGURE 6.33 Q-switched fiber laser with a PZT.

A simpler approach consists of using a Fabry-Perot linear cavity with two identical fiber Bragg gratings as mirrors. One of the FBG is fixed on a mechanical stretcher driven by a piezoelectric transducer (PZT) to tune its spectral response as illustrated in Figure 6.33. By applying a variable periodic voltage to the PZT, the spectrum of the grating can be moved back and forth.

When the spectra of the two gratings are misaligned, the loss in the cavity is high and energy is stored inside the cavity. On the other hand, when the spectra of the two gratings overlap, the cavity loss drop and energy is released in a large pulse. Using this technique, Russo et al. successfully operated an active Q-switched fiber laser with an output peak power of 530 mW and 3 kHz repetition rate (Russo et al. 2002). A similar cavity was used by Cheng et al. In addition they investigated the influence of fiber Bragg grating side lobes on the characteristics of the laser (Cheng et al. 2007). The technique was also used in linear cavity (Wu et al. 2016) and ring cavity (Cheng et al. 2008) along with a π-phase shifted fiber Bragg grating to narrow the line width of the laser as illustrated in Figure 6.34.

Using a FBG combined with a piezoelectric stretcher is a simple technique of building all fiber Q-switch fiber laser. However, this technique shows a limitation on the frequency and waveform of the modulation signal because of the mechanical resonance of the stretcher. An improvement of the method consists of replacing the PZT controlled stretcher by a magnetostrictive rod. When a FBG is fixed to such a rod, it is possible to tune its spectral response by applying to it a magnetic field

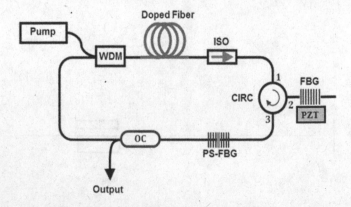

FIGURE 6.34 All fiber Q-switched fiber laser with PS-FBG.

(Mora et al. 2002; Pérez-Millán et al. 2005). Magnetostrictive transducers have the advantage of relatively flat frequency response permitting a continuous tuning of the repetition frequency and using a rectangular waveform to control the Q-value of the cavity. Using this technique tunable Q-switched fiber laser with tuning range in the range of 200 kHz have been demonstrated in both Erbium and Ytterbium. When pumped with 180 mW, pulse peaks of several watts with a pulse width of about 100 ns were obtained (Andersen et al. 2006; Pérez-Millán et al. 2005). Other techniques of all fiber actively Q-switched fiber lasers include using the interaction of longitudinal acoustic waves with a FBG (Cuadrado-Laborde et al. 2007; D. W. Huang, Liu, and Yang 2000; Delgado-Pinar et al. 2006) or overlap between a tunable Fabry-Perot fiber filter and fiber Bragg grating spectra (Monga et al. 2019).

6.10 MATLAB CODE

6.10.1 ACTIVE Q-SWITCH FIBER LASER FUNCTION

```
function
[n1,n2,ppf,psf,gain,xc,N1,N2,Psf,Ppf,Gain,time,
reflectivity ] =…
  ActiveQswitch(Pump Power,L,Core_D,Rise_Time)
% Function ActiveQswitch implemente a finite difference
method to compute
% output characteristics of an Erbium doped active Q-
switched fiber laser
% using an acousto-optic modulator as switching element
%
% Inputs:
% - Pump_Power: Input pump power
% - L:Length of the Erbium doped fiber
% - Core_D: Core diameter of the doped fiber
% - Rise_Time: Rise time of the acousto-optic modulator
%
% Outputs:
%   - n1: Ground level population density distribution
along the cavity
%   - n2: Excited state population density distribution
along the cavity
%    - N1: Ground level population density variation
with time
%    - N2: excited state population density variation
with time
% - gain: gain distribution along the cavity
% - Gain variation as a function of time
```

```
%   - ppf: Pump power distribution along the cavity
%   - psf: laser power distribution along the cavity
%   - Gain: Gain variation as a function of time
%   - Ppf: Residual pump
%   - Psf: Q-switched fiber laser output power
%   - xc: sections number along the cavity
%   - time: computation time
% Comments:
%   - Erbium doped gain medium is considered as a two level
system and the
% resulting rate equations are solved with simple ana-
lytical method using
% substitution
%   -Initial conditions of the finite difference scheme are
obtained by
%   solving the rate equations first in steady state
%
% References:
%
% Written by Justice Sompo, University of Johannesburg,
South Africa
%
%
%
%===
===============
=========
=========
===============
=========
=====
======
% Values for the variables in simulation
%================
=========
=========
===============
=========
==============
close all
clc
% Check inputs
if nargin <4
  Rise_Time = 40e-9;
  if nargin < 3
    Core_D = 2.3e-6;
```

```
      if nargin < 2
        L = 5;
        if nargin < 1
                    error(message('MATLAB:ActiveQswitch:
NotEnoughInputs'));
        end
      end
  end
end
lamddas = 1550e-9;        % Signal wavelength in meters
sigmaas = 3.15e-25;        % Absorption cross section at
signal wavelength m^2
sigmaes = 4.44e-25;        % Emission cross section at signal
wavelength m^2
NA = 0.2;
dcore = Core_D;            % Fiber core diameter
alfap = 0.005;            % Attenuation coefficient at pump
wavelength
alfas = 0.005;            % Attenuation coefficient at signal
wavelength
n0 = 4e24;                % Doping concentration of the fiber
from ions
c0 = 3e8;                % Speed of light in vacuum
gamas = 0.7;
h = 6.626e-34;            % Planck constant
nfiber = 1.45;            % Refractive index of the fiber
vg = c0/nfiber;            % Group velocity in doped fiber
step_number = 50;
deltax = L/step_number;        % Longitudianl section in
doped fiber
deltat = L/(step_number*vg);    % Sections in the time
domain
deltalambda = 3e-6;
Aco = pi*(dcore/2)^2;        % Transverse section of the
doped fiber
lambdap = 980e-9;            % Pump wavelength
gamap = 0.83;
sigmaep = 1.8e-25;            % Absorption cross section at pump
wavelength
sigmaap = 2.44e-24;            % Emission cross section at pump
wavelength
tao = 10e-3;                % Erbium excited state lifetime
Rmax = 0.9;
total_loss = 0.2032;        % 3 dB filter, 3 dB circulator, 0.8
dB WDM, 0.1 dB Coupler
leng = L;
```

```
coupler = Rmax;
Nt = n0;
sections = step_number+1;
round_trip_time = L/vg;
% compute steady state values distribution
pumppower = Pump_Power; % Pump power in watts
Pump = pumppower;
[PP,Ps,N2x] = LaserOutput(Pump,leng,coupler,Nt,sec-
tions);
% Initial conditions for doped fiber
%-------------------
------------------------------------------------------
-----------------------
s = 1:step_number+1;
n2(s,1) = N2x;
n1(s,1) = n0-N2x;
ppf(s,1) = PP;
psf(s,1) = Ps;
gain(s,1) = 0.0;
xc(s,1) = 0.0;
%=======================
========================
============================
% Finite difference scheme implementation
%============================
==================================================
ppf(step_number,1) = pumppower;
temps = deltat*1e6;
Time = deltat:deltat:temps;
t = 0;            % very important to be used in the boundary
conditions function
tr = Rise_Time;
to = 5e-6;
for k = 1:length(Time)
  ppf(1,2) = pumppower;
  for s = 2:step_number+1
          n2(s,2) = n2(s,1)+deltat*((gamap*lambdap/
(h*c0*Aco))*(sigmaap*(n0-n2(s,1))...
          -sigmaep*n2(s,1))*ppf(s,1)+(gamas*lamddas/
(h*c0*Aco))*(sigmaas*(n0-n2(s,1))...
      -sigmaes*n2(s,1))*psf(s,1)-n2(s,1)/tao);
    ppf(s,2) = ppf(s-1,1)+deltax*(gamap*(sigmaep*n2(s-
1,1)-sigmaap*(n0-n2(s-1,1)))*ppf(s-1,1)...
      -alfap*ppf(s-1,1));
```

```
    psf(s,2) = psf(s-1,1)+deltax*(gamas*(sigmaes*n2(s-
1,1)-sigmaas*(n0-n2(s-1,1)))*psf(s-1,1)…
      +2*gamas*sigmaes*n2(s-1,1)*h*c0*c0*deltalambda/
(lamddas)^3-alfas*psf(s-1,1));
    n1(s,2) = n0-n2(s,1);
    gain(s,2) = gamas*(sigmaes*n2(s-1,1)-sigmaas*(n0-
n2(s-1,1)));
  end
        n2(1,2)   =   n2(1,1)+deltat*((gamap*lambdap/
(h*c0*Aco))*(sigmaap*(n0-n2(1,1))…
          -sigmaep*n2(1,1))*ppf(1,1)+(gamas*lamddas/
(h*c0*Aco))*(sigmaas* (n0-n2(1,1))…
    -sigmaes*n2(1,1))*psf(1,1)-n2(1,1)/tao);
  n1(1,2) = n0-n2(1,1);
  % Boundary conditions)
  %=====================
  t = t + deltat;
  R = feval('boundary',Rmax,tr,to,t);
  psf(1,2) = psf(step_number+1,1)*R*total_loss;
  for s = 1:step_number+1;
    n2(s,1) = n2(s,2);
    n1(s,1) = n1(s,2);
    ppf(s,1) = ppf(s,2);
    psf(s,1) = psf(s,2);
    gain(s,1) = gain(s,2);
    xc(s) = (s-1)*deltax;
  end
  N2(k) = n2(step_number+1);
  N1(k) = n1(step_number+1);
  Psf(k) = psf(step_number+1)*(1-R);
  Ppf(k) = ppf(step_number+1);
  Gain(k) = gain(step_number+1);
  z(k) = k;
  time(k) = t*1e4;
  reflectivity(k) = R;
end
end
```

6.10.2   DISTRIBUTION FUNCTION

```
% This script plot differents values resulting from the
main Active Q
```

```
% switched fiber laser function
clear all
close all
clc
tic
Pump_Power0 = 40e-3;
[n10,n20,ppf0,psf0,gain0,xc0,N10,N20,Psf0,Ppf0,
Gain0,time0,reflectivity0 ]…
  = ActiveQswitch(Pump_Power0);
Pump_Power1 = 80e-3;
[n11,n21,ppf1,psf1,gain1,xc1,N11,N21,Psf1,Ppf1,
Gain1,time1,reflectivity1 ]…
  = ActiveQswitch(Pump_Power1);
Pump_Power2 = 120e-3;
[n12,n22,ppf2,psf2,gain2,xc2,N12,N22,Psf2,Ppf2,
Gain2,time2,reflectivity2 ]…
  = ActiveQswitch(Pump_Power2);
Pump_Power3 = 160e-3;
[n13,n23,ppf3,psf3,gain3,xc3,N13,N23,Psf3,Ppf3,
Gain3,time3,reflectivity3 ]…
  = ActiveQswitch(Pump_Power3);
figure(1)
subplot(2,2,1)
plot(xc0,ppf0,'b','Linewidth',2)
xlabel('Position (m)')
ylabel('Pump Power (W)')
subplot(2,2,2)
plot(xc0,psf0,'Linewidth',2)
xlabel('Position (m)')
ylabel('Laser Power (W)')
subplot(2,2,3)
plot(xc0,n10,'m','Linewidth',2)
hold on
plot(xc0,n20,'r','Linewidth',2)
xlabel('Position (m)')
ylabel('Population Density (ions/m^3)')
subplot(2,2,4)
plot(xc0,gain0,'k','Linewidth',2)
xlabel('Position (m)')
ylabel('Gain (m^-1)')
figure(2)
subplot(2,2,1)
plot(time0,Gain0,'k','Linewidth',2)
xlabel('Time (Microseconds)')
ylabel('Gain (m^-1)')
```

```
subplot(2,2,2)
plot(time0,Psf0,'Linewidth',2)
xlabel('Time (Microseconds)')
ylabel('Laser Power (W)')
subplot(2,2,3)
plot(time0,N10,'m','Linewidth',2)
xlabel('Time (Microseconds)')
ylabel('N1 (ions/m^3)')
subplot(2,2,4)
plot(time0,N20,'r','Linewidth',2)
xlabel('Time (Microseconds)')
ylabel('N2 (ions/m^3)')
figure(3)
subplot(2,2,1)
plot(xc2,ppf2,'b','Linewidth',2)
xlabel('Position (m)')
ylabel('Pump Power (W)')
subplot(2,2,2)
plot(xc2,psf2,'Linewidth',2)
xlabel('Position (m)')
ylabel('Laser Power (W)')
subplot(2,2,3)
plot(xc2,n12,'m','Linewidth',2)
hold on
plot(xc2,n22,'r','Linewidth',2)
xlabel('Position (m)')
ylabel('Population Density (ions/m^3)')
subplot(2,2,4)
plot(xc2,gain2,'k','Linewidth',2)
xlabel('Position (m)')
ylabel('gain (m^-1)')
figure(4)
subplot(2,2,1)
plot(time2,Gain2,'b','Linewidth',2)
xlabel('Time (Microseconds)')
ylabel('Gain (m^-1)')
subplot(2,2,2)
plot(time2,Psf2,'Linewidth',2)
xlabel('Time (Microseconds)')
ylabel('Laser Power (W)')
subplot(2,2,3)
plot(time2,N12,'m','Linewidth',2)
xlabel('Time (Microseconds)')
ylabel('N1 (ions/m^3)')
subplot(2,2,4)
plot(time2,N22,'r','Linewidth',2)
```

```
xlabel('Time (Microseconds)')
ylabel('N2 (ions/m^3)')
toc
```

6.10.3   MULTIPLE PULSE ACTIVE Q-SWITCHED FUNCTION

```
function [N1,N2,Psf,Ppf,Gain,time] =…
                        MultiPulseActiveQswitch
(Pump_Power,RepRate,Rise_Time,FiberLength,Core_D,Ner)
% Function MultipleActiveQswitch implement a finite dif-
ference method to compute
% output characteristics of an Erbium-doped active Q-
switched fiber laser
% using an acousto-optic modulator as a switching element.
A pulse train
% which number depends on the repetition frequency is
obtained
%
% Inputs:
% - Pump_Power: Input pump power
% - L: Length of the Erbium-doped fiber
% - Core_D: Core diameter of the doped fiber
% - Rise_Time: Rise time of the acousto-optic modulator
%
% Outputs:
%    - n1: Ground level population density distribution
along the cavity
%    - n2: Excited-state population density distribution
along the cavity
%     - N1: Ground level population density variation
with time
%     - N2: excited state population density variation
with time
% - Gain variation as a function of time
% - Gain: Gain variation as a function of time
% - Ppf: Residual pump
% - Psf: Q-switched fiber laser output power
% - time: computation time
% Comments:
% - Erbium-doped gain medium is considered as a two-level
system and the
```

```
%       resulting rate equations are solved with a simple
analytical method using
%    substitution
% -Initial conditions of the finite difference scheme are
obtained by
%  solving the rate equations first in steady-state with a
shooting
%  algorithm
%
% References:
%
% Written by Justice Sompo, University of Johannesburg,
South Africa
%
%
%
close all
clc
% Check inputs
if nargin < 6
  Ner = 4e24;
  if nargin < 5
    Core_D = 2.3e-6;
    if nargin < 4
      FiberLength = 3;
      if nargin < 3
        Rise_Time = 40e-9;
        if nargin < 2
          RepRate = 20e3;
          if nargin < 1
                                                error(message
('MATLAB:MultiPulseActiveQswitch:NotEnoughInputs'));
          end
        end
      end
    end
  end
end
% Values for the variables in simulation
%==================
===============
===============
===============
===========
lambdas = 1550e-9;       % signal wavelength
```

```
sigmaas = 3.15e-25;        % Absorption cross section at
signal wavelength
sigmaes = 4.44e-25;      % Emission cross section at signal
wavelength
NA = 0.2;
dcore = Core_D;         % fiber core diameter
L = FiberLength;        % Fiber length
alfap = 0.005;          % Attenuation coefficient at pump
wavelength
alfas = 0.005;          % Attenuation coefficient at signal
wavelength
n0 = Ner;            % Doping concentration of the fiber from
MORASSE
c0 = 3e8;              % Speed of light in vacuum
gamas = 0.7;
h = 6.626e-34;          % Planck constant
nfiber = 1.45;          % Refractive index of the fiber
vg = c0/nfiber;         % Group velocity in doped fiber
step_number = 50;       % Number of sections
deltax = L/step_number;        % Longitudianl section in
doped fiber
deltat = L/(step_number*vg);     % Sections in the time
domain
deltalambda = 3e-6;
Aco = pi*(dcore/2)^2;        % Transverse section of the
doped fiber
lambdap = 980e-9;       % Pump wavelength
gamap = 0.83;
sigmaep = 1.8e-25;       % Absorption cross section at pump
wavelength
sigmaap = 2.44e-24;       % Emission cross section at pump
wavelength
tao = 10e-3;            % Erbium metastable state lifetime
Rmax = 0.9;
total_loss = 0.2032;     % 3 dB filter, 3 dB circulator, 0.8
dB WDM, 0.1 dB Coupler
leng = L;
coupler = Rmax;
Nt = n0;
sections = step_number+1;
round_trip_time = L/vg;
% compute steady state values distribution
pumppower = Pump_Power;       % Pump power in watt
Pump = pumppower;
[PP,Ps,N2x]                  =                  LaserOutput
(Pump,leng,coupler,Nt,sections); % The function
```

```
% LaserOuput is called to solve the rate equations in
steady state and
% provide initial conditions for finite difference scheme
% Initial conditions for doped fiber
%--------------
------------------
--------------------
----------------------
s = 1:step_number+1;
n2(s,1) = N2x;
n1(s,1) = n0-N2x;
ppf(s,1) = PP;
psf(s,1) = Ps;
gain(s,1) = 0.0;
xc(s,1) = 0.0;
%================================
============================================
% Finite difference scheme implementation
%====================================
====================================
ppf(step_number,1) = pumppower;
temps = deltat*1e6;
Time = deltat:deltat:temps;
t = 0;  % very important to be used in the boundary condi-
tions function
tr = Rise_Time;
to = 2e-6;
ti = 50e-6;
rep_rate = RepRate;
per = 1/rep_rate;
for k = 1:length(Time)
  ppf(1,2) = pumppower;
  for s = 2:step_number+1
          n2(s,2)  =  n2(s,1)+deltat*((gamap*lambdap/
(h*c0*Aco))*(sigmaap*(n0-n2(s,1))...
          -sigmaep*n2(s,1))*ppf(s,1)+(gamas*lambdas/
(h*c0*Aco))*(sigmaas*(n0-n2(s,1))...
      -sigmaes*n2(s,1))*psf(s,1)-n2(s,1)/tao);
    ppf(s,2) = ppf(s-1,1)+deltax*(gamap*(sigmaep*n2(s-
1,1)...
        -sigmaap*(n0-n2(s-1,1)))*ppf(s-1,1)-alfap*ppf
(s-1,1));
    psf(s,2) = psf(s-1,1)+deltax*(gamas*(sigmaes*n2(s-
1,1)...
      -sigmaas*(n0-n2(s-1,1)))*psf(s-1,1)...
```

```
       +2*gamas*sigmaes*n2(s-1,1)*h*c0*c0*deltalambda/
(lambdas)^3-alfas*psf(s-1,1));
    n1(s,2) = n0-n2(s,1);
     gain(s,2) = gamas*(sigmaes*n2(s-1,1)-sigmaas*(n0-
n2(s-1,1)));
  end
        n2(1,2)    =    n2(1,1)+deltat*((gamap*lambdap/
(h*c0*Aco))*(sigmaap*(n0-n2(1,1))...
          -sigmaep*n2(1,1))*ppf(1,1)+(gamas*lambdas/
(h*c0*Aco))*(sigmaas* (n0-n2(1,1))...
    -sigmaes*n2(1,1))*psf(1,1)-n2(1,1)/tao);
  n1(1,2) = n0-n2(1,1);
  % Boundary conditions)
  %=====================
  t = t + deltat;
  if (t < ti)
    R = 0;
  elseif (t > ti && t < ti+tr)
    R = (Rmax/tr)*(t-ti);
  elseif (t > (ti+tr) && t < (ti+tr+to)) || (t > per+ti+tr &&
t < per+ti+tr+to)...
       || (t > 2*per+ti+tr && t < 2*per+ti+tr+to) ||...
       (t > 3*per+ti+tr && t < 3*per+ti+tr+to)
    R = Rmax;
  elseif (t > ti+per && t < ti+per+tr)
    R = (Rmax/tr)*(t-(ti+per));
  elseif (t > ti+2*per && t < ti+2*per+tr)
    R = (Rmax/tr)*(t-(ti+2*per));
  elseif (t > ti+3*per && t < ti+3*per+tr)
    R = (Rmax/tr)*(t-(ti+3*per));
  else
    R = 0;
  end
  psf(1,2) = psf(step_number+1,1)*R*total_loss;
  for s = 1:step_number+1;
    n2(s,1) = n2(s,2);
    n1(s,1) = n1(s,2);
    ppf(s,1) = ppf(s,2);
    psf(s,1) = psf(s,2);
    gain(s,1) = gain(s,2);
    xc(s) = (s-1)*deltax;
  end
  N2(k) = n2(step_number+1);
  N1(k) = n1(step_number+1);
  Psf(k) = psf(step_number+1)*(1-R);
```

```
  Ppf(k) = ppf(step_number+1);
  Gain(k) = gain(step_number+1)*5;
  z(k) = k;
  time(k) = t*1e4;
  reflectivity(k) = R;
end
figure
subplot(2,2,1)
plot(time,Psf,'Linewidth',2);
xlabel('Time (Microseconds)')
ylabel('Pulse Power (W)')
subplot(2,2,2)
plot(time,Gain,'r','Linewidth',2)
xlabel('Time (Microseconds)')
ylabel('Gain (m^-1)')
subplot(2,2,3)
plot(time,N2,'m','Linewidth',2).
hold on
plot(time,N1,'k','Linewidth',2)
xlabel('Time (Microseconds)')
ylabel('N1,N2 (ions/m^3)')
subplot(2,2,4)
plot(time,Ppf,'g','Linewidth',2)
xlabel('Time (Microseconds)')
ylabel('Pump Power (W)')
```

6.10.4 LONG CAVITY ACTIVE Q-SWITCHED FUNCTION

```
function [N1,N2,Psf,Ppf,Gain,time] =…
                                LongCavityActiveQswitch
(Pump_Power,UndopedFL,Rise_Time,FiberLength,Core_D,Ner)
%=====================================================
============================
%Part I: Provide initial values for the variables in
simulation
%=========================================
================================
% Function LongCavityActiveQswitch implements a finite dif-
ference method to compute
% output characteristics of an Erbium doped active Q-
switched fiber laser
```

```
% using an acousto-optic modulator as switching element.
This function
% include a length of undoped fiber in the cavity
% Inputs:
% - Pump_Power: Input pump power
% - UndopedFL: Length of undoped fiber
% - Rise_Time: Rise time of the acousto-optic modulator
% - FiberLength:Length of the Erbium doped fiber
% - Core_D: Core diameter of the doped fiber
% - Ner: Erbium doped fiber concentration
%
%
% Outputs:
%  - N1: Ground level population density variation with time
%  - N2: excited state population density variation with time
%  - Gain variation as a function of time
%  - Gain: Gain variation as a function of time
%  - Ppf: Residual pump
%  - Psf: Q-switched fiber laser output power
%  - time: computation time
% Comments:
%  - Erbium doped gain medium is considered as a two level
system and the
%  resulting rate equations are solved with simple analy-
tical method using
%  substitution
%  -Initial conditions of the finite difference scheme are
obtained by
%  solving the rate equations first in steady state with a
shooting
%  algorithm
%
% References:
%
% Written by Justice Sompo, University of Johannesburg,
South Africa
%
%
%
close all
clc
% Check inputs
if nargin < 6
  Ner = 4e24;
  if nargin < 5
    Core_D = 2.3e-6;
```

```
      if nargin < 4
         FiberLength = 3;
         if nargin < 3
            Rise_Time = 40e-9;
            if nargin < 2
               UndopedFL = 12;
               if nargin < 1
                                                     error(message
('MATLAB:LongCavityActiveQswitch:NotEnoughInputs'));
               end
            end
         end
      end
   end
end
% Values for the variables in simulation
%==================================================
============================
lambdas = 1550e-9;      % Signal wavelength
sigmaas = 3.15e-25;     % Absorption cross section at signal
wavelength
sigmaes = 4.44e-25;     % Emission cross section at signal
wavelength
NA = 0.2;
dcore = Core_D;         % Fiber core diameter
L = FiberLength;        % Fiber length
L1 = UndopedFL;         % Length of undoped fiber
alfap = 0.005;              % attenuation coefficient at pump
wavelength
alfas = 0.005;              % Attenuation coefficient at signal
wavelength
n0 = Ner;               % Doping concentration of the fiber
c0 = 3e8;               % Speed of light in vacuum
gamas = 0.7;
h = 6.626e-34;          % Planck constant
nfiber = 1.45;          % Refractive index of the fiber
vg = c0/nfiber;         % Group velocity in doped fiber
step_number = 50;
deltax = L/step_number;               % Longitudianl section in
doped fiber
deltat = L/(step_number*vg);    % Sections in the time domain
step_number_1 = 100;
deltax1 = L1/step_number_1;
deltat1 = L1/(step_number*vg);
deltalamda = 3e-6;
```

```
Aco = pi*(dcore/2)^2;      % Area of the doped fiber core
lambdap = 980e-9;          % Pump wavelength
gamap = 0.83;
sigmaep = 1.8e-25;         % Absorption cross section at pump
wavelength
sigmaap = 2.44e-24;        % Emission cross section at pump
wavelength
tao = 10e-3;               % Erbium excited metastable state
lifetime
Rmax = 0.9;
% Adding this to generate initial condition
leng = L;
coupler = Rmax;
Nt = n0;
sections = step_number+1;
% Compute steady state values distribution
pumppower = Pump_Power;    % Pump power in watts
Pump = pumppower;
[PP,Ps,N2x]                        =                 LaserOutput
(Pump,leng,coupler,Nt,sections);
% Initial conditions for doped fiber
%-----------------------------------------------
----------------------------
s = 1:step_number+1;
n2(s,1) = N2x;
n1(s,1) = n0-N2x;
ppf(s,1) = PP;
psf(s,1) = Ps;
gain(s,1) = 0.0;
xc(s,1) = 0.0;
%Initial conditions for undoped fiber
%--------------------
----------------------
----------------------------
----
s1 = 1:step_number_1+1;
ppfun(s1,1) = 0;
psfun(s1,1) = 0;
%=========================================
================================
%Part II: Iterative part
%=========================================
================================
ppf(1,1) = pumppower;
temps = deltat*400000;
Time = deltat:deltat:temps;
```

```
t = 0;  % very important to be used in the boundary conditions
function
tr = Rise_Time;
to = 5e-6;
longueur = 0;
for k = 1:length(Time)
  ppf(1,2) = pumppower;
  for s = 2:step_number+1
            n2(s,2)  =  n2(s,1)+deltat*((gamap*lambdap/
(h*c0*Aco))*(sigmaap*(n0-n2(s,1))...
              -sigmaep*n2(s,1))*ppf(s,1)+(gamas*lambdas/
(h*c0*Aco))*(sigmaas*(n0-n2(s,1))...
      -sigmaes*n2(s,1))*psf(s,1)-n2(s,1)/tao);
      ppf(s,2) = ppf(s-1,1)+deltax*(gamap*(sigmaep*n2(s-
1,1)...
            -sigmaap*(n0-n2(s-1,1)))*ppf(s-1,1)-alfap*ppf
(s-1,1));
      psf(s,2) = psf(s-1,1)+deltax*(gamas*(sigmaes*n2(s-
1,1)...
      -sigmaas*(n0-n2(s-1,1)))*psf(s-1,1)...
        +2*gamas*sigmaes*n2(s-1,1)*h*c0*c0* deltalamda/
(lambdas)^3)-alfas*psf(s-1,1);
    n1(s,2) = n0-n2(s,1);
      gain(s,2) = gamas*(sigmaes*n2(s-1,1)-sigmaas*(n0-
n2(s-1,1)));
  end
          n2(1,2)   =    n2(1,1)+deltat*((gamap*lambdap/
(h*c0*Aco))*(sigmaap*(n0-n2(1,1))...
              -sigmaep*n2(1,1))*ppf(1,1)+(gamas*lambdas/
(h*c0*Aco))*(sigmaas* (n0-n2(1,1))...
    -sigmaes*n2(1,1))*psf(1,1)-n2(1,1)/tao);
  n1(1,2) = n0-n2(1,1);
  ppfun(1,1) = ppf(step_number+1,2);
  psfun(1,1) = psf(step_number+1,2);
  for s1 = 2:step_number_1+1
    longueur = longueur+deltax1;
      psfun(s1,2) = psfun(s1-1,1)+deltax1*(-alfas*psfun
(s1-1,1));
  end
  % Boundary conditions)
  %====================
  t = t + deltat;
  R = feval('boundary',Rmax,tr,to,t);
  psf(1,2) = psfun(step_number_1+1,2)*R;
  % Updating initial conditions
```

```
%==============================
for s1 = 1:step_number_1+1
   psfun(s1,1) = psfun(s1,2);
end
for s = 1:step_number+1;
   n2(s,1) = n2(s,2);
   n1(s,1) = n1(s,2);
   ppf(s,1) = ppf(s,2);
   psf(s,1) = psf(s,2);
   gain(s,1) = gain(s,2);
   xc(s) = (s-1)*deltax;
end
N2(k) = n2(step_number+1);
N1(k) = n1(step_number+1);
Psf(k) = psf(step_number+1)*(1-R);
Ppf(k) = ppf(step_number+1);
Gain(k) = gain(step_number+1);
z(k) = k;
time(k) = t*1e4;
reflectivity(k) = R;
end
figure
subplot(2,2,1)
plot(time,Psf,'Linewidth',2);
xlabel('Time (Microseconds)')
ylabel('Pulse Power (W)')
subplot(2,2,2)
plot(time,Gain,'r','Linewidth',2)
xlabel('Time (Microseconds)')
ylabel('Gain (m^-1)')
subplot(2,2,3)
plot(time,N2,'m','Linewidth',2)
xlabel('Time (Microseconds)')
ylabel('N2 (ions/m^3)')
subplot(2,2,4)
plot(time,N1,'k','Linewidth',2)
xlabel('Time (Microseconds)')
ylabel('N1 (ions/m^3)')
toc
```

6.10.5 LASER OUTPUT FUNCTION

```
function [PP,Ps,N2,N1,N,t] = LaserOutput(Pump,leng,
coupler,Nt,sections)
% This function solves the rate equations and
propagation
% equations for a ring cavity fiber laser using the MATLAB
ODE 45 routine
% and a shooting algorithm
%
%
% Inputs:
% -Pump: Pump power
% -leng: Length of the doped fiber
% -coupler: Coupling ratio of the output coupler
%
% Outputs:
% -PP : Pump power distribition
% -Ps: Laser power
% -N2: Population density at the metastable level
% -N1: Population density at ground level
% -N : Total population density of the doped fiber
%
% Comments:
% - Computation is done in the ideal case of a two level
energy system
% equations analytically.
% References:
%
%
% Written by Justice Sompo, University of Johannesburg,
South Africa
close all
clc
format long
% Check inputs
if nargin < 5
  sections = 200;
  if nargin <4
    %   Nt = 1.26e25;
    Nt = 8e24;
    if nargin < 3
      coupler = 0.1;
      if nargin < 2
        leng = 5;
        if nargin < 1
```

```
                                          error(message
('MATLAB:LaserOutput:NotEnoughInputs'));
        end
      end
    end
  end
end
%=============================================
==============================
%Rate equations parameters
%==========================
====================================================
sigma12S = 3.15e-25;
sigma21S = 4.44e-25;
sigma12P = 1.8e-25;
% Lifetimes
t21 = 10e-3;
A21 = 1/t21;
% Concentrations
Ner = Nt;
% Overlap factors and background losses
Gammap = 0.83;
Gammas = 0.7;
lambdap = 980e-9; % Pumb wavelength
lambdas = 1550e-9; %signal wavelength
% Other physical constants
r = 2.3e-6;
h = 6.626e-34;
cel = 3e8;
% Calculated parameters
A = pi*r^2;
F_s = cel/lambdas;
den1 = A*h*F_s;
F_p = cel/lambdap;
den2 = A*h*F_p;
%===================================================
============================
% Implementation of the shooting secant algorithm to
solve propagation
% equations in steady state
%======================================
====================================
epsilon = 1e-6; % tolerance for computation
count = 0; % iniltialise iterations before the while loop
RL = coupler;  % reflectivity of mirror at L.
Pp = Pump;
% Default values of initial guess
% x1 = 10e-3; % first signal power guess
```

```
% x2 = 100e-3; % second signal power guess
x1 = 10e-3; % first signal power guess
x2 = 100e-3; % second signal power guess
L = leng;
%tspan = [0 6];
tspan = linspace(0,L,sections); % to obtain solutions at
each position for 100 positions one can write linspace
(0,L,100)
[t1,y1]=ode45(@propa,tspan,[Pp x1]); %y1 will have the
values of Pp in its first column and the values of Ps in its
second column.
[t2,y2]=ode45(@propa,tspan,[Pp x2]);
i = 1;
Psl1=y1(end,2); % Ps1 power at z = L
Psl2=y2(end,2); % Ps2 power at z = L
P01 = Psl1*RL;
P02 = Psl2*RL;
m1 = P01-x1;
m2 = P02-x2;
while (abs(x2-x1) > epsilon)
  tmp=x2;
  count = count+1;
  x2 = x1-(x1-x2)/(m1-m2)*m1;
  x1 = tmp;
  [t1,y1]=ode45(@propa,tspan,[Pp x1]);
  [t2,y2]=ode45(@propa,tspan,[Pp x2]);
  Psl1=y1(end,2); % Ps power at z = L
  Psl2=y2(end,2); % Ps power at z = L
  P01 = Psl1*RL;
  P02 = Psl2*RL;
  m1 = (P01-x1);
  m2 = (P02-x2);
  i = i+1;
  t = t2;
  PP = y2(:,1); % Pump power distribution
  Ps = y2(:,2); % Signal power distribution
  %Population distribution
  W12 = ((sigma12S*Gammas)*Ps)/den1;
  W21 = ((sigma21S*Gammas)*Ps)/den1;
  R12 = ((sigma12P*Gammap)*PP)/den2;
  N1 = Ner*(W21+A21)./(W12+R12+W21+A21);
  N2 = Ner*(W12+R12)./(W21+A21+W12+R12);
  N = N2./Ner;
end
%==========================================================
=========================
```

```
% Plotting
%
=======================================================
=======================
% figure(1)
% plot(t,N2,'m','linewidth',2)
% hold on
% plot(t,N1,'g','linewidth',2)
% figure(2)
% plot(t,Ps,'r-','linewidth',2)
% hold on
% plot(t,PP,'linewidth',2)
```

6.10.6 PROPA FUNCTION

```
function dpdz = propa(~,p)
% PROPA function represents the system of pump and laser
propagation equations
%
% Inputs:
% -The function takes the pump power and initial guesses of
output powers
%
% Outputs:
%   dpdz(1): Represents the pump propagating power
%   dpdz(2): Represents the laser propagating power
%
% Comments:
%   - Computation is done in the ideal case of a two level
energy system
%
% References:
%
% Written by Justice Sompo, University of Johannesburg,
South Africa
close all
clc
%
format longe
% Rate equation Parameters
sigma12S = 3.15e-25;   % Laser absorption cross section
```

```
sigma21S = 4.44e-25;  % Laser emission cross section
sigma12P = 1.8e-25;   % Pump absorption cross section
sigma21P = 0;         % Pump emission cross section
% Lifetimes
t21 = 10e-3;       % Lifetime of the metastable energy level
of Erbium
A21 = 1/t21;
% Concentrations
%Ner = 1.2e25;        % Erbium ions concentration
Ner = 8e24;
% Overlap factors and background losses
Gammap = 0.83;       % Overlap factor at pump wavelength
Gammas = 0.7;        % Overlap factor at signal wavelength
alphap = 0.005;      % Background loss at pump wavelength
alphas = 0.005;      % Background loss at laser wavelength
lambdap = 980e-9; % Pumb wavelenght
lambdas = 1550e-9; %signal wavelength
% Other physical constants
r = 2.3e-6;        % Doped fiber core
h = 6.626e-34;       % Planck's constant
cel = 3e8;         % speed of light in free space
% Calculated parameters
A = pi*r^2;
F_s = cel/lambdas;den1 = A*h*F_s;
F_p = cel/lambdap; den2 = A*h*F_p;
% Fiber Bragg Grating Parameters
W12 = ((sigma12S*Gammas)*(p(2)))/den1;
W21 = ((sigma21S*Gammas)*(p(2)))/den1;
R12 = ((sigma12P*Gammap)*(p(1)))/den2;
%Rate equations solving
N1  = Ner*(W21+A21)./(W12+R12+W21+A21); %  Population
density of the ground energy level
N2  = Ner*(W12+R12)./(W21+A21+W12+R12); %  Population
density of themetastable energy level
dpdz = zeros(2,1);
dpdz(1)=  Gammap*(sigma21P*N2-sigma12P*N1)*p(1)-alph-
ap*p(1);
dpdz(2)=  Gammas*(sigma21S*N2-sigma12S*N1)*p(2)-alph-
as*p(2);
```

6.10.7 BOUNDARY FUNCTION

```
function R2 = boundary(Rmax,tr,to,t)
% This function calculate the time varying boundary
condition for fiber
% laser BVP problem
% Rmax is the maximum reflectivity of the outpout mirror,
% tr is the rise time of the AOM switching function
% to is the open time of the AOM switching function
tstart = 100e-6; % time to switch on the acousto optic
modulator
tstart_palier = tstart+tr;
%t = t+deltat;
if (t < tstart)
  R2 = 0;
elseif (t > tstart)&&(t < tstart_palier)  % this corre-
sponds to the rise time of the modulator
  R2 = (Rmax/tr)*(t-tstart);
elseif (t > tstart_palier)&& (t <= tstart_palier+to)
  R2 = Rmax;
else
  R2 = 0;
end
```

6.10.8 BOUNDARY2 FUNCTION

```
function R2 = boundary2(Rmax,tr,to,t)
% This function calculate the time varying boundary
condition for fiber
% laser BVP problem
% Rmax is the maximum reflectivity of the outpout mirror,
% tr is the rise time of the AOM switching function
% to is the open time of the AOM switching function
tstart = 100e-6; % time to switch on the acousto optic
modulator
tstart_palier = tstart+tr;
%t = t+deltat;
if (t < tstart)
  R2 = 0;
elseif (t > tstart)&&(t < tstart_palier)  % this corre-
sponds to the rise time of the modulator
  R2 = 0.5*(1+sin(pi*((t-tstart)-0.5*tr))/tr);
```

```
elseif (t > tstart_palier)&& (t <= tstart_palier+to)
  R2 = Rmax;
else
  R2 = 0;
end
```

6.10.9 PASSIVE Q-SWITCHED FIBER LASER

```
function    [Psf,time,N2,Nsa2]    =    PassiveQswitch
(Pump,Leng,Nt)
% Function PassiveQswitch implementes a finite differ-
ence method to compute
% output characteristics of an Ytterbiium doped passive
Q-switched fiber laser
% using Cr4+:YAG saturable absorber
%
% Inputs:
% - Pump: Input pump power
% - Leng:Length of the Ytterbium doped fiber
% - Nt: doped fiber concentration
%
%
% Outputs:
%
%   - N2: excited state population density variation
with time
%   - Nsa2 : excited state
%   - Psf: Q-switched fiber laser output power
%   - time: computation time
% Comments:
%   - Ytterbium doped gain medium is a two level system
%
% References:
%
% Written by Justice Sompo, University of Johannesburg,
South Africa
%
%
%
close all
clc
```

```matlab
% Check inputs
if nargin < 3
  Nt = 9e25;
  if nargin < 2
    Leng = 5;
    if nargin < 1
                                                        error(message
('MATLAB:PassiveQswitch:NotEnoughInputs'));
    end
  end
end
lambdas = 1064e-9;            % Laser signal waveength
sigmaas = 1.16e-27;          % Absorption cross section at
signal wavelength
sigmaes = 2.34e-25;           % Emission cross section at
signal wavelength
sigmagsa = 3.74e-22;         % Absorption cross section of
the saturable absorber
sigmaesa = 0.935e-22;         % Emission cross section of
the saturable absorber
NA = 0.15;
dcore = 5.4e-6;              % fiber core diameter
dclad = 125e-6;              % Fiber cladding diameter
Step_number = 50;            % Number of section
L = Leng;                   % Length of the doped fiber
R = 0.04;
alfap = 0.005;                % Background loss at pump
wavelength
alfas = 0.005;                % Background loss at laser
wavelength
alfasa = 20;               % Background loss of the saturable
absorber
nfiber = 1.45;               % Refractive index of the fiber
n0 = Nt;                    % Concentration of doped fiber
c0 = 3e8;
V = 2*pi*dcore*NA/(2*lambdas);
w = (dcore/2)*(0.65+1.619*V^(-1.5)+2.879*V^(-6));
gamas = 1-exp(-(2*dcore^2/(4*w^2)));
wsa0 = 7.2e-6;                % Focused beam waist in SA
nsa = 6.22e23;
nSA = 1.8;
taosa = 1e-6;
lsa = 0.0025;
Step_number1 = 25;
z0 = pi*wsa0*wsa0*nSA/(lambdas);
```

```
%============================================
===================================
q = 1:Step_number1+1;
z(q) = (lsa/(Step_number1-1))*(q-(2+Step_number1)/2);
wsa(q) = wsa0*sqrt(1+(z(q)/z0).^2);
Asa(q) = pi*wsa(q).^2;
%============================================
===================================
h = 6.626e-34;              %Planck constant
vg = c0/nfiber;
deltax1 = lsa/Step_number1;
deltax = L/Step_number;
deltat = L/(Step_number*vg);
deltalamda = 3e-6;
Aco = pi*(dcore/2)^2;       % Doped fiber core diameter
lamdap = 976e-9;            % Pump wavelength
gamap = dcore^2/(dclad^2);  % Overlap factor between the
pump and doped area
sigmaep = 2.44e-24;         % Emission cross section at pump
power
sigmaap = 2.5e-24;          % Absorption cross section at
pump power
tao = 850e-6;
%============================================
===================================
% Initial conditions of the saturable absorber
ssa = 1:Step_number1+1;
nsa2(ssa,1) = 100;
psfSA(ssa,1) = 0;
psbSA(ssa,1) = 0;
%============================================
===================================
% Initial condition for the Ytterbum doped fiber
s = 1:Step_number+1;
n2(s,1) = 100;
ppb(s,1) = 0.0;
psf(s,1) = 0.0;
psb(s,1) = 0.0;
%============================================
===============================
Pp = Pump;             %(Pump power in watt)
temps = deltat*450000;  % Simulation time value can be
choosen arbitrary
Time = deltat:deltat:temps;
ppb(Step_number,1) = Pp;
```

```
t = 0;
for ii = 1:length(Time)
  ppb(Step_number+1,2) = Pp;
  for s = 1:Step_number
            n2(s,2)  = n2(s,1)+deltat*((gamap*lamdap/
(h*c0*Aco))*(sigmaap*(n0-n2(s,1))…
           -sigmaep*n2(s,1))*ppb(s,1)+(gamas*lambdas/
(h*c0*Aco))*(sigmaas*(n0-n2(s,1))…
              -sigmaes*n2(s,1))*(psf(s,1)+psb(s,1))-
n2(s,1)/tao);
      ppb(s,2) = ppb(s+1,1)+deltax*(gamap*(sigmaep*n2(s
+1,1)…
        -sigmaap*(n0-n2(s+1,1)))*ppb(s+1,1)-alfap*ppb
(s+1,1));
     psb(s,2) = psb(s+1,1)+deltax*(gamas*(sigmaes*n2(s
+1,1)…
                  -sigmaas*(n0-n2(s+1,1)))*psb(s+1,1)
+2*gamas*sigmaes*n2(s+1,1)…
           *h*c0*c0*deltalamda/(lambdas)^3-alfas*psb
(s+1,1));
                          psf(s+1,2)    =    psf(s,1)
+deltax*(gamas*(sigmaes*n2(s,1)-sigmaas*(n0-
n2(s,1)))…
                                    *psf(s,1)
+2*gamas*sigmaes*n2(s,1)*h*c0*c0*deltalamda/
(lambdas)^3-alfas*psf(s,1));
  end
    n2(Step_number+1,2)  = n2(Step_number+1,1)+del-
tat*((gamap*lamdap/(h*c0*Aco))…
               *(sigmaap*(n0-n2(Step_number+1,1))-
sigmaep*n2(Step_number+1,1))*ppb(Step_number+1,1)…
          +(gamas*lambdas/(h*c0*Aco))*(sigmaas*(n0-
n2(Step_number+1,1))…
      -sigmaes*n2(Step_number+1,1))*(psf(Step_number
+1,1)+psb(Step_number+1,1))…
    -n2(Step_number+1,1)/tao);
  for s1=1:Step_number1
      nsa2(s1,2) = nsa2(s1,1)+deltat*(((psfSA(s1,1)
+psbSA(s1,1))*sigmagsa*lambdas)…
        *(nsa-nsa2(s1,1))/(Asa(s1)*h*c0)-nsa2(s1,1)/
taosa);
                      psbSA(s1,2)    =    psbSA(s1+1,1)-
deltax1*(sigmaesa*nsa2(s1+1,1)…
                        +sigmagsa*(nsa-nsa2(s1+1,1))-
alfasa)*psbSA(s1+1,1);
```

```
                    psfSA(s1+1,2)    =    psfSA(s1,1)-
deltax1*(sigmaesa*nsa2(s1,1)…
      +sigmagsa*(nsa-nsa2(s1,1))-alfasa)*psfSA(s1,1);
  end
    nsa2(Step_number1+1,2)  = nsa2(Step_number1+1,1)
+deltat*(((psfSA(Step_number1+1,1)…
    +psbSA(Step_number1+1,1))*sigmagsa*lambdas)*(nsa-
nsa2(Step_number1+1,1))/(Asa(Step_number1+1)…
    *h*c0)-nsa2(Step_number1+1,1)/taosa);
  psbSA(Step_number1+1,2) = 0.85*psb(1,1);
  psfSA(1,2) = psbSA(1,1);
  psf(1,2) = 0.85*psfSA(Step_number1+1,1);
  psb(Step_number+1,2) = psf(Step_number+1,1)*R*0.85;
  for s1 = 1:Step_number1+1
    nsa2(s1,1) = nsa2(s1,2);
    psbSA(s1,1) = psbSA(s1,2);
    psfSA(s1,1) = psfSA(s1,2);
  end
  for s = 1:Step_number+1
    n2(s,1) = n2(s,2);
    ppb(s,1) = ppb(s,2);
    psb(s,1) = psb(s,2);
    psf(s,1) = psf(s,2);
  end
  t = t + deltat;
  Psf(ii) = psf(Step_number+1);
  Psb(ii) = psb(Step_number+1);
  Ppb(ii) = ppb(Step_number+1);
  time(ii) = t;
  N2(ii) = n2(Step_number+1);
  Nsa2(ii) = nsa2(Step_number1+1);
end
```

6.10.10 Script to Plot the Dynamics of the Output Characteristics of Passive Q-switched Fiber Laser

```
clear all
close all
clc
tic
```

```
% This script plots differents values of the out put of
the function
% PassiveQswitch
Pump1 = 2.5; % Pump power in watts
Pump2 = 10;
[Psp1,time1,N21,Nsa21] = PassiveQswitch(Pump1);
[Psp2,time2,N22,Nsa22] = PassiveQswitch(Pump2);
% The plottings
figure(1)
plot(time1,Psp1,'Linewidth',2);
hold on
plot(time1,Psp2,'r','Linewidth',2);
figure(2)
plot(time1,N21,'Linewidth',2);
hold on
plot(time1,N22,'r','Linewidth',2);
figure(3)
plot(time1,Nsa21,'Linewidth',2);
hold on
plot(time1,Nsa22,'r','Linewidth',2);
toc
```

REFERENCES

Adachi, Shoji and Yahei Koyamada. 2002. "Analysis and Design of Q-switched Erbium-Doped Fiber Lasers and Their Application to OTDR." *Journal of Lightwave Technology* 20 (8): 1506–1511. 10.1109/JLT.2002.800293.

Adel, Panssee, M. Auerbach, Carsten Fallnich, S. Unger, H.R. Müller, and J. Kirchhof. 2003. "Passive Q-Switching by Tm3+ Co-Doping of a Yb3+-Fiber Laser." *Optics Express* 11 (November): 2730–2735. 10.1364/OE.11.002730.

Alvarez-Chavez, J.A., H.L. Offerhaus, J. Nilsson, P.W. Turner, W.A. Clarkson, and D.J. Richardson. 2000. "High-Energy, High-Power Ytterbium-Doped Q-switched Fiber Laser." *Optics Letters* 25 (1): 37. 10.1364/ol.25.000037.

Álvarez-Tamayo, R. Iván, Manuel Durán-Sánchez, Olivier Pottiez, Baldemar Ibarra-Escamilla, Evgeny A. Kuzin, and M. Espinosa-Martínez. 2016. "Active Q-switched Fiber Lasers with Single and Dualwavelength Operation." *Fiber Laser*, no. March. 10.5772/61571.

Andersen, T.V., P. Pérez-Millán, S.R. Keiding, S. Agger, R. Duchowicz, and M.V. Andrés. 2006. "All-Fiber Actively Q-switched Yb-Doped Laser." *Optics Communications* 260 (1): 251–256. 10.1016/j.optcom.2005.10.036.

Andrés, M.V., J.L. Cruz, A. Díez, P. Pérez-Millán, and M. Delgado-Pinar. 2008. "Actively Q-switched All-Fiber Lasers." *Laser Physics Letters* 5 (2): 93–99. 10.1002/lapl. 200710104.

åWIDERSKI, J., A. Zaj, Masilan Karunanithi, and P. Konieczny. 2005. "Q-switched Nd-Doped Double-Clad Fiber Laser." *Opto-Electronics Review* 13 (January).

Barmenkov, Yuri O., Alexander V. Kirryanov, Jose L. Cruz, and Miguel V. Andres. 2014. "Pulsed Regimes of Erbium-Doped Fiber Laser Q-switched Using Acousto-Optical Modulator." *IEEE Journal of Selected Topics in Quantum Electronics* 20 (5). 10.1109/JSTQE.2014.2304423.

Boni, Leonardo, Erin Wood, Carlos Toro, and Florencio Hernandez. 2008. "Optical Saturable Absorption in Gold Nanoparticles." *Plasmonics*, October.

Chakravarty, Usha, P.K. Mukhopadhyay, A. Kuruvilla, B.N. Upadhyaya, and K.S. Bindra. 2017. "Narrow-Linewidth Broadly Tunable Yb-Doped Q-switched Fiber Laser Using Multimode Interference Filter." *Applied Optics* 56 (13): 3783. 10.1364/ao.56.003783.

Chandonnet, Alain and Gilles Larose. 1993. "High-Power Q-switched Erbium Fiber Laser Using an All-Fiber Intensity Modulator." *Optical Engineering* 32 (9): 2031. 10.1117/12.143949.

Chen, Bohua, Hao Wang, X. Zhang, J. Wang, K. Wu, and J. Chen. 2015. "Q-switched Ring-Cavity Erbium-Doped Fiber Laser Based on Tungsten Disulfide (WS2)." In *2015 Opto-Electronics and Communications Conference (OECC)*, 1–3. 10.1109/OECC.2015.7340169.

Chen, Y.-C., N.R. Raravikar, L.S. Schadler, P.M. Ajayan, Y.-P. Zhao, T.-M. Lu, G.-C. Wang, and X.-C. Zhang. 2002. "Ultrafast Optical Switching Properties of Single-Wall Carbon Nanotube Polymer Composites at 1.55 Mm." *Applied Physics Letters* 81 (6): 975–977. 10.1063/1.1498007.

Chen, Y., C. Zhao, H. Huang, S. Chen, P. Tang, Z. Wang, S. Lu, H. Zhang, S. Wen, and D. Tang. 2013. "Self-Assembled Topological Insulator: $Bi_{2}Se_{3}$ Membrane as a Passive Q-Switcher in an Erbium-Doped Fiber Laser." *Journal of Lightwave Technology* 31 (17): 2857–2863. 10.1109/JLT.2013.2273493.

Chen, Yu, Guobao Jiang, Shuqing Chen, Zhinan Guo, Xuefeng Yu, Chujun Zhao, and Han Zhang, et al. 2015. "Mechanically Exfoliated Black Phosphorus as a New Saturable Absorber for Both Q-Switching and Mode-Locking Laser Operation." *Optics Express* 23 (10): 12823–12833. 10.1364/OE.23.012823.

Chen, Yung-Fu, T.M. Huang, and C.L. Wang. 1997. "Passively Q-switched Diode-Pumped Nd: YVO4/Cr4+: YAG Single-Frequency Microchip Laser." *Electronics Letters* 33 (22): 1880–1881.

Cheng, X.P., C.H. Tse, P. Shum, R.F. Wu, M. Tang, W.C. Tan, and J. Zhang. 2008. "All-Fiber Q-switched Erbium-Doped Fiber Ring Laser Using Phase-Shifted Fiber Bragg Grating" 26 (8): 945–951.

Cheng, X.P., J. Zhang, P. Shum, M. Tang, and R.F. Wu. 2007. "Influence of Sidelobes on Fiber-Bragg-Grating-Based Q-switched Fiber Laser." *IEEE Photonics Technology Letters* 19 (20): 1646–1648. 10.1109/LPT.2007.904920.

Cuadrado-Laborde, C., M. Delgado-Pinar, S. Torres-Peiró, A. Díez, and M.V. Andrés. 2007. "Q-switched All-Fiber Laser Using a Fiber-Optic Resonant Acousto-Optic Modulator." *Optics Communications* 274 (2): 407–411. 10.1016/j.optcom.2007.02.032.

Dawlaty, Jahan M, Shriram Shivaraman, Jared Strait, Paul George, Mvs Chandrashekhar, Farhan Rana, Michael G Spencer, Dmitry Veksler, and Yunqing Chen. 2008. "Measurement of the Optical Absorption Spectra of Epitaxial Graphene from Terahertz to Visible." *Applied Physics Letters* 93 (13): 131905. 10.1063/1.2990753.

Delgado-Pinar, M., D. Zalvidea, A. Díez, P. Perez-Millan, and M. Andres. 2006. "Q-Switching of an All-Fiber Laser by Acousto-Optic Modulation of a Fiber Bragg Grating." *Optics Express* 14 (3): 1106. 10.1364/OE.14.001106.

Eilers, Hergen, Uwe Hömmerich, Stuart M Jacobsen, William M Yen, K.R. Hoffman, and W. Jia. 1994. "Spectroscopy and Dynamics of Cr 4+: Y 3 Al 5 O 12." *Physical Review B* 49 (22): 15505.

El-Sherif, Ashraf F. and Terence A. King. 2003. "High-Energy, High-Brightness Q-switched Tm3+-Doped Fiber Laser Using an Electro-Optic Modulator." *Optics Communications* 218 (4–6): 337–344. 10.1016/S0030-4018(03)01200-8.

Feldman, R., Y. Shimony, and Z. Burshtein. 2003. "Passive Q-Switching in Nd:YAG/ Cr4+:YAG Monolithic Microchip Laser." *Optical Materials* 24 (1): 393–399. 10.1016/ S0925-3467(03)00153-8.

Filippov, Valery, A. Kir'yanov, and S. Unger. 2004. "Advanced Configuration of Erbium Fiber Passively Q-switched Laser with Co2+:ZnSe Crystal as Saturable Absorber." *Photonics Technology Letters, IEEE* 16 (February): 57–59. 10.1109/LPT.2003.819397.

Fluck, R., R. Häring, R. Paschotta, E. Gini, H. Melchior, and U. Keller. 1998. "Eyesafe Pulsed Microchip Laser Using Semiconductor Saturable Absorber Mirrors." *Applied Physics Letters* 72 (25): 3273–3275. 10.1063/1.121621.

Fluck, R., U. Keller, E. Gini, and H.E.D. Bosenberg Melchior W., and M. Fejer 1998. "Eyesafe Pulsed Microchip Laser." In *Advanced Solid State Lasers*, 19:LS1. OSA Trends in Optics and Photonics Series. Coeur d'Alene, Idaho: Optical Society of America. 10.1364/ASSL.1998.LS1.

Gao, Yachen, Xueru Zhang, Yuliang Li, Hanfan Liu, Yuxiao Wang, Qing Chang, Weiyan Jiao, and Yinglin Song. 2005. "Saturable Absorption and Reverse Saturable Absorption in Platinum Nanoparticles." *Optics Communications* 251 (4): 429–433. 10.1016/j.optcom.2005.03.003.

Geng, Jihong, Qing Wang, Jake Smith, Tao Luo, Farzin Amzajerdian, and Shibin Jiang. 2009. "All-Fiber Q-switched Single-Frequency Tm-Doped Laser near 2?M." *Optics Letters* 34 (23): 3713–3715. 10.1364/OL.34.003713.

González-García, A., B. Ibarra-Escamilla, E.A. Kuzin, M. Durán-Sánchez, O. Pottiez, and F. Maya-Ordoñez. 2013. "Tunable Actively Q-switched Fiber Laser Based on Fiber Bragg Grating." In *Proc. SPIE*. Vol. 8601. 10.1117/12.2002854.

González García, Andrés, Baldemar Ibarra-Escamilla, E. Kuzin, M. Durán-Sánchez, Olivier Pottiez, and F. Maya-Ordoñez. 2013. "Tunable Actively Q-switched Fiber Laser Based on Fiber Bragg Grating." *Proceedings of SPIE - The International Society for Optical Engineering* 8601 (February). 10.1117/12.2002854.

González García, Andrés, Baldemar Ibarra-Escamilla, Olivier Pottiez, E. Kuzin, F. Maya-Ordoñez, and M. Durán-Sánchez. 2015. "Compact Wavelength-Tunable Actively Q-switched Fiber Laser in CW and Pulsed Operation Based on a Fiber Bragg Grating." *Laser Physics* 25 (April): 45104. 10.1088/1054-660X/25/4/045104.

Gurudas, Ullas, Elijah Brooks, Daniel M Bubb, Sebastian Heiroth, Thomas Lippert, and Alexander Wokaun. 2008. "Saturable and Reverse Saturable Absorption in Silver Nanodots at 532 Nm Using Picosecond Laser Pulses." *Journal of Applied Physics* 104 (7): 73107. 10.1063/1.2990056.

Haitao, Huang, Min Li, Li Wang, Xuan Liu, Deyuan Shen, and Dingyuan Tang. 2015. "Gold Nanorods as Single and Combined Saturable Absorbers for a High-Energy $Q $-switched Nd:YAG Solid-State Laser." *Photonics Journal, IEEE* 7 (August): 1–10. 10.1109/JPHOT.2015.2460552.

Häring, R., R. Paschotta, R. Fluck, E. Gini, H. Melchior, and U. Keller. 2001. "Passively Q-switched Microchip Laser at 1.5 Mm." *Journal of the Optical Society of America B* 18 (12): 1805–1812. 10.1364/JOSAB.18.001805.

He, Xiaoying, Zhi-Bo Liu, Dongning Wang, Minwei Yang, C.R. Liao, and Xin Zhao. 2012. "Passively Mode-Locked Fiber Laser Based on Reduced Graphene Oxide on Microfiber for Ultra-Wide-Band Doublet Pulse Generation." *Journal of Lightwave Technology* 30 (7): 984–989.

Huang, Ding Wei, Wen Fung Liu, and C.C. Yang. 2000. "Q-switched All-Fiber Laser with an Acoustically Modulated Fiber Attenuator." *IEEE Photonics Technology Letters* 12 (9): 1153–1155. 10.1109/68.874219.

Huang, J.Y., H.C. Liang, K.W. Su, and Yung-Fu Chen. 2007. "High Power Passively Q-switched Ytterbium Fiber Laser with Cr^4+:YAG as a Saturable Absorber." *Optics Express* 15 (2): 473. 10.1364/oe.15.000473.

Huang, J.Y., W.C. Huang, W.Z. Zhuang, K.W. Su, Y.F. Chen, and K.F. Huang. 2009. "High-Pulse-Energy, Passively Q-switched Yb-Doped Fiber Laser with AlGaInAs Quantum Wells as a Saturable Absorber." *Optics Letters* 34 (15): 2360–2362. 10.1364/OL. 34.002360.

Huo, Yanming, Robert T Brown, George G King, and Peter K Cheo. 2004. "Kinetic Modeling of Q-switched High-Power Ytterbium-Doped Fiber Lasers." *Applied Optics* 43 (6): 1404–1411. 10.1364/AO.43.001404.

Jedrzejewski, K.P., E.R. Taylor, and D.N. Payne. 1992. "Short-Pulse, High-Power Q-switched Fiber Laser." *IEEE Photonics Technology Letters* 4 (6): 545–547. 10.1109/ 68.141962.

Jeon, Jinwoo, Junsu Lee, and Ju Han Lee. 2013. "Theoretical Analysis of Impact of Q-Switch Rise Time on Output Pulse Performance in an Ytterbium-Doped Actively Q-switched Fiber Laser." *Korean Journal of Optics and Photonics* 24 (2): 58–63. 10.3807/kjop. 2013.24.2.058.

Kaneda, Yushi, Yongdan Hu, Christine Spiegelberg, Jihong Geng, and Shibin Jiang. 2004. "Single-Frequency, All-Fiber Q-switched Laser at 1550 Nm." *OSA Trends in Optics and Photonics Series* 94: 126–130.

Keller, U., D.A.B. Miller, G.D. Boyd, T.H. Chiu, J.F. Ferguson, and M.T. Asom. 1992. "Solid-State Low-Loss Intracavity Saturable Absorber for Nd:YLF Lasers: An Antiresonant Semiconductor Fabry–Perot Saturable Absorber." *Optics Letters* 17 (7): 505–507. 10.1364/OL.17.000505.

Keller, U., K.J. Weingarten, F.X. Kartner, D. Kopf, B. Braun, I.D. Jung, R. Fluck, C. Honninger, N. Matuschek, and J. Aus der Au. 1996. "Semiconductor Saturable Absorber Mirrors (SESAM's) for Femtosecond to Nanosecond Pulse Generation in Solid-State Lasers." *IEEE Journal of Selected Topics in Quantum Electronics* 2 (3): 435–453. 10.1109/2944.571743.

Kieu, Khanh and Masud Mansuripur. 2006. "Active Q Switching of a Fiber Laser with a Microsphere Resonator." *Optics Letters* 31 (24): 3568–3570. 10.1364/OL.31.003568.

Kim, Jun Wan, Sun Young Choi, Shanmugam Aravazhi, Markus Pollnau, Uwe Griebner, Valentin Petrov, Sukang Bae, Kwang Jun Ahn, Dong-Il Yeom, and Fabian Rotermund. 2015. "Graphene Q-switched Yb:KYW Planar Waveguide Laser." *AIP Advances* 5 (1): 17110. 10.1063/1.4905785.

Klimov, Igor, M. Nikol'skiĭ, Vladimir Tsvetkov, and Ivan Shcherbakov. 2007. "Passive Q Switching of Pulsed Nd3+ Lasers Using YSGG:Cr4+ Crystal Switches Exhibiting Phototropic Properties." *Soviet Journal of Quantum Electronics* 22 (October): 603. 10.1070/QE1992v022n07ABEH003552.

Kolpakov, Stanislav A., Yuri O. Barmenkov, Ana Dinora Guzman-Chavez, Alexander V. Kir'Yanov, José Luis Cruz, Antonio Diez, and Miguel V. Andrés. 2011. "Distributed Model for Actively Q-switched Erbium-Doped Fiber Lasers." *IEEE Journal of Quantum Electronics* 47 (7): 928–934. 10.1109/JQE.2011.2143695.

Kolpakov, Stanislav A., Sergey Sergeyev, Chengbo Mou, Neil T. Gordon, and Kaiming Zhou. 2014. "Optimization of Erbium-Doped Actively Q-switched Fiber Laser Implemented in Symmetric Configuration." *IEEE Journal on Selected Topics in Quantum Electronics* 20 (5). 10.1109/JSTQE.2014.2301015.

Kurkov, A.s. 2011. "Q-switched All-fiber Lasers with Saturable Absorbers." *Laser Physics Letters* 8 (May): 335–342. 10.1002/lapl.201010142.

Kurkov, A.s, E. Sholokhov, A. Marakulin, and L. Minashina. 2010. "Dynamic Behavior of Laser Based on the Heavily Holmium Doped Fiber." *Laser Physics Letters - LASER PHYS LETT* 7 (August): 587–590. 10.1002/lapl.201010024.

Lan, Ruijun, Lei Pan, Ilya Utkin, Quan Ren, Huaijin Zhang, Zhengping Wang, and Robert Fedosejevs. 2010. "Passively Q-switched Yb3+:NaY(WO4)(2) Laser with GaAs Saturable Absorber." *Optics Express* 18 (March): 4000–4005. 10.1364/OE.18.004000.

Laroche, Mathieu, H. Gilles, Sylvain Girard, Nicolas Passilly, and Kamel Aït-Ameur. 2006. "Nanosecond Pulse Generation in a Passively Q-switched Yb-Doped Fiber Laser by Cr4+:YAG Saturable Absorber." *Photonics Technology Letters, IEEE* 18 (February): 764–766. 10.1109/LPT.2006.871678.

Lecourt, Jean-Bernard, Gilles Martel, Maud Guezo, C. Labbé, and S. Loualiche. 2006. "Erbium-Doped Fiber Laser Passively Q-switched by an InGaAs/InP Multiple Quantum Well Saturable Absorber." *Optics Communications* 263 (July): 71–83. 10.1016/j.optcom.2006.01.009.

Lee, J., Minwan Jung, Joonhoi Koo, C. Chi, and Jong Hyeon Lee. 2015. "'Passively Q-switched 1.89-Mm Fiber Laser Using a Bulk-Structured Bi2Te3 Topological Insulator.'" *IEEE Journal of Selected Topics in Quantum Electronics* 21 (January): 1–6.

Lees, G.P. and T.P. Newson. 1996. "Diode Pumped High Power Simultaneously Q-switched and Self Mode-Locked Erbium Doped Fiber Laser." *Electronics Letters* 32 (4): 332–333. 10.1049/el:19960226.

Li, Wencai, Haowei Liu, Ji Zhang, Bo Yao, Sujuan Feng, Li Wei, and Qinghe Mao. 2017. "Mode-Hopping-Free Single-Longitudinal-Mode Actively Q-switched Ring Cavity Fiber Laser with an Injection Seeding Technique." *IEEE Photonics Journal* 9 (1): 1–7. 10.1109/JPHOT.2017.2654999.

Lim, E., S. Alam, and D.J. Richardson. 2011. "The Multipeak Phenomena and Nonlinear Effects in ${Q}$-switched Fiber Lasers." *IEEE Photonics Technology Letters* 23 (23): 1763–1765. 10.1109/LPT.2011.2169395.

Liu, Chun, Chenchun Ye, Zhengqian Luo, Huihui Cheng, Duanduan Wu, Yonglong Zheng, Zhen Liu, and Biao Qu. 2013. "High-Energy Passively Q-switched 2 Mm Tm3+-Doped Double-Clad Fiber Laser Using Graphene-Oxide-Deposited Fiber Taper." *Optics Express* 21 (1): 204–209. 10.1364/OE.21.000204.

Luo, Z., Y. Huang, M. Zhong, Y. Li, J. Wu, B. Xu, H. Xu, Z. Cai, J. Peng, and J. Weng. 2014. "1-, 1.5-, and 2-Mm Fiber Lasers Q-switched by a Broadband Few-Layer MoS2 Saturable Absorber." *Journal of Lightwave Technology* 32 (24): 4679–4686. 10.1109/JLT.2014.2362147.

Mears, R.J., L. Reekie, S.B. Poole, and D.N. Payne. 1986. "Low-Threshold Tunable CW and Q-switched Fiber Laser Operating at 1.55 Mm." *Electronics Letters* 22 (3): 159–160.

Mgharaz, Driss, Marc Brunel, and Abdelkader Boulezhar. 2009. "Design of a Bi-Pulse Fiber Laser for Particle Image Velocimetry Applications." *The Open Optics Journal* 3 (1): 56–62. 10.2174/1874328500903010056.

Monga, Kaboko Jean Jacques, Rodolfo Martinez-Manuel, Faouzi Bahloul, Sompo Mpoyo Justice, Mikhail G. Shlyagin, and Johan Meyer. 2019. "Numerical Analysis of Ring Erbium-Doped Fiber Laser with Q-Switching Based on Dynamic Overlapping of Narrowband Filters." *Optical Engineering* 58 (10): 1. 10.1117/1.oe.58.10.106102.

Mora, José, B. Ortega, M.V. Andres, J. Capmany, D. Pastor, and José Cruz. 2002. "Tunable Chirped Fiber Bragg Grating Device Controlled by Variable Magnetic Fields." *Electronics Letters* 38 (March): 118–119. 10.1049/el:20020086.

Myslinski, P., J. Chrostowski, J.A. Koningstein, and J.R. Simpson. 1992. "High Power Q-switched Erbium Doped Fiber Laser." *IEEE Journal of Quantum Electronics* 28 (1): 371–377. 10.1109/3.119537.

Okhrimchuk, Andrey G, Vladimir K Mezentsev, Vladislav V Dvoyrin, Andrey S Kurkov, Evgeny M Sholokhov, Sergey K Turitsyn, Alexander V Shestakov, and Ian Bennion. 2009. "Waveguide-Saturable Absorber Fabricated by Femtosecond Pulses in YAG:Cr4+ Crystal for Q-switched Operation of Yb-Fiber Laser." *Optics Letters* 34 (24): 3881–3883. 10.1364/OL.34.003881.

Pan, Lei, Ilya Utkin, and Robert Fedosejevs. 2010. "Experiment and Numerical Modeling of High-Power Passively Q-switched Ytterbium-Doped Double-Clad Fiber Lasers." *Quantum Electronics, IEEE Journal Of* 46 (February): 68–75. 10.1109/JQE.2009. 2028031.

Pan, Lei, Ilya Utkin, Robert Fedosejevs, and Senior Member. 2007. "Passively Q -switched Ytterbium-Doped Double-Clad Fiber Laser With a Cr 4 +: YAG Saturable Absorber." *Technology* 19 (24): 1979–1981.

Pérez-Millán, P., A. Díez, M.V. Andrés, D. Zalvidea, and R. Duchowicz. 2005. "Q-switched All-Fiber Laser Based on Magnetostriction Modulation of a Bragg Grating." *Optics Express* 13 (13): 5046–5051. 10.1364/OPEX.13.005046.

Philip, Reji, G. Ravindrakumar, Sandhyarani N., and Pradeep Thalappil. 2000. "Picosecond Optical Nonlinearity in Monolayer-Protected Gold, Silver, and Gold-Silver Alloy Nanoclusters." *Physical Review B* 62 (July): 13160. 10.1103/PhysRevB.62.13160.

Renaud, C.C., R.J. Selvas-Aguilar, J. Nilsson, P.W. Turner, and A.B. Grudinin. 1999. "Compact High-Energy Q-switched Cladding-Pumped Fiber Laser with a Tuning Range over 40 Nm." *IEEE Photonics Technology Letters* 11 (8): 976–978. 10.1109/68.775318.

Renaud, Cyril C., H.L. Offerhaus, J.A. Alvarez-Chavez, J. Nilsson, W.A. Clarkson, P.W. Turner, D.J. Richardson, and A.B. Grudinin. 2001. "Characteristics of Q-switched Cladding-Pumped Ytterbium-Doped Fiber Lasers with Different High-Energy Fiber Designs." *IEEE Journal of Quantum Electronics* 37 (2): 199–206. 10.1109/3.903069.

Russo, N.A., R. Duchowicz, J. Mora, J.L. Cruz, and M.V. Andrés. 2002. "High-Efficiency Q-switched Erbium Fiber Laser Using a Bragg Grating-Based Modulator." *Optics Communications* 210 (3–6): 361–366. 10.1016/S0030-4018(02)01815-1.

Sakakibara, Youichi, Satoshi Tatsuura, Hiromichi Kataura, Madoka Tokumoto, and Yohji Achiba. 2003. "Near-Infrared Saturable Absorption of Single-Wall Carbon Nanotubes Prepared by Laser Ablation Method." *Japanese Journal of Applied Physics* 42 (Part 2, No. 5A): L494–L496. 10.1143/jjap.42.l494.

Seguin, Francois and Tanya K Oleskevich. 1993. "Diode-Pumped Q-switched Fiber Laser." *Optical Engineering* 32 (9): 2036–2041. 10.1117/12.143955.

Sejka, M., C.V. Poulsen, and Yuan Shi. 1993. "High Repetition Rate Q-switched Er/Sup 3+/- Doped Fiber Ring Laser." In *Proceedings of LEOS '93*, 710–711. 10.1109/LEOS.1993. 379388.

Siegman, Anthony E.. 1986. "Siegman,_Lasers,1986.Pdf."

Sejka, Milan, Christian V. Poulsen, Jørn Hedegaard Povlsen, Yuan Shi, and Ove Poulsen. 1995. "High Repetition Rate Q-switched Ring Laser in Er3+-Doped Fiber." *Optical Fiber Technology*. 10.1006/ofte.1995.1007.

Sobon, Grzegorz, Jaroslaw Sotor, Joanna Jagiełło, Rafal Kozinski, Krzysztof Librant, Mariusz Zdrojek, Ludwika Lipińska, and Krzysztof Abramski. 2012. "Linearly Polarized, Q-switched Er-Doped Fiber Laser Based on Reduced Graphene Oxide Saturable Absorber." *Applied Physics Letters* 101 (December). 10.1063/1.4770373.

Sotor, J., G. Sobon, K. Grodecki, and K.M. Abramski. 2014. "Mode-Locked Erbium-Doped Fiber Laser Based on Evanescent Field Interaction with Sb2Te3 Topological Insulator." *Applied Physics Letters* 104 (25): 251112. 10.1063/1.4885371.

Spuehler, Gabriel, Rüdiger Paschotta, M.P. Kullberg, Martin Graf, Mohammed Moser, E. Mix, Guenter Huber, C. Harder, and U. Keller. 2001. "A Passively Q-switched Yb:YAG Microchip Laser." *Applied Physics B* 72 (February): 285–287. 10.1007/ s003400100507.

Spühler, G.J., R. Paschotta, R. Fluck, B. Braun, M. Moser, G. Zhang, E. Gini, and U. Keller. 1999. "Experimentally Confirmed Design Guidelines for Passively Q-switched Microchip Lasers Using Semiconductor Saturable Absorbers." *Journal of the Optical Society of America B* 16 (3): 376–388. 10.1364/JOSAB.16.000376.

Stutzki, Fabian, Florian Jansen, Andreas Liem, Cesar Jauregui, Jens Limpert, and Andreas Tünnermann. 2012. "26 MJ, 130 W Q-switched Fiber-Laser System with near-Diffraction-Limited Beam Quality." *Optics Letters* 37 (6): 1073–1075. 10.1364/OL.37.001073.

Sun, Yi-Jian, Chao-Kuei Lee, Jin-Long Xu, Zhao-Jie Zhu, Ye-Qing Wang, Shu-Fang Gao, Hou-Ping Xia, Zhen-Yu You, and Chao-Yang Tu. 2015. "Passively Q-switched Tri-Wavelength Yb3+:GdAl3(BO3)4 Solid-State Laser with Topological Insulator Bi2Te3 as Saturable Absorber." *Photonics Research* 3 (3): A97–A101. 10.1364/PRJ.3.000A97.

Sun, Zhipei, Tawfique Hasan, Felice Torrisi, Daniel Popa, Giulia Privitera, Fengqiu Wang, Francesco Bonaccorso, Denis M. Basko, and Andrea C. Ferrari. 2010. "Graphene Mode-Locked Ultrafast Laser." *ACS Nano* 4 (2): 803–810. 10.1021/nn901703e.

Tan, Yang, Chujun Zhao, Shavkat Akhmadaliev, Shengqiang Zhou, and Feng Chen. 2015. "Bi2Se3 Q-switched Nd:YAG Ceramic Waveguide Laser." *Optics Letters* 40 (February). 10.1364/OL.40.000637.

Wang, Yong, Hou Ren Chen, Xiaoming Wen, Wen Hsieh, and Jau Tang. 2011. "A Highly Efficient Graphene Oxide Absorber for Q-switched Nd:GdVO4 Lasers." *Nanotechnology* 22 (November): 455203. 10.1088/0957-4484/22/45/455203.

Wang, Yong, Alejandro Martinez-Rios, and Hong Po. 2003. "Analysis of a Q-switched Ytterbium-Doped Double-Clad Fiber Laser with Simultaneous Mode Locking." *Optics Communications* 224 (1–3): 113–123. 10.1016/S0030-4018(03)01722-X.

Wang, Yong and Chang-Qing Xu. 2004. "Switching-Induced Perturbation and Influence on Actively Q-switched Fiber Lasers." *IEEE Journal of Quantum Electronics* 40 (11): 1583–1596. 10.1109/JQE.2004.835212.

Wang, Yong and Chang-Qing Xu. 2006. "Modeling and Optimization of Q-switched Double-Clad Fiber Lasers." *Applied Optics* 45 (9): 2058–2071. 10.1364/AO.45.002058.

Wang, Yong and Chang Qing Xu. 2007. "Actively Q-switched Fiber Lasers: Switching Dynamics and Nonlinear Processes." *Progress in Quantum Electronics* 31 (3–5): 131–216. 10.1016/j.pquantelec.2007.06.001.

Wu, Liangying, Li Pei, Jianshuai Wang, Jing Li, Tigang Ning, and Shuo Liu. 2016. "Q-switched Erbium-Doped Fiber Ring Laser with Piezoelectric Transducer-Based PS-CFBG." *Laser Physics Letters* 13 (9): 095101. 10.1088/1612-2011/13/9/095101.

Wurtz, P. 1966. "Etude de Lemission Dun Laser au Neodyme Declenche Par Effet Pockels." *Philips Research Reports* 21 (4): 213.

Yariv, Amnon. 1991. *Optical Electronics*. Saunders College Publ.

Zhang, Jiaojiao, Zuxing Zhang, Yu Cai, Hongdan Wan, Zhiqiang Wang, and Lin Zhang. 2018. "An Actively Q-switched Fiber Laser with Cylindrical Vector Beam Generation." *Laser Physics Letters* 15 (3). 10.1088/1612-202X/aaa335.

Zhang, Qinduan, Jun Chang, Qiang Wang, Zongliang Wang, Wang Fupeng, and Zengguang Qin. 2017. "Acousto-Optic Q-switched Fiber Laser-Based Intra-Cavity Photoacoustic Spectroscopy for Trace Gas Detection." *Sensors* 18 (December): 42. 10.3390/s18010042.

7 Narrow Linewidth Fiber Lasers

7.1 INTRODUCTION

Narrow linewidth fiber lasers are highly desirable in several applications such as optical communication and spectroscopy. In this chapter, the main theories as well as techniques for achieving narrow linewidth, single longitudinal mode, and tunability of fiber lasers are presented.

Narrow linewidth operation is fundamentally easily achievable in fiber lasers. The core diameter of the fiber is usually comprised between 3 μm and 10 μm. Hence, a relatively low input power often results in significantly high intensities. Because of the inherent waveguiding property of the fiber, the high-intensity optical wave interacts with the rare-earth ions over a long distance producing a high gain. This gain allows the use of intra-cavity linewidth narrowing devices such as filters without significantly increasing the intracavity loss and deteriorating the efficiency of the fiber laser (Morkel et al. 1990; Iwatsuki et al. 1990; Maeda et al. 1990).

The theoretical limit of the linewidth of a laser widely known as the Schawlow-Townes limit was established in the late 1950s and is expressed as (Schawlow and Townes 1958):

$$\Delta\nu_{laser} = \frac{\pi h\nu\,(\Delta\nu_0)^2}{P_{out}} \tag{7.1}$$

where ν is the lasing frequency, $\Delta\nu_0$ is the resonator bandwidth, h is the Planck constant, and P_{out} is the output power. The Schawlow-Townes limit results from the contribution of the spontaneous emission into the lasing mode. This contribution is weak in fiber lasers because of the large transition cross-section of rare-earth ions; therefore, rare-earth-doped lasers tend to have narrower linewidth than semiconductor laser, for example. Also, the linewidth of the laser is inversely proportional to the cavity length. Because of the fiber geometry, it is possible to build fiber laser cavities several meters long. The ombination of all the reasons mentioned in the preceding makes fiber lasers excellent narrow linewidth lasers.

7.2 FUNDAMENTAL CONCEPTS OF NARROW LINEWIDTH FIBER LASERS

A photon trapped inside a laser cavity travels a distance corresponding to the optical path length. The intracavity optical field is a function of the optical path length and the beam frequency given by (Siegman 1986):

DOI: 10.1201/9781003256380-7

$$E_{cav}(\nu) = \frac{E_{in}(\nu)}{1 - r_1 r_2 \exp(i\phi - \alpha l)} \tag{7.2}$$

$$\phi = \frac{2\pi \nu l}{c} \tag{7.3}$$

where E_m is the input field, ν is the frequency, r_1 and r_2 are the reflection coefficient for mirrors 1 and 2, respectively, and ϕ is the intra-cavity phase shift. It can be shown from equation (7.2) that the intra-cavity field reaches its maximum when the phase shift ϕ is an integer multiple of 2π. These maxima of the field correspond to the allowed longitudinal modes of the cavity. These longitudinal modes are periodic and have between them a frequency separation of:

$$\Delta \nu = \frac{c}{l} \tag{7.4}$$

where c is the speed of light in free space and l is the round trip optical path length.

Typical fiber lasers have round-trip optical path lengths comprised between 1 cm and 50 m corresponding to a longitudinal mode spacing in the range from 30 GHz to 6 MHz. This mode spacing combined with the fluorescence of rare-earth ion spectra which can potentially extend over hundreds of terahertz makes it possible for the fiber laser to support several cavity longitudinal modes.

As already discussed in Chapter 3, in an amorphous environment such as glass, the spectral broadening of the transition linewidth is the result of homogeneous and inhomogeneous broadening effects (Zyskind et al. 1990). Assuming a purely homogeneously broadened amplifying medium, one can say that the first long-itudinal mode to oscillate will be the one experiencing the highest small-signal gain. This longitudinal mode will then saturate the gain preventing the other mode to oscillate because gain saturation will depopulate upper energy levels. Therefore, theoretically, the output of a homogeneously broadened gain medium will always be a single mode. The linewidth limit as we already said in such a system will be imposed by the Schawlow-Townes limit. In practice, however, multiple simulta-neous longitudinal mode outputs have been observed. Several reasons are at the origin of such behaviour. The frequency behaviour in a homogeneously broadened medium is the same for the excited ions and the intra-cavity signal. This behaviour is broken by the effect of inhomogeneous-broadening mechanisms. The local en-vironment experienced by the ion will alter the frequency response, making dif-ferent groups of ions respond differently to the intra-cavity field. In this scenario, a given longitudinal mode may saturate the gain associated with a certain group of ions, but there can still be an available gain for a different group that will trigger oscillation of another longitudinal mode. As a result, several longitudinal modes may oscillate simultaneously in an inhomogeneously broadened gain medium. This behaviour has been taken advantage of to achieve multiple wavelength operation of a rare-earth-doped fiber laser (Chow et al. 1996).

Another effect that prevents single longitudinal mode operation of linear cavity fiber lasers is spatial hole burning. In linear cavity fiber lasers, the interaction between forwards and backwards propagating waves result in the formation of standing wave patterns which in turn generate a periodic spatial modulation of the gain with a periodicity of half the oscillating wavelength. This periodicity of the gain modulation creates a grating effect. Because the period is half the propagating wavelength, the counter-propagating waves will be 180 degrees out of phase to each other and will destructively interfere. The consequence of this effect is to lower the intensity of the propagating waves, thus the grating strength which in turn produce a more uniformly distributed gain across the gain medium and favours incoherent multi-mode operation rather than coherent single-mode operation.

Inhomogeneous broadening and spatial hole burning remain the most detrimental effects of the single-mode operation of fiber lasers. To overcome these limitations, several techniques have been adopted. For example, spatial hole burning can be eliminated by making sure that standing wave patterns do not develop within the gain medium. One way to do it is to make sure that the frequency spacing between neighbouring longitudinal modes is larger than the gain spectrum of the rare-earth ion of interest; therefore, only one longitudinal mode will experience gain. Using this technique, the effects of inhomogeneous broadening can be eliminated too.

The other method to eliminate the multi-longitudinal mode operation of fiber lasers is to use an intracavity frequency control element to discriminate undesirable frequencies. Such an element can be a wavelength-selective mirror (Bragg grating or diffraction grating), a tunable filter such as Fabry-Perot filter or optical etalon. The two methods, wavelength-selective mirrors, and filters have the same results but work differently. The wavelength-selective mirror reflects into the cavity only light beams whose frequencies are comprised within the reflection bandwidth. The filters, on the other hand, transmit certain wavelengths only, thus limiting the number of modes available for the gain.

If the frequency selective element limits the output to only a few longitudinal modes, the resultant laser is termed as narrow linewidth fiber laser, whereas if the frequency selective element allows for only a single longitudinal mode to oscillate, the laser is referred to as a single-frequency or single longitudinal mode fiber laser. Single longitudinal mode operation combined with the broad fluorescence spectra of the trivalent rare-earth ions present a remarkable potential to build broadly tunable fiber lasers.

7.3 NARROW LINEWIDTH FIBER LASERS

As mentioned in section 7.2, the free-running laser may support thousands of modes. Narrow linewidth operation of a fiber laser is often achieved using a bandwidth restriction element in the cavity of a free-running laser to restrict the allowed modes. The progress in fiber component technologies has made possible the use of fiber compatible selective devices, which has resulted in the construction of robust all-fiber narrow linewidth lasers. Narrow linewidth fiber lasers most of the time use fiber Bragg grating as the frequency selective device. The central wavelength of the laser, in this case, corresponds to the Bragg wavelength of the fiber

Bragg grating and the linewidth of the laser corresponds to the reflection bandwidth of the Bragg grating which can be very narrow (<0.5 nm) resulting in a narrow linewidth laser. In the early stage of development of narrow linewidth fiber lasers, the fiber Bragg gratings were either etched close to the core of a polished fiber or in the case of a photosensitive fiber, directly printed into the fiber core. Nowadays, most of the FBG is obtained by directly printing into the core of photosensitive fiber using UV light from other laser sources.

The first reported narrow linewidth fiber laser used an Nd^{3+}-doped fiber. The configuration of this laser is shown in Figure 7.1. A 6-m long Nd-doped fiber was pumped with a GaAlAs semiconductor laser at 830 nm through an input dichroic mirror with 99% reflectivity and 85% transmission at the oscillating wavelength. The cavity was closed by an FBG etched on the core of the fiber with a reflectivity of around 75% at 1084 nm. The fiber laser oscillated at 1084 nm and the measured linewidth was 16 GHz (Jauncey et al. 1986).

Among the rare-earth ions, Erbium is desirable because of its emission spectrum which corresponds to the maximum transmission of the third telecommunication window; therefore, an effort was made to build Erbium-doped fiber lasers transmitting in this band. A design similar to one proposed in the previous section using Erbium-doped fiber was constructed. In this case, the laser oscillated at 1551 nm with a linewidth of 4.9 GHz (Jauncey et al. 1987). However, the linewidth of this type of structure is usually limited by the interaction length between the evanescent field from the fiber core and the grating structure. One way to increase this interaction is to use a diffraction grating in contact with the D-shape fiber a linewidth of 13 GHz was achieved using this technique (Yennadhiou and Cassidy 1990) Fabrication of both etched grating and modular D shape fiber and diffraction grating proved to be very complicated and lead to fragile designs. A more reliable method was printing fiber Bragg grating directly into the core of the photosensitive fiber. This technique was used in a 30-m cavity long fiber laser. The reflectivity of the fiber Bragg grating was 0.5% at the wavelength of 1538 nm. The other reflector of the cavity was made using a high reflective mirror. The laser oscillated at 1537 nm with a linewidth of approximately 1 GHz (Kashyap et al. 1990). The previous type of fiber lasers both used a cavity composed of a grating and a reflecting mirror. This situation resulted in cumbersome devices. To solve this problem, fiber lasers with two FBG printed directly on the core of the fiber were proposed. The first report of this type of configuration was by Ball et al. They used two Bragg gratings printed

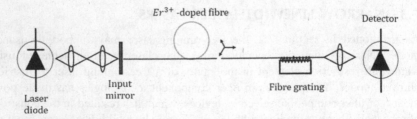

FIGURE 7.1 Experimental configuration of a narrow linewidth fiber laser with intracavity fiber Bragg grating.

into a Nd^{3+}-doped fiber. The two Bragg grating had a reflectivity of 94%. This 3-m long Nd^{3+}-doped fiber laser was pumped by a CW dye laser emitting at 830 nm. The laser oscillated at 1085 nm which corresponds to the peak of the fiber Bragg grating. The output power of 11.7 mW was measured for 265 mW pump power and a linewidth of 30 GHz ("Tunable Nd/Sup 3+/-Doped Fiber Ring Laser - Electronics Letters", n.d.). One of the biggest advantages of using a fiber Bragg grating resonator is the possibility to obtain a tunable fiber laser. Tunability can be achieved by either increasing the temperature (Toebben 1993) of the FBG to exploit its thermal expansion or by stretching (Ball and Morey 1992) or compressing the FBG (Ball and Morey 1992; Babin et al. 2007).

Although using fiber Bragg gratings is the most reliable way of defining the lasing wavelength of a rare-earth doped fiber laser, there exist alternative methods too. One of these methods consists of using an etalon inside the cavity of the fiber laser. The spectral transmission profile of the etalon superimposes the mode distribution. Modes that are perfectly aligned with the maximum transmission wavelength will experience minimum loss, whereas modes that are not tuned with the maximum transmission of the etalon will experience maximum loss (Sullivan et al. 1989). It is worth mentioning that lossy elements such as bulk Fabry-Perot cavities can only be used because of high gains developed in fiber lasers.

7.4 LINEAR CAVITY SINGLE LONGITUDINAL MODE FIBER LASERS

It is essential that narrow linewidth lasers also be single longitudinal mode fiber lasers. In a linear cavity fiber laser also known as standing wave fiber lasers, spatial hole burning is a major drawback that prevents the occurrence of single longitudinal mode operation. Several approaches to eliminate the effect of standing waves have been demonstrated and presented since the first years of research on fiber single-frequency lasers (Glassner and Esman 1997; Zhang et al. 2018; Amouzou and Bisson 2019; Lux et al. 2016; Kishi and Yazaki 1999; Sabert et al. 1991).

The simplest and straightforward approach is to increase the frequency spacing between the cavity longitudinal modes by reducing the cavity length. When the cavity length is reduced continuously, one reaches a point where the mode spacing becomes wider than the cavity length. At this critical length, the fiber laser ceases to be multi-mode as only one mode fits inside the cavity. Excellent results were achieved using this approach, for example, by Jauncey et al. Starting with a 50-cm long cavity where 10 longitudinal modes were found to oscillate, gradually reduced the cavity length and single-mode operation was achieved at a cavity length of 2.7 cm (Jauncey et al. 1988).

Gilbert et al. followed a similar approach to achieve a single longitudinal mode operation in Erbium-doped fiber. However, they used in addition to the 2.8 cm a 1-cm non-reflection coated fiber gain medium which acts as etalons and a Littrow-mounted diffraction grating with a reflection bandwidth of 5 GHz as shown in Figure 7.2. Coarse tuning was achieved by rotating the diffraction grating and fine-tuning by stretching the fiber with piezoelectric transitions as well as tilting the grating with another piezoelectric transition (Gilbert 1991). All these resulted in a

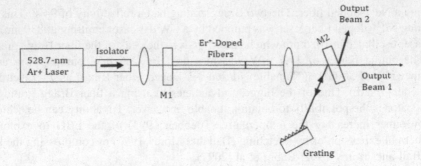

FIGURE 7.2 Schematic diagram of a fiber laser using a diffraction grating as a tuning element.

very cumbersome device whose continuous tuning was almost impossible to achieve.

Also, a perfect resonance condition must always be respected between the transmission maxima of the various etalons in the cavity. Not respecting this condition will result in mode hopping created from the net loss of the oscillating mode being higher than that experienced by a nearby mode.

Larose et al. came up with an elegant solution to the preceding problem by replacing the two pieces of Erbium-doped fiber that made the etalon with a unique fiber (Têtu et al. 1994). To achieve tuning, they use a grating-mirror combination as shown in Figure 7.3. This device acted like a tunable filter with a transmission bandwidth of 7.6 GHz. An etalon was introduced in the cavity for single-frequency operation. Tunability of the device was achieved by varying the incident angle of the light on the grating. Subsequent frequency stabilization relative to the R(10) line of acetylene resulted in a resolution-limited linewidth of 70 kHz and an RMS jitter of 2 MHz.

The work of Gilbert (1991) and Têtu et al. (1994) described in the foregoing is a clear demonstration that single longitudinal mode operation can be achieved in fiber lasers by introducing additional spectral filtering with an intra-cavity Etalon.

Another way to achieve a single longitudinal mode operation of fiber lasers is by using a Fox-Smith resonator. A Fox-Smith resonator is obtained by splicing one

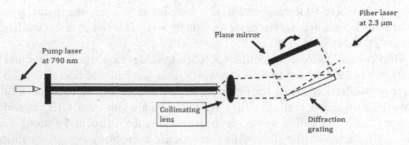

FIGURE 7.3 Diagram of a linear cavity used to obtain a narrow linewidth operation of a Tm^{3+}-doped Fluorozirconate fiber laser.

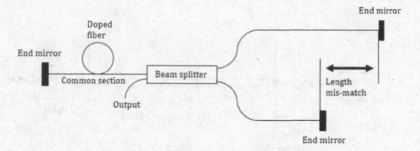

FIGURE 7.4 Schematic diagram of the Fox-Smith resonator comprising 2 arms with different lengths.

arm of a beam splitter to the end of the doped fiber. The two opposite arms have a fiber of different lengths terminated with a mirror connected to each one of them and the last arm is taken as the output as shown in Figure 7.4 Modern technology allows the construction of low loss fiber compatible beam splitters.

The mismatch in the length between the two arms of the beam splitter provides additional spectral filtering in such a way that single longitudinal mode operation can be achieved. A cavity like this one was used by Barnsley et al., except that, in their configuration, the end mirror on the left was replaced by a diffraction grating to allow tunability of the laser. The laser oscillated in single longitudinal mode with a resolution limited linewidth of 8.5 MHz.

All of the previous configurations for single longitudinal operation have in common a rather complicated design and if the single-frequency operation was achieved, implementations of such configuration outside of laboratory conditions or mass production is difficult. The simplest and more robust way of achieving a single longitudinal mode of fiber lasers remains to combine a short cavity length with a narrow bandwidth frequency-selective element. The discovery of Fiber Bragg Gratings printed in the core of photosensitive fiber using a UV interference pattern provided an excellent opportunity to solve this problem.

Ball et al. reported the first robust fiber laser using fiber Bragg gratings printed in the core of an Erbium-doped fiber as reflective mirrors as illustrated in Figure 7.5. The two fiber Bragg gratings had a reflectivity of 72% and 80%, respectively. The fiber laser cavity length was 50 cm and each one of the two Bragg gratings was 12.5 mm long. The fiber laser was end-pumped through the 80% fiber Bragg grating using a Ti:Saphire laser emitting at 980 nm and oscillated at 1548 nm corresponding to the peak reflectivity of the fiber Bragg gratings. Using a commercially available Fabry-Perot Etalon with a 15 GHz free spectral range and finesse of 1000 a 180 MHz free spectral range was measured for the fiber laser (Ball et al. 1991; Chernikov et al. 1993).

Also, a fiber Michelson interferometer was used to measure the coherence length of the laser. A coherence length well above 4.4 km was measured, which corresponds to a linewidth of less than 47 kHz at an assumed refractive index of 1.46.

Fiber lasers using fiber Bragg grating as reflective mirrors are an elegant solution but their single longitudinal mode operation is not stable and they are prone to mode

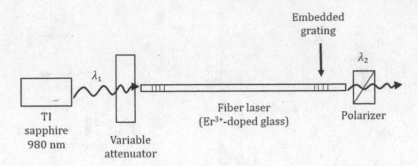

FIGURE 7.5 Schematic representation of the first fiber laser using fiber Bragg grating as reflecting mirrors.

hopping. The simplest way to overcome mode hopping is by shortening the cavity as mentioned early. Fiber lasers with cavity length in the range of centimetres have been reported by Zyskind et al. (1992). In their experiments, they found that mode-hop free single longitudinal mode operation was achieved for a cavity length of less than 2 cm at both 1480 nm and 980 nm pumping. For a 1-cm long gain medium, the threshold power was found to be 15 mW at 980 nm and the maximum output power obtained was 57 μW. These rather poor performances can be attributed to the available gain which depends on the length of the cavity. In such a situation of a short cavity fiber laser, increasing output power often correspond to increasing the doping concentration of rare-earth ions. However, high doping concentration inevitably results in ion-ion interaction problems such as cooperative up-conversion (Masuda et al. 1992), which leads to luminosity quenching (Nilsson et al. 1993) and self-pulsing (Leboudec et al. 1993) these deleterious effects negatively affect the performance of the fiber laser. Cooperation up-conversion results in an increased threshold and reduced output power of the laser whereas self-pulsing lead to laser output power fluctuation. Attempts to reduce these amplitude fluctuations have been made possible using an appropriate feedback signal applied to the pump source (Park et al. 1993).

A more effective way of inhibiting mode hopping is making sure that the loss experienced by the dominant cavity mode is ways below the neighbouring modes such that the dominant mode alone is allowed to oscillate. This can be achieved by carefully tailoring the fiber Bragg grating reflectors. Zyskind et al. provided full modelling and reported accurate results. In their analysis, it was found that a single longitudinal mode operation of a short-cavity fiber laser can only be achieved using fiber Bragg grating with near 100% reflectivity (Ball and Glenn 1992).

7.4.1 TUNABLE SHORT CAVITY LASERS

Short length fiber lasers can be effectively tuned by applying strain to them. A change in laser wavelength of 0.72 nm was reported for a 90 μm strain applied to the cavity using a piezoelectric transducer (Ball and Morey 1992). However, this system is limited by the maximum tension applicable to the laser before the breaking point. Silica glass is more resistant to compression than tension, therefore

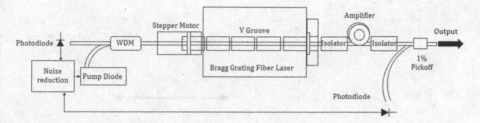

FIGURE 7.6 Schematic diagram of the compression tuned Brag grating fiber laser.

a large tunability range can be obtained with compression rather than tension. For example, using compressive strain, tuning of up to 32 nm of a 4-cm long Erbium-doped fiber laser was obtained (Ball and Morey 1994). This interesting tuning range is very close to what can be achieved using a semiconductor laser; therefore, short cavity single-frequency fiber lasers are a serious candidate for applications where wavelength tunability is required (Figure 7.6).

7.4.2 Efficiency Enhancement Yb^{3+} Co-Doping

As already mentioned, the use of short cavity fiber lasers requires high doping concentration to develop enough gain to allow lasing. As already mentioned, increasing rare-earth ions concentration will always result in relaxation oscillation due to ion-ion interactions. This problem can be solved by using alternative pumping wavelengths rather than the traditional 980 nm and 1480 nm in Erbium, for example. It was proved that wavelengths of 528 or 650 nm have significantly higher absorption cross-section, and are free from excited absorption influence, which improve the overall absorption of the laser device hence improved efficiency. A more interesting solution, however, is co-doping the rare-earth ion in the gain medium with another specie of rare-earth ion. Ytterbium is the preferred co-doping ion for Erbium-doped gain medium because of its inherent properties such as a total absence of cooperative upconversion which is the consequence of its two energy level structures. Also, Ytterbium ions can resonantly transfer their energy to Erbium ions. By carefully engineering the concentrations of the two species, this energy transfer can be made close to 100%. The first short cavity Erbium-Ytterbium co-doped fiber laser was demonstrated by Kringlebotn et al. (1994). They constructed a 10-cm long fiber laser with a cavity. This cavity was made of a high reflective mirror butt-coupled at one end and a fiber Bragg grating printed in a piece of photosensitive fiber spliced to the doped fiber at the other end as illustrated in Figure 7.7. When pumped at 980 nm the laser oscillated at 1544.8 nm. The threshold pump power was reported to be 7 mW and the output power was 7.6 mW for a pump power around 100 mW.

The Erbium-Ytterbium co-doped fiber laser was presented by Kringlebotn et al. (1994). They presented two major drawbacks. First, it used a bulk mirror making its construction technologically challenging. Also, they used non-photosensitive aluminosilicate fiber as the gain medium which made it impossible to write the grating directly into its core. Therefore, the fiber Bragg grating was instead printed in the

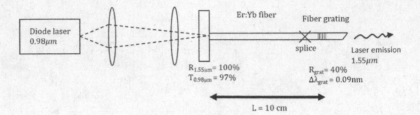

FIGURE 7.7 Schematic representation of a short cavity Erbium-Ytterbium doped fiber laser.

core of a separate photosensitive fiber which was then spliced to the active fiber. Given the dimension of the available splicing machines, fiber laser with a cavity length of less than 5 cm is very difficult to realize using this technique. Increasing the photosensitivity of the aluminosilicate fiber can be done by loading it with hydrogen. Hydrogen loading was used by Archambault et al. to print fiber Bragg grating with reflectivity up to 100% in alumina silicate fibers (Archambault et al. 1994).

In an attempt to produce the first distributed feedback fiber laser Kringlebotn et al. use the same method of hydrogen loading to print gratings directly in the photosensitive fiber (Kringlebotn et al. 1994). As described in Chapter 5, a distributed feedback fiber laser constructed using a uniform fiber Bragg grating printed into a doped fiber will oscillate in two longitudinal modes with wavelengths corresponding to the Bragg wavelength of the FBG. The output of this kind of lasers is emitted from both ends of the device. This behaviour is explained by the round-trip phase criterion associated with a distributed feedback laser (Kogelnik and Shank 1972). To force single-frequency operation from a DFB fiber laser π-phase shift is often required inside the cavity. The phase-shift can be obtained by either perturbing the cavity Bragg structure or changing the cavity round trip by using an end reflector (Kringlebotn et al. 1994). However, a more reliable way of forcing a single frequency operation is by introducing a π-phase shift into the Bragg grating. Physically this means to permanently modify the refractive index of the grating at a given position. If the position is exactly in the middle of the grating then the DFB fiber laser will emit light with the same intensity from both ends. It was shown that if the phase shift is situated asymmetrically from the extremities of the DFB fiber laser, more power will be emitted from the end closer to the phase shift location. However, there is an optimum position of the phase shift corresponding to the highest power emitted from one side (Yelen et al. 2005). The first phase-shift was obtained by heating a point on the fiber Bragg grating with a resistance wire as shown in Figure 7.8. Using this technique a single-frequency DFB fiber laser was fabricated by Kringlebotn and Payne (1994).

The demonstration of this DFB fiber laser was an important milestone in the field of single-frequency fiber lasers. However, after the demonstration of the concept, it became clear that the DFB fiber laser developed by Payne et al. was not suitable for real-life applications. More robust ways of introducing the phase shift on DFB fiber lasers were therefore needed. One of these techniques is to use a π phase-shifted

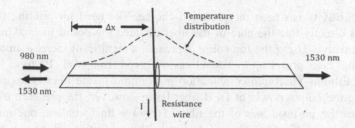

FIGURE 7.8 Distributed feedback fiber laser with phase-shift introduced by heating up a point on the fiber Bragg grating using an electric wire.

phase mask to produce the DFB fiber laser directly by exposing the core of the fiber to UV light like in any standard phase mask technique. In addition to its simplicity, this technique allows one to control the length of the grating. The relatively long gratings obtained with this technique increases the value of achievable gain.

Another technique to produce DFB fiber lasers uses a post-processing method to introduce phase shift after the fiber Bragg grating has been printed. More precisely post exposition consists of altering the refractive index of a specific point on the body of the grating by a further exposition to UV light. The first distributed feedback fiber laser using a permanent phase shift was reported by Asseh et al. They printed a 10-cm long grating into a Ytterbium-doped fiber. Phase shift was obtained using the technique of post exposition. The reflectivity of the grating was 99% and the Bragg wavelength was 1047 nm. The laser oscillated emitting an output power of 7 mW for 18 mW pump power at 974 nm (Asseh et al. 1995). Another DFB fiber laser was constructed by Loh and Laming (1995). The DFB fiber laser was 10 cm long with a $\pi/2$ phase shift. When pumped the laser did not as expected oscillate in a single longitudinal mode but in two modes. When these two modes were analysed, they seemed to be perfectly orthogonal, which is a clear indication that they originated from the birefringence in the fiber. Using a polarization discriminator the two modes could be separated and each one of them oscillated with a linewidth of 13 kHz. However, the intensities of the two modes fluctuated with time as a consequence of energy exchange between the two orthogonally polarized modes induced by external perturbations. To counteract this mutual influence of the two orthogonally polarized modes, a technique of twisting the fiber was applied and result in the suppression of one of the two polarization modes (Harutjunian et al. 1996). The exact reason why one of the polarization modes was eliminated by simply twisting the fiber is not yet fully understood. A robust way of eliminating two-mode operation resulting from polarization is by performing UV post exposition. By doing so, Storoy et al. (1997) achieved inside the fiber Bragg grating the quarter-wave phase shift required for single-frequency operation.

As already discussed in a previous section, the short cavity single-frequency fiber laser of DFB type has a major drawback which is the limited output power due to essentially low absorption and ion-ion detrimental effects at high concentrations. The most popular way of tackling this problem has been so far to use co-doping with another rare-earth species. In the case of Erbium-doped fiber, co-doping with Ytterbium in an aluminosilicate fiber loaded with hydrogen to increase its photosensitivity has been the preferred choice. The need for printing fiber Bragg gratings directly into the core of the fiber dictated the desire for making the fiber photosensitive. Using the foregoing approach, a significant increase in the Erbium DFB fiber laser power up to 1 mW was observed (Loh et al. 1996).

The Erbium-Ytterbium combination in aluminosilicate fiber is excellent in increasing the output power of DFB fiber laser; however, the presence of hydrogen increases the intrinsic loss of the fiber. To solve this problem, one must choose germanosilicate fiber which is inherently photosensitive. However, the energy transfer between Ytterbium and Erbium ions has low efficiency in germanosilicate fibers. One way of solving this problem consists of pumping Erbium in its strongest absorption region around 520 nm. Such pumping is possible now because of the availability of laser diodes emitting in this range. Using this approach power levels up to 16 mW was obtained directly from a laser pumped at 523 nm (Loh et al. 1996). However, colour centres were reported to develop in Erbium-doped fiber laser when pumped with the green light. These colour centre absorb the pump power resulting in the severe gradual decrease in output power. Therefore, Ytterbium co-doping remains the best technique for increasing the output power of DFB fiber lasers.

Shortening the cavity length is not the only way of preventing issues resulting from spatial hole burning in linear cavity fiber lasers. Other methods such as the modified Sagnac cavity geometry have been successfully used. In this configuration, the two output fibers from a 50% fused fiber coupler are spliced forming a Sagnac cavity as illustrated in Figure 7.9. Light launched from one input propagate in both directions and recombines at the coupler. If one considers the ideal case of the lossless coupler with zero birefringences in the loop, all the light launched is collected at the coupler. Therefore, the loop acts like a 100% reflectivity mirror and is called for this reason a "loop mirror" (Mortimore 1988). If a length of the doped fiber and an isolator is placed inside the loop, the light will travel in a single and unique direction resulting in the cancellation of spatial hole burning. If a length of standard fiber terminated by a reflective mirror such as a bulk-optic mirror, a bulk diffraction grating or a fiber Bragg grating complete the cavity, a sort of linear cavity with a travelling wave in the amplifying medium is obtained. In this resulting configuration, spatial hole burning is eliminated. The free port of the coupler is then used as the output of the cavity. Cowle et al. used a similar configuration with a polished Bragg reflector as reflector mirror (Cowle et al. 1991).

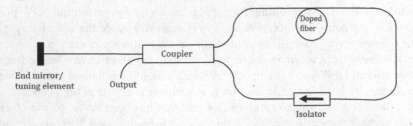

FIGURE 7.9 Schematic representation of the modified Sagnac cavity geometry.

Using a bulk diffraction grating, O'Cochlain and Mears (1992) were able to tune the laser from 1525 nm to 1568 nm (43 nm). In another report, the diffraction grating was replaced by a fiber Bragg grating and tunability from 1518 to 1535 nm was achieved by compression of the fiber Bragg grating.

Using a fiber Bragg grating as the wavelength selective element, Guy et al. (1995) found that the laser oscillated on multiple longitudinal modes. A phase-shifted fiber Bragg grating was then introduced into the loop to get rid of this problem. Another technique designed to counteract the tendency to mode hopping of this configuration was introduced by Guy 1995) who used a mirror to the free port of the cavity reflecting light into the cavity as shown in Figure 7.10.

In addition to the bulk diffraction grating and fiber Bragg grating, other tuning methods such as using reflection Mach-Zehnder interferometer has been reported (Millar et al. 1989). The difference in path length of the two arms of the Mach-Zehnder interferometers results in wavelength-dependent transmission maxima. Therefore, the device behaves as a filter, tunable by varying the path length difference. Using this technique, tunability of around 40 nm was achieved by Chieng and Minasian (1994).

The modified loop mirror configuration is very effective in achieving single frequency, mode hops free operation of a linear cavity fiber laser. However, it possesses a major drawback which is the additional loss introduced by the isolator used to force unidirectional propagation in the loop section of the laser. Pan and Shi (1994) proposed an elegant solution to get rid of this problem. In their configuration, the coupler was removed and a circulator was used instead. This circulator prevents a standing wave from forming in the cavity without the inconvenience of supplementary loss introduced by the isolator. This resulted in a neat improvement in output power where the authors reported an output power of 22 mW at 80 mW pump power, a 2 mW power gained from the 20 mW at the same pump power achieved with a 50% coupler.

Another way to prevent spatial hole burning in a linear cavity fiber laser takes advantage of the twisted mode technique already applied successfully to bulk lasers (Kane and Byer 1985). It consists of engineering the cavity in such a way that the circulating polarized light counter propagates inside the amplifying medium. In fiber, this is achieved by taking advantage of the weak birefringence of the rare-earth-doped fiber. Applying appropriate strain to such fiber permits to generate any

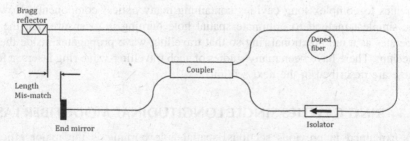

FIGURE 7.10 Schematic diagram of the modified Sagnac cavity with an end mirror to the free port.

state of polarization at a given point along with the fiber (Ghera et al. 1992). If using the technique previously described, circular polarization can be achieved, then the twisted mode technique can be easily implemented in a fiber laser. Applying this technique, Chang et al. (1996) reported an output power of 0.6 mW with a linewidth of 10 kHz.

Both modified Sagnac loop lasers and twisted-mode lasers have in common their sensitivity to environmental perturbations. A consequence of these perturbations is the random frequency shifting and mode hopping. Taking advantage of the standing wave in saturable absorbers, Horowitz et al. (1994) developed a relatively simple approach to suppress mode hopping. In an absorber, the lights transmitted and scattered by the grating are in phase, and constructively interfere. This means that the lasing frequency experiences the lowest loss and single-longitudinal-mode operation is promoted. In their experiment, Horowitz et al. used a piece of un-pumped Erbium-doped fiber as the saturable absorber (Horowitz et al. 1994).

The reason why spatial hole burning in a laser is a problem is because the grating induced in the gain medium reduces the coherence of the light circulating in the cavity and destabilizes the cavity. As already mentioned, the effect of spatial hole burning can be compensated by using a saturable absorber. However, this stabilizing effect only takes place at the condition that the pump transition rate is greater than the signal transition rate (Paschotta et al. 1997). The pump and signal rates are proportional to absorption and emission cross-section, respectively (Siegman 1986). Therefore, a higher absorption cross-section at pump wavelength than emission cross-section at signal wavelength results in a minimum influence of spatial hole burning. This was demonstrated in a Ytterbium-doped fiber linear cavity fiber laser. At 975 nm, the measured absorption cross-section in silica glass was 2.6×10^{-20} cm^2 and the measured emission cross-section was 5.13×10^{-21} cm^2 at 1040 nm. The cavity was 6 m long, and at such a high concentration, pump power was completely absorbed in the first few meters and the remaining portion of fiber acted as a saturable absorber. With a fiber Bragg grating defining oscillating wavelength, single longitudinal mode operation was observed from 70 mW pump power. Also, at that specific pump power, the laser was tuned over 5Ghz without mode hopping (Paschotta et al. 1997).

In this section, several narrow linewidth linear-cavity fiber lasers have been described. One of the major problems associated with a linear-cavity laser is spatial holes burning, which can preclude single-frequency operation. Various methods to overcome this effect have been discussed ranging from simple short two-mirror cavities to complex long cavities containing many optical components. However, the simplest method to eliminate spatial hole burning is to ensure that the laser operates as a unidirectional ring so that travelling wave propagates inside the gain medium. There have been many studies of such travelling-wave ring lasers, a few of these are described in the next section.

7.5   RING CAVITIES SINGLE LONGITUDINAL MODE FIBER LASERS

As explained in previous sections, spatial hole burning is the major efficiency limiting factor of linear cavity fiber lasers. Various methods such as shortening the cavity length or using modified Sagnac cavities were used to suppress this deleterious effect. However, shortening the cavity often results in reducing absorption capability of a fiber laser thus reducing output power. On the other hand, using a modified Sagnac cavity increases intracavity loss. Spatial holes burning can be reduced more effectively by using an optical isolator inside a ring cavity configuration to unsure unidirectional travelling of the optical field. With modern technology, it is possible to manufacture isolators with insertion losses of around 1 dB and isolation of up to 35 dB; hence, inserting an isolator inside the cavity does not significantly affect the efficiency of the fiber laser while completely blocking counter-propagating wave.

Ring cavity fiber laser is also excellent in making tunable lasers. The principle often consists of introducing a tuning element inside the cavity of a ring cavity fiber laser. Reported configurations used Fabry-Perot filters (Miller and Janniello 1990), integrated interference filters (Selvas et al. 2005), and continuous-filters overlay filters (McCallion et al. 1994). The weakness of this type of fiber lasers often originates from a polarization mode competition. This type of problems can often be avoided by using polarization-maintaining (PM) fiber as shown by Iwatsuki et al. (1990). The PM fiber they used in the cavity eliminated potential problems associated with polarization mode competition. They observed a single longitudinal mode with a linewidth of less than 28 kHz. Another problem associated with this type of fiber lasers results from the influence of the output coupling ratio. Numerical modelling showed that there is a relationship between coupling ratio and the signal to ASE ratio. The output coupling ratio was optimized to 90%. However, at this coupling ratio, the signal level to ASE was only 30 dB, while reducing the ratio to 10% yields 50 dB signal level to ASE ratio (Pfeiffer et al. 1992). In addition to reducing the signal level to ASE ratio, the variation of output coupling ratio also has a notable influence in the linewidth of the output laser beam. An output coupling of 10% resulted in a linewidth of 20 kHz, which is in disagreement with Schawlow-Townes limit. This anomalous behaviour was thought to originate from the laser sensitivity to environmental fluctuations. These fluctuations induce random phase variations which broaden the linewidth of the laser. The relation between the perturbation and the coupling ratio lies in the fact that a higher coupling ratio means higher cavity losses hence, higher sensitivity to perturbations.

The laser presented by Iwatsuki et al. was not tunable over the full range of the gain profile of the Erbium ions in the glass. This limit seems to have originated from the filter used in their design. Therefore, to address this issue, Maeda et al. (1990) used a different type of filter to broaden the tunability range up to 45 nm. They used as a filter, a Fabry-Perot Etalon constructed with a 10-μm thick birefringent liquid crystal layer sandwiched in two high-reflecting dielectric coated glass plates. An applied alternating voltage to the film, modified its refractive index in one of the birefringence axes, resulting in the modification of the optical path length. Varying the amplitude of the alternating voltage will result in a change in wavelength and the filter can be tuned in this way. The tunability observed was not continuous but rather jumped in intervals of 1 nm. This was attributed to the fact that the isolator showed some birefringence which combined with the intra-cavity polarizer acted like a birefringent filter with 1 nm separation between transmission maxima. Other tuning techniques include the use of an integrated acoustically tunable filter. Such an approach was successfully used by Wysocki et al. (1990). The cavity was to all extents similar to that of Maeda et al. (1990), except that the liquid crystal glass was replaced by the acoustic filter with a transmission bandwidth of 1.5 nm and insertion loss of 6 dB. The recorded tuning ratio was reported to 8.8 nm/MHz when controlling the filter using an RF signal. Using this filter, they were able to tune the laser over 12 nm from 1560 to 1548 nm. Similarly, to the design presented by Maeda et al., the laser was only continuously tunable after the compensation of birefringence induced by the isolator.

Another tuning mechanism was presented by Schmuk et al. and was based on tuning of an "air-space" Fabry-Perot fiber filter by varying the separation between the fiber ends (Schmuck et al. 1991). Fabry-Perot fiber filters are desirable due to their minimum reflectivity-polarization dependence. Therefore, their finesse is determined primarily by the reflectivities of the dielectric coatings applied to the fiber end. The laser showed a tendency to randomly shift to multimode operation making it difficult to be used in practical applications.

One of the most robust ways of preventing the multimode operation of single-mode fiber lasers is to increase the spectral selectivity of the cavity. In practice, this can be achieved by introducing in the cavity two Fabry-Perot cavities with different free spectral ranges. This approach was successfully used by Park et al. (1991). In their design, they used two Fabry-Perot filters in a ring cavity Erbium-doped fiber laser as shown in Figure 7.11. The two filters were used in tandem to tune the laser.

The first filter (BB) had a wider bandwidth with a free spectral range of 4 THz and 26.1 GHz of transmission bandwidth. This filter permitted coarse tuning of the laser. The second filter (NB) possessed a much narrower free spectral width of 100 GHz and a bandwidth of 1.39 GHz. To prevent back reflection from forming a cavity between the two filters, an isolator was used between them. This laser was tunable over 30 nm from 1530 to 1560 nm. Using the 100 GHz filter inside the cavity

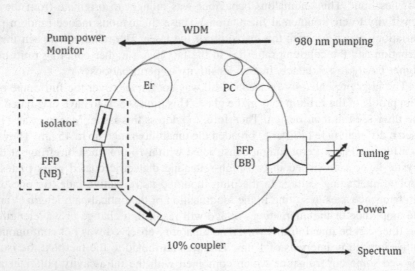

FIGURE 7.11 Erbium-doped unidirectional ring fiber laser using two fiber Fabry-Perot filters with different FSRs, one is narrowband (NB) and the other is broadband (BB).

prevented the laser from multi longitudinal mode operation. However, the laser frequency was not stable. This instability was attributed to the perturbations induced by the surrounding environment. The laser linewidth was found to be 1.4 kHz.

Spontaneous emission induced a higher level of noise affecting the output of the fiber laser. Optimizing the output coupling and also the position of the intra-cavity filters significantly reduce the influence of the spontaneous emission induced noise. It was found experimentally that placing the narrowband filter between the gain medium and the output coupler while adjusting the output coupler to 10% provided the lowest noise intensity as reported by Sanders et al. (1992) and Pfeiffer et al. (1992). The experiment was conducted by Sanders et al. who measured the excess intensity noise with the narrowband filter between the doped fiber and output coupler and then broadband filter between the doped fiber and output coupler. The results of the experiment clearly showed that a reduction in excess intensity noise was recorded with the narrowband filter between doped fiber and an output coupler (Figure 7.12).

The laser constructed by Schmuck et al. (1991) had negligible excess intensity noise. However, it was still affected by random longitudinal mode hop induced by environmental changes. Park et al. (1993) reported that the use of a variation of the Pound-Drever method (Hamilton 1989) could be used to overcome this limitation. In this method, the laser frequency was stabilized to both a transmission maximum of one of the intra-cavity mode-selecting Fabry-Perot fiber filters and to an external Fabry-Perot fiber cavity. In the Pound-Drever method, an optical field from a source such as a laser is phase modulated by an external source at a frequency ω. The resultant field consists of a carrier frequency and two sidebands. This field is directed onto an optical cavity. The signal reflected from the cavity is then monitored using a phase-sensitive detection system. An error signal depending on how much the frequency has shifted from resonance with reference frequency is generated by the phase-sensitive detector. This signal is then used to compensate the drift in the laser frequency and keep the two frequency in resonance. In the case of Park et al. (1993), an intra-cavity integrated phase modulator was used as illustrated in Figure 7.13. The error signal generated by the phase-sensitive detector was used to control either the Fabry-Perot filter in the cavity or a length of the metal-coated fiber.

The design resulted in a stable, mode hop-free operation for hours. A similar approach for fiber laser stabilization was adopted by Sabert in Nd-doped ring cavity fiber laser with two intracavity Fabry-Perot etalons. In this case, however, the error signal was obtained by a simple modulation of the optical path length of the Fabry-Perot etalons (Schmuck et al. 1991).

The use of two Fabry-Perot filters to achieve stable single longitudinal mode operation of a ring cavity rare-earth-doped fiber laser presented in the previous sections have been a very successful technique. However, the design possesses a series of drawbacks. The insertion loss associated with the use of a Fabry-Perot fiber filter is comprised between 2 and 3 dB. Also, there is the insertion loss resulting from the isolator which can be in the range of 1 dB added to this the cost associated with all these extra components, one finds that there is enough room for improvement left. Also, introducing the Fabry-Perot filters in the cavity requires that the fiber is broken, coated and realigned several times which is time-consuming

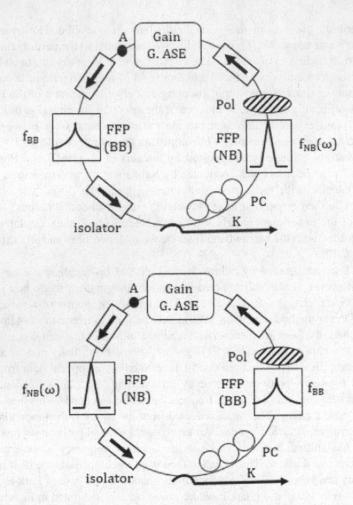

FIGURE 7.12 Schematic representation of uni-directional ring cavity fiber laser with the position of the spectral filters. BB is the broadband Fabry-Perot filter and NB is the narrowband Fabry-Perot filter.

and must be accomplished with special care. The ideal solution to this is to introduce the additional spectral filtering without light leaving the fiber. The most viable solutions to the problem were found by using either 2×2 or 3×3 fused fiber couplers (Zhang et al. 1996; Gloag et al. 1996; 1997a; 1997b).

The design using the 2×2 coupler is shown in Figure 7.14. In this design, a dual coupler fiber ring with two fused fiber couplers is used along with another fiber ring containing the doped fiber and the cavity components. The two rings assembled form a dual loop resonator. It was proved that the dual coupler fiber ring acted like a Fabry-Perot cavity in terms of its frequency response (Fraile-Peláez et al. 1991). In the configuration shown in Figure 7.14, the transmission frequency of the dual coupler fiber ring modulates the spacing between the longitudinal modes of the

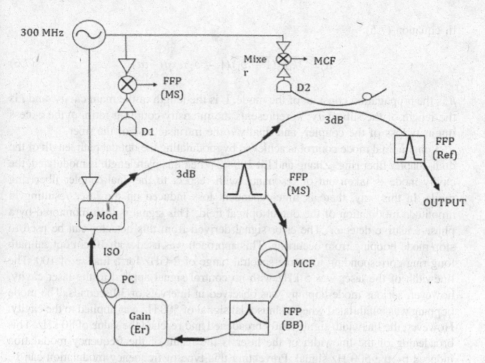

FIGURE 7.13 Schematic representation of the experimental arrangement used to lock the oscillating frequency of a unidirectional ring fiber laser to an external reference etalon. FFP, fiber Fabry-Perot; NB, narrowband; BB, broadband; PC, polarization controller; WDM, wavelength division multiplexer; Er, Erbium-doped fiber; ISO, isolator; MOD, phase modulator; MCF, metal-clad fiber; MS, the mode selecting the filter.

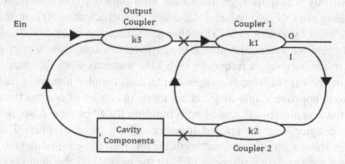

FIGURE 7.14 Schematic representation of a ring cavity unidirectional fiber laser using a dual coupler fiber ring for longitudinal mode control.

cavity. This oscillation frequency was expressed by Zhang and Lit (1994) with the expression:

$$\beta L - tan^{-1}\left(\frac{Rsin\beta l}{1 + Rcos\beta l}\right) = 2m\pi \quad (m = 1, 2 \dots) \tag{7.5}$$

In equation (7.5):

$$R = \sqrt{k_1 k_2 (1 - \delta_1)}(1 - \delta_2) exp(-\alpha l) \tag{7.6}$$

β is the propagation constant of the mode, L is the length of the main cavity, and l is the length of the subcavity; k_i represents the intensity coupling ratio, δ_i the excess intensity loss of the coupler, and finally α the intrinsic loss of the fiber.

Longitudinal mode control is achieved by modulating the optical path length of the dual-coupler fiber ring (Zhang and Lit 1994). When this path length is modulated, the cavity mode is taken out of resonance with respect to the dual-coupler fiber ring mode. In this way, there is time-dependent loss induced on the laser resulting in amplitude modulation of the output optical field. This signal is then monitored by a phase-sensitive detector. The error signal derived from this detector can be used to stop mode hopping from occurring. This approach was used with 10 cm optical path long ring corresponding to a free spectral range of 2 GHz for a finesse of 100. The linewidth of the laser was 5 kHz with no control signal applied to the laser cavity, however, serious mode-hopping was observed in intervals of 30 seconds. The mode hopping was annihilated when a sinusoidal signal of 500 Hz was applied to the cavity. However, the linewidth significantly broadened and reached the value of 80 kHz. This broadening of the linewidth of the laser is the result of the frequency modulation induced by the 500 Hz signal. Preventing this type of frequency modulation can be achieved by applying the modulation signal to both the cavity optical path length and the dual-coupler fiber ring (Zhang et al. 1996).

Variants of the system were presented to eliminate other problems related to the dual coupler ring with two fused fiber couplers such as the sensitivity of the throughput and finesse of the loop to the excess loss. In Zhang et al.'s configuration, this sensitivity was quite high due to the high-intensity coupling ratio of the couplers. Gloag et al. (1996) reduced the coupling coefficient to 50% instead of the 97 and 98% used by Zhang et al. This change reduces the finesse of the dual-coupler fiber ring and the system became less sensitive to excess loss within the ring. To prevent mode hopping, a frequency of 8 kHz was necessary this time.

One of the biggest disadvantages of the dual coupler fiber ring method to stabilize mode hop-free single longitudinal mode operation of unidirectional ring fiber laser is the complexity associated with building the fiber ring. Also, the cavity and the dual-coupler ring are spliced together with a piece of fiber. Therefore, any change in the length of this fiber resulting for example environmental fluctuations will affect the frequency response of both the cavity and the dual-coupler to some extent. To overcome the foregoing limitations, an alternative approach using a 3 × 3 fused fiber coupler is preferred. Such a cavity is presented in Figure 7.15.

Two cavities were formed closing the input and output of two of the three ports of the 3 × 3 coupler. One of the cavities is the main cavity containing the cavity elements and the other is the sub-cavity playing the role of additional spectral filtering. Numerical analysis showed that when the two sub-cavities constituting the composite resonator are in resonance, the signal coupled out of the monitor port of the 3 × 3 coupler is minimum. On the other hand, the field coupled out of the

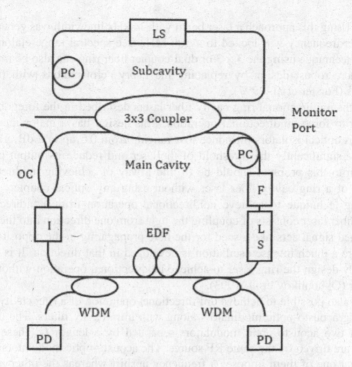

FIGURE 7.15 Unidirectional ring cavity fiber laser using a 3 × 3 fused fiber coupler.

monitor port increases when the two cavities drift out of resonance with each other. The signal generated is, therefore, frequency sensitive and can be used as an error signal for stabilizing cavity frequency change. Using this approach a linewidth of 14 kHz was reached for a laser frequency stabilized to a 300 free-spectral-range confocal etalon (Gloag et al. 1997).

The 3 × 3 coupler approach described in the foregoing section is easy to implement. However, it is also affected by the problem of frequency modulation (Sabert 1994; Park et al. 1993; Zhang et al. 1996). As seen previously, the problem could have been solved by using cavity length modulation to the main cavity but the approach suffers from a complicated implementation and the drive signals that control the cavity must be carefully controlled. A simpler and more elegant solution is to passively derive the error signal. In this approach, the change in the intra-cavity signal is monitored and provide the required signal. For example, change in polarization can be exploited to prevent mode hopping in a uni-directional ring fiber laser (Forster and Langford 1996; 1997). Due to the weak birefringence of the fiber-forming the cavity, light circulating inside it takes an arbitrary polarization, which however repeats itself at any point after a complete round trip. The amount of light leaving the main cavity to the sub-cavity, the amount of light at the monitor port is dependent on the birefringence of the sub-cavity. Therefore if the frequency changes, the intensity of light but also its polarization state changes. These changes of polarization can be sensed and used to generate an error signal to stabilize the cavity. Using this approach a laser beam with 14 kHz linewidth was generated when the laser frequency was locked to a 300 MHz free-spectral range etalon.

The technique using the 3 × 3 or dual-coupler fiber ring can also be implemented in a more robust design by replacing the Fabry-Perot etalons with fiber Bragg gratings (Forster et al. 1997).

All the unidirectional ring cavity fiber lasers described in the foregoing used an isolator to force unidirectional operation. This design has a major inconvenience which is optical isolators introduce loss ranging from 0.5 up to 2 dB. These losses increase significantly the threshold of the laser and reduce its output power. The solution to this problem should be to find a way of achieving unidirectional operation of a ring cavity fiber laser without using any optical coupler. A very interesting technique to achieve unidirectional operation from a bidirectional ring cavity fiber laser consists of coupling the light from one direction into the other. The reinjected signal acts like a seed for the field propagating in the opposite direction, therefore a much intense oscillation is developed in that direction. It is possible to carefully design the ring laser to achieve unidirectional operation without using an isolator (Okhotnikov et al. 1993).

It is also possible to achieve uni-directional operation of a ring cavity fiber laser by using acousto-optic modulators along with intra-cavity filters. The design consists of two acousto-optic modulators separated by a length L. These two modulators are driven by the same RF source. The acousto-optic modulators are chosen such that one of them imposes a frequency upshift whereas the other imposes the frequency downshift on the optical field propagating inside the fiber laser cavity. If the driving source has a frequency F and the laser has a frequency f, then the optical field propagating in the region between the two acousto-optic modulators have frequency $f - F$ and $f + F$, respectively; hence, a phase delay $\Delta\phi$ is generated between the two optical fields according to:

$$\Delta\phi = 4\pi F T_g \tag{7.7}$$

where

$$T_g = \frac{N_g L}{c} \tag{7.8}$$

N_g is the group velocity refractive index. To compensate the initial phase delay between the two optical fields, the drive signal to one of the acousto-optic modulators is delayed by $\Delta\phi$ which allows continuous tuning of the fiber laser. The unidirectional operation of the fiber laser is possible because of the tuning element placed between the two acousto-optic modulators. The working principle of this system is as follow: first, the central frequency of the filter is detuned from the laser frequency. This implies that the optical field passing through this filter will have a different frequency and will experience greater loss than the field not passing through this filter. Therefore, a preferred direction can be chosen just by varying the centre frequency of the filter. When this central frequency corresponds to the laser frequency, the two counter-propagating optical fields experience the same loss and the laser works in a bidirectional way. Using two modulators driven at 80 MHz, a narrow linewidth Nd-doped ring cavity fiber laser was obtained. The unidirectional operation was obtained with a linewidth of 10 kHz. Other simpler cavities using Faraday rotator were also presented (Kiyan et al. 1996).

REFERENCES

Amouzou, K.N. and J.-F. Bisson. 2019. "Elimination of Spatial Hole Burning in Solid-State Lasers Using Anisotropic Nanostructured Thin Films." In *Optical Interference Coatings Conference (OIC) 2019*, TB.5. OSA Technical Digest. Santa Ana Pueblo, New Mexico, Optical Society of America. doi:10.1364/OIC.2019.TB.5.

Archambault, J.-L., L. Reekie, L. Dong, and P. S. J. Russell. 1994. "High Reflectivity Photorefractive Bragg Gratings in Germania-Free Optical Fibers." In *Conference on Lasers and Electro-Optics*, 8:CWK3. OSA Technical Digest. Anaheim, CA, Optical Society of America.

Asseh, A., H. Storoy, J.T. Kringlebotn, W. Margulis, B. Sahlgren, S. Sandgren, R. Stubbe, and G. Edwall. 1995. "10 cm Yb^{3+} DFB Fiber Laser with Permanent Phase Shifted Grating." *Electronics Letters* 31 (July): 969–970. doi:10.1049/el:19950672.

Babin, S., S. Kablukov, and A.A. Vlasov. 2007. "Tunable Fiber Bragg Gratings for Application in Tunable Fiber Lasers." *Laser Physics* 17 (November): 1323–1326. doi:10.1134/S1054660X07110096.

Ball, G.A. and W.H. Glenn. 1992. "Design of a Single-Mode Linear-Cavity Erbium Fiber Laser Utilizing Bragg Reflectors." *Journal of Lightwave Technology* 10 (10): 1338–1343. doi:10.1109/50.166773.

Ball, G.A. and W.W. Morey. 1992. "Continuously Tunable Single-Mode Erbium Fiber Laser." *Optics Letters* 17 (6): 420–422. doi:10.1364/OL.17.000420.

Ball, G.A. and W.W. Morey. 1994. "Compression-Tuned Single-Frequency Bragg Grating Fiber Laser." *Optics Letters* 19 (23): 1979–1981. doi:10.1364/OL.19.001979.

Ball, G.A., W.W. Morey, and W.H. Glen. 1991. "Standing Wave Monomode Erbium Fiber Laser." *Ieee Photonics Technology Letters* 3 (7): 613–615. doi:10.1109/68.87930.

Chaoyu, Y., Jiangde, P., and Bingkun, Z. 1989. "Tunable Nd^{3+}-Doped Fiber Ring Laser - Electronics Letters." 25 (2): 101–102.

Chang, D.I., M.J. Guy, S.V. Chernikov, J.R. Taylor, and H.J. Kong. 1996. "Single-Frequency Erbium Fiber Laser Using the Twisted-Mode Technique." *Electronics Letters* 32 (19): 1786–1787. doi:10.1049/el:19961194.

Chernikov, S.V., R. Kashyap, P.F. McKee, and J.R. Taylor. 1993. "Dual Frequency All Fiber Grating Laser Source." *Electronics Letters* 29 (12): 1089–1091. doi:10.1049/el:1993 0727.

Chieng, Y.T. and R.A. Minasian. 1994. "Tunable Erbium-Doped Fiber Laser with a Reflection Mach-Zehnder Interferometer." *IEEE Photonics Technology Letters* 6 (2): 153–156. doi:10.1109/68.275413.

Chow, J., G. Town, B. Eggleton, M. Ibsen, K. Sugden, and I. Bennion. 1996. "Multiwavelength Generation in an Erbium-Doped Fiber Laser Using in-Fiber Comb Filters." *IEEE Photonics Technology Letters* 8 (1): 60–62. doi:10.1109/68.475778.

Cowle, G.J., D.N. Payne, and D. Reid. 1991. "Single-Frequency Travelling-Wave Erbium-Doped Fiber Loop Laser." *Electronics Letters* 27 (3): 229–230. doi:10.1049/el:1991 0148.

Forster, R.J. and N. Langford. 1996. "Longitudinal Mode Control of a Narrow-Linewidth Fiber Laser by Use of the Intrinsic Birefringence of the Fiber Laser." *Optics Letters* 21 (20): 1679–1681. doi:10.1364/OL.21.001679.

Forster, R.J. and N. Langford. 1997. "Polarization Spectroscopy Applied to the Frequency Stabilization of Rare-Earth-Doped Fiber Lasers: A Numerical and Experimental Demonstration." *Journal of the Optical Society of America B* 14 (8): 2083–2090. doi: 10.1364/JOSAB.14.002083.

Forster, R.J., N. Langford, A. Gloag, L. Zhang, J.A.R. Williams, and I. Bennion. 1997. "Narrow Linewidth Operation of an Erbium Fiber Laser Containing a Chirped Bragg Grating Etalon." *Journal of Lightwave Technology* 15 (11): 2130–2136. doi: 10.1109/50.641533.

Fraile-Peláez, F.J., J. Capmany, and M.A. Muriel. 1991. "Transmission Bistability in a Double-Coupler Fiber Ring Resonator." *Optics Letters* 16 (12): 907–909. doi: 10.1364/OL.16.000907.

Ghera, U., N. Konforti, and M. Tur. 1992. "Wavelength Tunability in a Nd-Doped Fiber Laser with an Intracavity Polarizer." *IEEE Photonics Technology Letters* 4 (1): 4–6. doi: 10.1109/68.124856.

Gilbert, S.L. 1991. "Frequency Stabilization of a Tunable Erbium-Doped Fiber Laser." *Optics Letters* 16 (3): 150–152. doi: 10.1364/OL.16.000150.

Glassner, D.S. and R.D. Esman. 1997. "Spatial Hole Burning in Erbium Fiber Lasers Using Faraday Rotator Mirrors." In *Proceedings of Optical Fiber Communication Conference,* 66–67. doi: 10.1109/OFC.1997.719700.

Gloag, A., R.J. Forster, and N. Langford. 1997a. "Theoretical Model and Experimental Demonstration of Frequency Control in Rare-Earth-Doped Fiber Lasers with a 3 × 3 Nonplanar Fused-Fiber Coupler." *Journal of the Optical Society of America B* 14 (4): 895–902. doi: 10.1364/JOSAB.14.000895.

Gloag, A., R.J. Forster, and N. Langford. 1997b. "Frequency Stabilization of Rare-Earth-Doped Fiber Lasers by Use of 3 × 3 Nonplanar Fused-Fiber Couplers." *Applied Optics* 36 (18): 4077–4080. doi: 10.1364/AO.36.004077.

Gloag, A., N. Langford, K. McCallion, and W. Johnstone. 1996. "Continuously Tunable Single-Frequency Erbium Ring Fiber Laser." *Journal of the Optical Society of America B* 13 (5): 921–925. doi: 10.1364/JOSAB.13.000921.

Guy, M.J., J.R. Taylor, and R. Kashyap. 1995. "Single-Frequency Erbium Fiber Ring Laser with Intracavity Phase-Shifted Fiber Bragg Grating Narrowband Filter." *Electronics Letters* 31 (22): 1924–1925. 10.1049/el:19951297.

Hamilton, M.W. 1989. "An Introduction to Stabilized Lasers." *Contemporary Physics* 30 (1): 21–33. doi: 10.1080/00107518908222588.

Harutjunian, Z.E., W.H. Loh, R.I. Laming, and D.N. Payne. 1996. "Single Polarisation Twisted Distributed Feedback Fiber Laser." *Electronics Letters* 32 (4): 346. doi: 10.1049/el:19960207.

Horowitz, M., R. Daisy, B. Fischer, and J. Zyskind. 1994. "Narrow-Linewidth, Singlemode Erbium-Doped Fiber Laser with Intracavity Wave Mixing in Saturable Absorber." *Electronics Letters* 30 (8): 648–649. doi: 10.1049/el:19940448.

Iwatsuki, K., Okamura, H. and Saruwatari, M. 1990. "Wavelength-Tunable Single Frequency and Single-Polarisation Er-Doped Fiber Ring-Laser with 1.4 KHz Linewidth." *Electronics Letters* 26: 2033–2035.

Jauncey, I.M., L. Reekie, R.J. Mears, D.N. Payne, C.J. Rowe, D.C.J. Reid, I. Bennion, and C. Edge. 1986. "Narrow-Linewidth Fiber Laser with Integral Fiber Grating." *Electronics Letters* 22 (19): 987. doi: 10.1049/el:19860675.

Jauncey, I.M., L. Reekie, R.J. Mears, and C.J. Rowe. 1987. "Narrow-Linewidth Fiber Laser Operating at 1.55 Mm." *Optics Letters* 12 (3): 164–165. doi: 10.1364/OL.12.000164.

Jauncey, I.M., L. Reekie, J.E. Towsend, D.N. Payne, and C.J. Rowe 1988. "Single-Longitudinal-Mode Operation of an Nd^{3+}-Doped Fiber Laser." *Electron Lett* 24 (1): 24–26

Kane, T.J. and R.L. Byer. 1985. "Monolithic, Unidirectional Single-Mode Nd:YAG Ring Laser." *Optics Letters* 10 (2): 65–67. doi:10.1364/OL.10.000065.

Kashyap, R., J.R. Armitage, R. Wyatt, S.T. Davey, and D.L. Williams. 1990. "All-Fiber Narrowband Reflection Gratings at 1500 Nm." *Electronics Letters* 26 (11): 730–732. doi:10.1049/el:19900476.

Kishi, N. and T. Yazaki. 1999. "Frequency Control of a Single-Frequency Fiber Laser by Cooperatively Induced Spatial-Hole Burning." *IEEE Photonics Technology Letters,* 11 (March): 182–184. doi:10.1109/68.740697.

Kiyan, R., S.K. Kim, and B.Y. Kim. 1996. "Bidirectional Single-Mode Er-Doped Fiber-Ring Laser." *IEEE Photonics Technology Letters* 8 (12): 1624–1626. doi:10.1109/68.544698.

Kogelnik, H. and C.V. Shank. 1972. "Coupled-Wave Theory of Distributed Feedback Lasers." *Journal of Applied Physics* 43 (5): 2327–2335. doi:10.1063/1.1661499.

Kringlebotn, J.T., J.-L. Archambault, L. Reekie, and D.N. Payne. 1994. "Er^{3+}:Yb^{3+}-Codoped Fiber Distributed-Feedback Laser." *Optics Letters* 19 (24): 2101–2103. doi:10.1364/OL.19.002101.

Kringlebotn, J.T. and D.N. Payne. 1994. "Single-Frequency Er^{3+}-Doped Fiber Lasers." In *Optical Amplifiers and Their Applications*, 14:ThB1. 1994 OSA Technical Digest Series. Breckenridge, CO, Optical Society of America. doi:10.1364/OAA.1994.ThB1.

Leboudec, P., P.L. Francois, E. Delevaque, J.F. Bayon, F. Sanchez, and G.M. Atephan. 1993. "Influence of Ion-Pairs on the Dynamical Behavior of Er^{3+}-Doped Fiber Lasers." *Optical and Quantum Electronics* 25 (8), pp. 501-507.

Loh, W.H., S.D. Butterworth, and W.A. Clarkson. 1996. "Efficient Distributed Feedback Erbium-Doped Germanosilicate Fiber Laser Pumped in 520 Nm Band." *Electronics Letters* 32 (22): 2088–2089.

Loh, W.H. and R.I. Laming. 1995. "1.55-Mm Phase-Shifted Distributed-Feedback Fiber Laser." *Electronics Letters* 31: 1440.

Loh, W.H., B.N. Samson, Z.E. Harutjunian, and R.I. Laming. 1996. "Intracavity Pumping for Increased Output Power from a Distributed Feedback Erbium Fiber Laser." *Electronics Letters* 32 (13): 1204–1205. doi:10.1049/el:19960778.

Lux, O., S. Sarang, O. Kitzler, D.J. Spence, and R.P. Mildren. 2016. "Intrinsically Stable High-Power Single Longitudinal Mode Laser Using Spatial Hole Burning Free Gain." *Optica* 3 (8): 876–881. doi:10.1364/OPTICA.3.000876.

Maeda, M.W., J.S. Patel, D.A. Smith, C. Lin, M.A. Saifi, and A. von Lehman. 1990. "An Electronically Tunable Fiber Laser with a Liquid-Crystal Etalon Filter as the Wavelength-Tuning Element." *IEEE Photonics Technology Letters.* doi:10.1109/68.63221.

Masuda, H., A. Takada, and K. Aida. 1992. "Modeling the Gain Degradation of High Concentration Erbium-Doped Fiber Amplifiers by Introducing Inhomogenous Cooperative Up-Conversion." *Journal of Lightwave Technology* 10 (12): 1789–1799.

McCallion, K., W. Johnstone, and G. Fawcett. 1994. "Tunable In-Line Fiber-Optic Bandpass Filter." *Optics Letters* 19 (8): 542–544. doi:10.1364/OL.19.000542.

Millar, C.A., D. Harvey, and P. Urquhart. 1989. "Fiber Reflection Mach-Zehnder Interferometer." *Optics Communications* 70 (4): 304–308. doi:10.1016/0030-4018(89)90324-6.

Miller, C.M. and F.J. Janniello. 1990. "Passively Temperature-Compensated Fiber Fabry-Perot Filter and Its Application in Wavelength Division Multiple Access Computer Network." *Electronics Letters* 26 (25): 2122–2123. doi:10.1049/el:19901365.

Morkel, P.R., G.J. Cowle, and D.N. Payne. 1990. "Travelling-Wave Erbium Fiber Ring Laser with 60 KHz Linewidth." *Electronics Letters* 26 (10): 632–634. doi:10.1049/el:19900414.

Mortimore, D.B. 1988. "Fiber Loop Reflectors." *Journal of Lightwave Technology* 6 (7): 1217–1224. doi:10.1109/50.4119.

Nilsson, J., B. Jaskorzynska, and P. Blixt. 1993. "Performance Reduction and Design Modification of Erbium-Doped Fiber Amplifiers Resulting from Pair-Induced Quenching." *IEEE Photonics Technology Letters* 5 (12): 1427–1429. doi:10.1109/68.262560.

O'Cochlain, C.R. and R.J. Mears. 1992. "Broadband Tunable Single Frequency Diode-Pumped Erbium Doped Fiber Laser." *Electronics Letters* 28 (2): 124–126. doi:10.1049/el:19920077.

Okhotnikov, O., A.L. Ribeiro, and J.R. Salcedo. 1993. "All-fiber Traveling-wave Laser with Nonreciprocal Ring Configuration." *Applied Physics Letters* 63 (December): 2726–2728. doi:10.1063/1.110341.

O'Sullivan, M.S., J. Chrostowski, E. Desurvire, and J.R. Simpson. 1989. "High-Power Narrow-Linewidth Er^{3+}-Doped Fiber Laser." *Optics Letters* 14: 438.

Pan, J.J. and Y. Shi. 1994. "Tunable Er^{3+}-Doped Fiber Ring Laser Using Fiber Grating Incorporated by Optical Circulator or Fiber Coupler." *Electronics Letters* 31: 1164–1165.

Park, N., J.W. Dawson, and K.J. Vahala. 1993. "Frequency Locking of an Erbium-Doped Fiber Ring Laser to an External Fiber Fabry–Perot Resonator." *Optics Letters* 18 (11): 879–881. doi:10.1364/OL.18.000879.

Park, N., J.W. Dawson, K.J. Vahala, and C. Miller. 1991. "All Fiber, Low Threshold, Widely Tunable Single-frequency, Erbium-doped Fiber Ring Laser with a Tandem Fiber Fabry–Perot Filter." *Applied Physics Letters* 59 (19): 2369–2371. doi:10.1063/1.106018.

Park, Y.K., J.M.P. Delavaux, R.M. Atkins, and S.G. Grubb. 1993. "Stable Single-Mode Erbium Fiber-Grating Laser For Digital Communication." *Journal of Lightwave Technology* 11 (12): 2021–2025. doi:10.1109/50.257965.

Paschotta, R., J. Nilsson, L. Reekie, A.C. Trooper, and D.C. Hanna. 1997. "Single-Frequency Ytterbium-Doped Fiber Laser Stabilized by Spatial Hole Burning." *Optics Letters* 22 (1): 40–42. doi:10.1364/OL.22.000040.

Pfeiffer, T., H. Schmuck, and H. Bulow. 1992. "Output Power Characteristics of Erbium-Doped Fiber Ring Lasers." *IEEE Photonics Technology Letters* 4 (8): 847–849. doi:10.1109/68.149883.

Sabert, H. 1994. "Suppression of Mode Jumps in a Single-Mode Fiber Laser." *Optics Letters* 19 (2): 111–113. doi:10.1364/OL.19.000111.

Sabert, H., A. Koch, and R. Ulrich. 1991. "Reduction of Spatial Hole Burning by Singlephase Modulator in Linear Nd^{3+} Fiber Laser." *Electronics Letters* 27 (23): 2176–2177. doi:10.1049/el:19911346.

Sanders, S., N. Park, J.W. Dawson, and K.J. Vahala. 1992. "Reduction of the Intensity Noise from an Erbium-Doped Fiber Laser to the Standard Quantum Limit by Intracavity Spectral Filtering." *Applied Physics Letters* 61 (16): 1889–1891. doi:10.1063/1.108379.

Schawlow, A.L. and C.H. Townes. 1958. "Infrared and Optical Masers." *Physical Review* 112 (6): 1940–1949. doi:10.1103/PhysRev.112.1940.

Schmuck, H., T. Pfeiffer, and G. Veith. 1991. "Widely Tunable Narrow Linewidth Erbium Doped Fiber Ring Laser." *Electronics Letters* 27 (23): 2117–2119. doi:10.1049/el:19911312.

Selvas, R., I. Torres-Gomez, A. Martinez-Rios, J.A. Alvarez-Chavez, D.A. May-Arrioja, P. LiKamWa, A. Mehta, and E.G. Johnson. 2005. "Wavelength Tuning of Fiber Lasers Using Multimode Interference Effects." *Optics Express* 13 (23): 9439–9445. doi:10.1364/OPEX.13.009439.

Siegman, A.E. 1986. Lasers. Mill Valley, California: University Science Books.

Storoy, H., B. Sahlgren, and R. Stubbe. 1997. "Single Polarisation Fiber DFB Laser." *Electronics Letters* 33 (1): 56–58. doi:10.1049/el:19970033.

Têtu, M., F. Ouellette, C. Latrasse, R. Larose, D. Stepanov, and M.A. Duguay. 1994. "Simple Frequency Tuning Technique for Locking a Singlemode Erbium-Doped Fiber Laser to the Centre of Molecular Resonances." *Electronics Letters* 30 (10): 791–793. doi:10. 1049/el:19940537.

Toebben, H. 1993. "Temperature-Tunable 3.5 Micron Fiber Laser." *Electronics Letters* 29(8): 667–669.

Wysocki, P.F., M.J.F. Digonnet, and B.Y. Kim. 1990. "Electronically Tunable, 1.55-Mm Erbium-Doped Fiber Laser." *Optics Letters* 15 (5): 273–275. doi:10.1364/OL.15. 000273.

Yelen, K., M.N. Zervas, and L.M.B. Hickey. 2005. "Fiber DFB Lasers with Ultimate Efficiency." *Journal of Lightwave Technology* 23 (1): 32–43. doi:10.1109/JLT.2004. 840037.

Yennadhiou, P. and S.A. Cassidy. 1990. "D-Fiber Grating Reflection Filters." In *Proceedings of the Conference on Optical Fiber Communications*. Optical Society of America.

Zhang, J. and J.W.Y. Lit. 1994. "All-Fiber Compound Ring Resonator with a Ring Filter." *Journal of Lightwave Technology* 12 (7): 1256–1262. doi:10.1109/50.301819.

Zhang, J., W.Y. Lit, and G.W. Schinn. 1996. "Cancellation of Associated Frequency Dithering in a Single-Frequency Compound-Ring Erbium-Doped Fiber Laser." *IEEE Photonics Technology Letters* 8 (12): 1621–1623.doi: 10.1109/68.544697.

Zhang, J., C.-Y. Yue, G.W. Schinn, W.R.L. Clements, and J.W.Y. Lit. 1996. "Stable Single-Mode Compound-Ring Erbium-Doped Fiber Laser." *Journal of Lightwave Technology* 14 (1): 104–109. doi:10.1109/50.476143.

Zhang, Z., P. Miao, J. Sun, S. Longhi, N.M. Litchinitser, and L. Feng. 2018. "Elimination of Spatial Hole Burning in Microlasers for Stability and Efficiency Enhancement." *ACS Photonics* 5 (8): 3016–3022. doi:10.1021/acsphotonics.8b00800.

Zyskind, J.L., E. Desurvire, J.W. Sulhoff, and D.J. Di Giovanni. 1990. "Determination of Homogeneous Linewidth by Spectral Gain Hole-Burning in an Erbium-Doped Fiber Amplifier with GeO/Sub 2/:SiO/Sub 2/ Core." *IEEE Photonics Technology Letters* 2 (12): 869–871. doi:10.1109/68.62013.

Zyskind, J.L., V. Mizrahi, D.J. DiGiovanni, and J.W. Sulhoff. 1992. "Short Single Frequency Erbium-Doped Fiber Laser." *Electronics Letters* 28 (15): 1385–1387. doi:10.1049/el: 19920881.

8 High-Power Fiber Lasers

8.1 INTRODUCTION

The development of high-power fiber lasers was fuelled by the maturation of active and passive fiber technologies combined with the availability of optical components such as bright laser diodes beam combiners, fiber Bragg gratings, isolators, etc. One of the main advantages of high-power fiber lasers over the other type of high-power lasers is their superior beam quality at high powers. For the same collimated beam size, higher beam quality radiation results in smaller spot size at the workpiece. Alternatively, for the same spot size at the workpiece, higher beam quality requires smaller and therefore lighter focusing beam optics.

High-power fiber lasers are not very different from other types of lasers in terms of their resonant cavities. The particularity of this type of lasers mostly comes from their pumping and doped fiber technologies. Therefore, much attention will be given to increasing pumping capacity as well as optimizing the design to improve light absorption, thus efficiency. The efficiency of high-power fiber lasers is often expressed in terms of their brightness; hence, parameters that affect brightness are described in detail.

Finally, at the powers of interest, non-linear effects in fibers are inevitable even if the fibers used as a gain medium are rather short. The limitations imposed by non-linear effects namely, stimulated Raman and Brillouin scattering, second harmonic generation, and four waves mixing are described and ways to mitigate them proposed. Other detrimental effects like photodarkening, optical damage, and transverse mode instabilities (TMI) also affect the fibers and must be understood for the design and operation of high-power fiber lasers.

8.2 HIGH-POWER FIBER LASERS DESIGN

8.2.1 Cavity Configurations

Similar to other types of fiber lasers, high-power fiber lasers can be of either linear or ring cavity configuration. What differentiates the high-power fiber lasers from other types of fiber lasers is the amplifying medium which in the case of high-power fiber laser is a rare-earth-doped fiber with a special geometry. The basic configuration is often a linear cavity using mirrors of which one is highly reflective (R > 99%) the other with lower reflectivity, constitutes the output coupler. The pump is launched through a dichroic mirror at one end which transmits the pump and reflects the signal or vice versa. An improved and wavelength selective version of this configuration replaces the bulk mirrors with fiber Bragg gratings. One has to note that in the latter configuration because of the order of powers involved, FBGs must have additional protection. Finally, the last configuration consists of intra-cavity

DOI: 10.1201/9781003256380-8

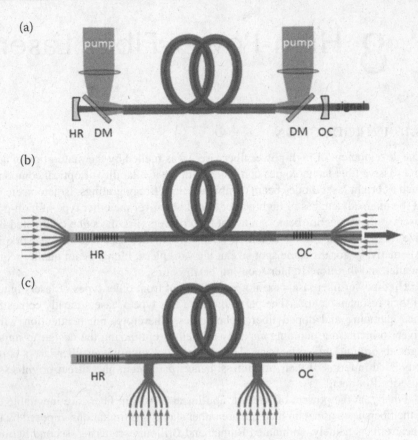

FIGURE 8.1 Cladding pumped fiber laser configurations: (a) hybrid end-pumped, (b) all fiber end-pumped, and (c) all fiber intra-pumped.

pumping which eliminates the requirement for additional protection on the fiber Bragg grating. Pumping is often done by launching a beam resulting from the combination of several diode lasers into the core or the cladding. The main configurations for linear cavity high-power fiber lasers are illustrated in Figure 8.1.

In the hybrid configuration, careful alignment of mirrors and excellent finishing of fiber end facet is required to reduce or eliminate undesirable back reflections as well as surface damage (Limpert et al. 2002; Böhme et al. 2012). These design concepts make this type of fiber laser more useful in lab demonstrations (Jeong et al. 2004; Liu et al. 2004; Nilsson et al. 2004) than in mass production. However, the device can be made less cumbersome by butt-coupling the mirror or directly depositing them onto the fiber facet (Minelly et al. 2011), the preceding approach is convenient for multi-mode operation. The all-fiber configuration, on the other hand, is preferable for the case where one wants to take advantage of the benefit of fiber technology, namely, the low cost of maintenance to improve the reliability of industrial systems (Gapontsev et al. 2005; Norman et al. 2006; Norman and Zervas 2007; Rath 2012; Kliner et al. 2011; Zervas 2010). Often written in single-mode fibers, fiber Bragg gratings offer the advantage of determining the output beam modality in addition to imposing the lasing wavelength.

8.3 INCREASING OUTPUT POWER: CLADDING PUMPING

The major disadvantage of fiber lasers is their poor light coupling. In fact, because of the small numerical aperture of the fiber due essentially to the tiny diameter of its core, only a fraction of launched light is coupled in the doped core. As a result, fiber lasers were for a long time low power devices with power ranges in milliwatts in the continuous-wave regime. To increase the output power, Snitzer proposed cladding pumping (Snitzer et al. 1988; Po et al. 1989) which has ever since been the most common power-enhancing technique. Cladding pumping has enabled fiber laser power scaling from few milliwatts to kilowatts in the continuous-wave regime propelling the technology from a laboratory interest to a reliable commercial product.

In cladding pumping scheme, low brightness pump power is launched in the cladding instead of core. Because the cladding possesses a bigger transverse section and bigger numerical aperture, energy propagate inside it in several modes. The multiple modes overlap with the doped core as they propagate down the fiber and energy is absorbed. The generated light, however, is obtained at a wavelength such that only the fundamental mode (FM) is excited; therefore, it is only trapped in the core and cannot propagate in the cladding. This result in a highly intense and brighter laser output light. In this aspect, cladding-pumped fiber lasers can be regarded as excellent brightness converters as explained by equation (8.1) giving the brightness or radiance of a beam.

$$B = \frac{P}{A\Omega} \quad (8.1)$$

where B is the beam brightness, P is the beam power, A is the area, and Ω is a solid angle. For the fiber lasers with a circular cross-section, the area is $A = \pi r^2$ and $\Omega = \pi NA^2$, where r is the radius of the core and NA is the numerical aperture.

The maximum pump power that can be launched in the fiber is given by equation (8.2) as follow:

$$P_p^{in} = B_p(\pi r_{cl}^2)(\pi NA_{cl}^2) = \frac{1}{2}B_p\lambda_p^2 N_p \quad (8.2)$$

where P_p^{in} is the launched power, B_p is the pump brightness, and λ_p and N_p are the pump wavelength and number of modes, respectively. The number of modes is given by $N_p = V_{cl}^2/2$, where V_{cl} is the cladding V-number. It can be seen from equation (8.2) that the maximum launched power is proportional to the number of propagating modes. Therefore, to scale-up fiber laser output power, one must engineer a way of increasing the number of supported modes. However, increasing output power cannot go beyond a certain limit imposed by the finite numerical aperture and non-linear effects that become important with increasing light intensity.

The other practical way of improving pump absorption efficiency, thus output power, is related to the shape of the core section. The overlap between propagating cladding modes and the doped core is not 100%. It can be seen from equation (8.2) that majority of cladding modes miss completely the doped core with only a fraction of 10% only overlapping with the core. These modes crossing the core correspond to LP_{0m} modes or meridional modes in the ray optics picture as well as a small fraction of the LP_{m0} modes corresponding to the skew rays in the ray optics approach. The majority of modes represented by ($m \neq n$) entirely miss the core. To improve mode absorption, several fiber section shapes have been proposed. The principle here is trying to breaking the cladding rotational symmetry and increase the fraction of cladding modes overlapping with the active core. Physically this can be understood as scattering skew modes towards the active core. Figure 8.2 illustrates the most common cladding-pumped fiber cross-sections.

The numerical aperture of the pump depends on the outer cladding material. To achieve an acceptable numerical aperture this material must be carefully chosen. In the most common cases, fluorinated material is used for outer cladding, resulting in NAs around 0.46. The drawback of using this material is the special cooling arrangement required to limit heating of the polymer. Such kind of cooling arrangement is not in the case of low index fluorosilicate glass where a much moderate cooling system can be used. However, the advantage of better thermal handling capability is compensated by the rather low numerical aperture of this material (~0.22–0.26), considerably reducing the amount of acceptable launched pump power. In such a case special care must be taken when coiling the fiber. Large coiling diameter is advised to minimize additional pump bending loss.

The main limitations to the effective operation of high-power fiber lasers were highlighted in the previous section and some solutions proposed. However, to achieve high power emission more compactly and robustly, new designs of fiber amplifying

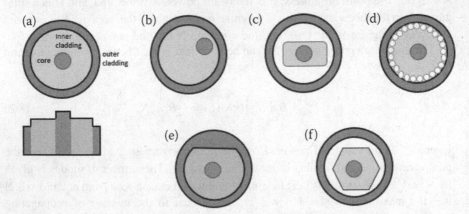

FIGURE 8.2 Most used double-clad-pumped fibers design: (a) centred core, (b) off-centred, (c) rectangular inner cladding, (d) cylindrically arranged air holes inner cladding, (e) D-shape inner cladding, and (f) hexagonal inner cladding.

mediums are required. One of these novel designs is the novel jacketed-air clad (JAC) type of fibers. This design relies on a row of cylindrically arranged air hole whose role is to provide a glass/air interface resulting in a numerical aperture greater than 0.8 (Sahu et al. 2001; Wadsworth et al. 2004). The same arrangement is also desirable to reduce the cladding diameter significantly, therefore, increasing the pumping rate and enable efficient three-level operation for 980 nm pumping wavelength, for example (Selvas et al. 2003).

The cladding perturbation must be chosen in such a way that mode scrambling is maximized while keeping scattering loss at an acceptable value (Åslund et al. 2006). For example, it was observed that the effective NA of circular fiber is bigger than that of rectangular fiber for the same material composition. The pumping loss was also reported to be lower in circular fiber than in rectangular cladding fibers (Liu and Ueda 1996). To maintain polarization, Boron-doped stress elements can be incorporated in the inner cladding. In the cladding-pumped high-birefringence fibers, the low refractive index of the borosilicate stress-applying element has an additional role. They prevent pump light from being trapped in them, and their presence combined with the applied stress results in a further scrambling of the helical modes within the inner cladding (Kliner et al. 2001; Pare 2003). Among the several inner-cladding shapes presented in Figure 8.2 the more efficient are the ones with multiple truncations.

Breaking the cladding rotational symmetry only, will not increase the effective absorption of the pump power over the entire length and for all wavelengths. Therefore, the propagating modes must be continuously mixed over the entire length of the fiber to maximize the pump absorption. This can be achieved by using periodic/quasi-periodic fiber bending along the fiber (Nilsson et al. 2003; Tünnermann et al. 2010) or by fiber tapering (Filippov et al. 2008).

In Figure 8.3, the pump absorption variation is plotted against fiber length for different wavelengths with and without mode mixing. The mode mixing of the reported example has been achieved by periodic bending. It can be seen that in the absence of mode mixing, pump absorption on the two wavelengths remains small and saturates quickly as the fiber length increases. In this case, there is no big advantage for pumping Ytterbium at its absorption peak of 976 nm over 940 nm. One consequence of this is the observed deformation of the absorption spectrum at all pumping schemes (Mortensen 2007). Mode mixing helps not only to increase absorption but also to restore the absorption linearity with length. However, it can be seen that the absorption still shows some saturation with length at 976 nm. This shows that stronger mixing is required for wavelengths with higher absorptions. The consequence of this non-linear pump absorption is the deterioration of signal quality and fiber lasers cavity efficiency (Cheng et al. 2006). However, this situation results in non-uniform heat generation along the fiber with most of the heat being generated over short length at the launching side.

8.3.1 BRIGHTNESS ENHANCEMENT IN CLADDING PUMPED FIBER LASERS

As mentioned, early brightness enhancement is the result of combinations of several factors including cladding pumping, excellent beam quality, and power scaling. To

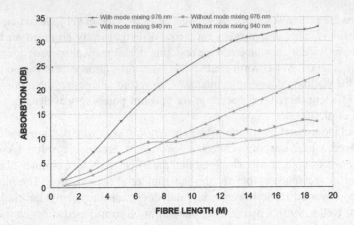

FIGURE 8.3 Effect of longitudinal mode mixing and wavelength on pump absorption in Ytterbium-doped alumina-silicate fiber (Mortensen 2007).

quantify the brightness enhancement, one define the brightness enhancement factor which can be written as (Zervas and Codemard 2014):

$$\eta_B = \frac{B_s^{out}}{B_p^{in}} = \eta_{oo} \left(\frac{\lambda_p}{\lambda_s} \right)^2 \frac{V_{cl}^2/2}{V_{co}^2/2} \approx \eta_{oo}^* \frac{N_{cl}}{N_{co}} \tag{8.3}$$

where B_s^{out} and B_p^{in} are the brightness of the output signal and input pump, respectively. $\eta_{oo} = P_s^{out}/P_p^{in}$ is the optical-to-optical power conversion efficiency, where P_s^{out} is the output laser power and P_p^{in} is the pump power. N_{cl} and N_{co} are the number of modes in the cladding and the core, respectively.

8.3.2 CLADDING-PUMPING SCHEME

Besides designing fiber with appropriate geometry, high-power fiber lasers also require the use of appropriate pumping schemes. These pumping schemes are separated between end pumping and side pumping. The schematic representations of the main end pumping and side pumping schemes are represented in Figure 8.4(a)–(c) and (d)–(f), respectively. Figure 8.4(a) illustrates a combination of pump modules using a focusing mirror, whereas Figure 8.4(b) is a scheme based on Tapered Fibers Bundle (TFBs; Kosterin et al. 2004; Tobergte and Curtis 2013). In this pumping method, several semiconductor lasers are combined and used to pump the fiber laser. The number of individual semiconductor diodes lasers used is only limited by the cladding diameter. In Figure 8.4(c), the end pumping scheme uses wavelength-multiplexed pump modules. The multiplexing is done using several wavelength division multiplexers (Liu et al. 2004). In the geometric combination, the resultant brightness of the combined pump does not exceed the brightness of the input pump modules taken individually because of the presence of the combiner component. On the other hand,

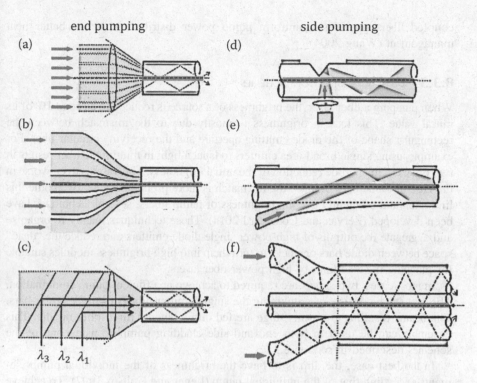

FIGURE 8.4 Main cladding-pumping schemes: (a)–(c) end pumping, (d)–(f) side pumping.

the brightness of the pumping module is increased in the case of wavelength-multiplexed schemes because multiple aligned beams are superimposed.

The side pumping scheme of Figure 8.4(e) relies on a total internal reflection effect. The pump beam is reflected on a V-groove milled in the cladding (Goldberg et al. 1996). The biggest disadvantage of such an approach is the difficulty of implementing it for power exceeding certain values, therefore scalability in the kilowatt levels is very hard to achieve. Another approach of side pumping consists of an angle-polished (Xiao et al. 2010) or tapered (Gapontsev and Samartsev 1999) pump fiber fused into the cladding of another fiber which constitutes the signal fiber. The biggest disadvantage of this method is that part of the launched light leaks out of the cladding, resulting in an efficiency loss of the laser (Ou et al. 2004). Finally, the approach in Figure 8.4(f) makes the use of a multi-fiber assembly of fibers held together by a low-index polymer cladding. The polymer ligand is applied as usual during the fiber drawing (Grudiain et al. 1999; Grudinin et al. 2000). If needed, the individual fibers can be accessed independently by removing the polymer. This type of fiber assembly offers the advantage of having multiple ports for launching pump power. In a variant form of this scheme, the multiple pumps and signal fibers are held together using an external heat-shrunk tube (Polynkin et al. 2004). The key advantage of the side pumping is the fact that the signal and pump beam paths are kept separated leading to a simplified and more robust design. Lastly, it is worth mentioning that the side-pumping scheme based on evanescently coupled fibers provides a uniform pump power distribution, hence better heat management (Wang 2004).

8.3.3 Pump Combination Schemes

When pumping a fiber laser, the brightness of a source is reduced to close to 1% of its initial value. This loss in brightness is mostly due to the mismatch between the rectangular shape of the diode emitting aperture and the receiving circular fiber. For example, using single broad-area emitters to launch light in multimode fiber results in two orders of magnitude reduction of the initial brightness. This ratio is even worse in the case of diode bars because the mismatch is more pronounced. To overcome this limitation and efficiently use the brightness of pump fibers, several techniques have been developed (Zervas and Codemard 2014). These techniques mainly re-organize and aggregate the outputs of high-power single-diode emitters and reduce the "dead" space between diode bars or stacks to turn them into high brightness modules suitable for pumping the claddings of high power fiber lasers.

In most cases, two stages are required to achieve an efficient pump combination. The first stage consists of combining the single-emitter diodes or diodes mini-bars. The beams resulting from this stage are fed into a tapered multi-fiber bundle. This ensemble can be used in both end and side cladding pumping using one of the schemes described in section 8.3.2.

In the best case, the aim is to have the brightness of the individual pump elements exceeding that of the individual pump (Leger and Goltsos 1992). To achieve this requirement, there must be mutual coherence between the different lasers constituting the pump. In this case, the output of the entire composed source behaves like a single spatial super mode. The resulting output beam brightness is, therefore, the sum of the contributing laser's brightness. Another way of increasing the brightness of a combined source consists of using pump devices with different Eigen-properties such as wavelength or polarization. The beams from such pumps are then multiplexed using passive bulk elements such as diffraction gratings or polarization beam splitters (Leger and Goltsos 1992). In the case of wavelength combination, however, the spectral brightness is reduced and such a configuration can only be beneficial in the case of rare-earth active ions with broad absorption spectra like Ytterbium at 940 nm.

Karlsen et al. reported the development of a high brightness diode module able to deliver over 100 W of optical power into a 105 μm, and 0.15 of NA fiber at 976 nm wavelength. The reported electrical to optical efficiency was above 40% (Karlsen et al. 2009). In another report, seven individual diodes with brightness around 0.21 W/(μm^2 sr) were combined using a 7:1 fused combiner resulting in an output of 500 W coupled into a 220 μm, 0.22 NA fibers. The resulting brightness was reduced to around 0.086 W/(μm^2 sr). Nowadays, there exist commercially available pump modules that use TFBs and geometrically combined single, large-area emitters providing up to 140 W in a 106.5 μm, 0.22 NA fibers. The resulting brightness, in this case, was 0.1 W/(μm^2 sr) ("JDSU Product Catalogue, JDS Uniphase Corporation" 2013). An example of wavelength-beam-combined pump modules using diode lasers has also been presented. This module provided 200 W at 91 nm in a 200 μm, 0.22 NA fiber (Xiao et al. 2012). The resulting brightness was 0.04 W/(μm^2 sr); this value is less than half of what was achieved with geometrically combined single emitters. Using a 7:1 tapered fused bundle, seven of such modules were combined to deliver around 1.5 kW pump power in a 400 μm diameter fiber.

A loss budget analysis shows that most of the pump insertion loss results from the brightness loss across the tapered fiber combiner. Therefore, by proper design and optical loss minimization in both directions, TFBs with kW- level power handling capabilities are easily achievable (Xiao et al. 2011; 2012; Wetter et al. 2007). A choice between different pump modules using several combination technologies is dictated by the resulting wall-plug efficiency, the lifetime as well as the final cost. These conditions are well fulfilled when using a large area, single emitters, and small-size diode bars.

So far, the TFBs have been used extensively to combine multimode pumps. However, a combination of SM to MM TFBs is also possible. Seo et al. demonstrated an efficient pump beam combining technique. In their work, up to 331 9/125 μm single-mode fibers with 0.12 NA were combined (Seo et al. 2011). Apart from being used to scale-up output power from single-mode fiber lasers into multimode output beams, this SM to MM TFBs technique can also be used to combine short-wavelength, SM fiber lasers for cladding in-band, also known as tandem pumping of other fiber lasers (Alam et al. 2008). This is the preferred pumping scheme for single-mode diffraction-limited fiber lasers with output power exceeding 3 kW (Alam et al. 2008; Zhu et al. 2011).

8.4 RARE-EARTH IONS FOR HIGH-POWER FIBER LASERS

Several rare-earth ions have been used as dopants for rare-earth doped fiber lasers including Neodymium, Erbium, Ytterbium, Thulium, and Holmium. However, only Ytterbium and Thulium are massively used to build high-power fiber lasers for their excellent power scaling capabilities with output power values over 1 kW.

Yb^{3+} energy level is very interesting because of its simple two energy levels and its lasing wavelength is in the range of 1 μm. A typical energy level diagram of Ytterbium in silica is illustrated in Figure 8.5(a). The figure shows that $^2F_{5/2}$ and $^2F_{7/2}$ energy levels are split into sub-levels because of Stark splitting. The exact sub-level splitting is a function of the glass composition and Ytterbium ions concentration (Barua et al. 2008).

Laser working in both three and four-level systems is only allowed by stark splitting. Whether the system will be three- or four-level will depend on the choice of pump and lasing wavelength. In Figure 8.5(b), the emission and absorption cross-section as a function of wavelength is represented in aluminosilicate and phosphosilicate fibers (Pask et al. 1995). The emission and absorption spectra, as well as metastable level lifetime, depend strongly on the host material. One of the most recently used hosts is phosphosilicate glass. Although it reduces significantly the absorption and emission cross-section, this type of glass possesses the advantage of allowing high doping concentration without clustering effects, thus, avoiding luminosity quenching as well as photodarkening (PD). As mentioned earlier, Ytterbium possesses a simplified energy level structure. This simplified structure

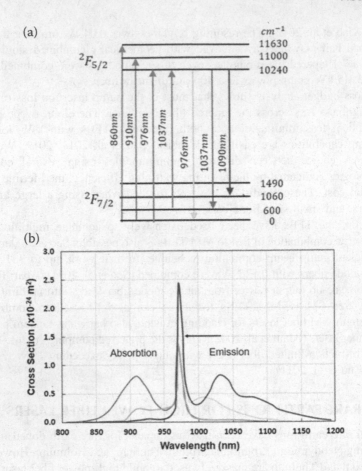

FIGURE 8.5 (a) Typical energy level diagram of Yb^{3+} ions in silica, (b) typical emission and absorption cross-sections in aluminosilicate (thicker lines) and phosphosilicate (thinner lines) fibers (the arrow shows the peak emission and absorption for phosphosilicate fibers).

prevents effect like cooperative up-conversion, excited-state absorption increasing the overall efficiency of the fiber laser.

The broadband absorption spectrum of Ytterbium, extending from ~850 to ~1080 nm enables pumping at various wavelengths even multi-wavelength pumping schemes resulting in easy power scaling. Another advantage of this broadband absorption spectrum is that it allows the use of un-stabilized low-cost pumps. This significantly simplifies the design, therefore the final cost of the high-power fiber laser. In addition to this, the small and finite absorption around the 1010–1020 nm band allows in-band, also known as tandem pumping using high-brightness fiber lasers which facilitate power scaling to multi-kilowatts levels (Gapontsev et al. 2005). On the other side, the broadband emission spectrum allows tunability on the wide range going from 980 nm to about 1100 nm as well as short pulse amplification.

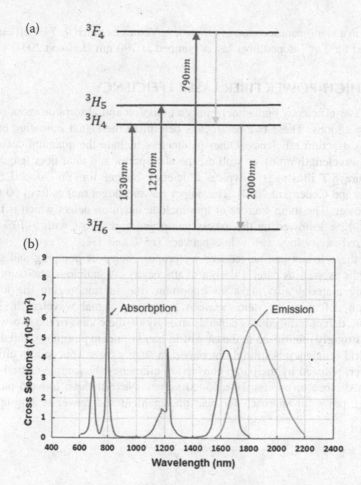

FIGURE 8.6 (a) Typical energy level diagram of Tm^{3+} (only lower levels are shown), and (b) typical emission and absorption cross-section in aluminosilicate fibers (Jackson and King 1999).

The other rare-earth ion of interest is the Tm^{3+}. Its energy level structure is much more complex than that of Ytterbium. In Figure 8.6(a), only the lower energy levels are represented. The diagram highlights the main ground state absorption transition as well as the interesting Tm^{3+} emission transition of Tm^{3+} ions at 2 μm. Figure 8.6(b) illustrates the absorption and emission cross-sections of Thulium in aluminosilicate fibers (Jackson and King 1999). The most interesting absorption band as far as high-power fiber lasers are concerned is the one lying around 1600 nm because it permits in-band pumping that can be easily achieved using high-power Erbium-doped fiber lasers, and the one at 790 nm because of the availably of powerful diodes pumps capable of pumping the fiber laser through cross-relaxation (Jackson 2004). By the mechanism of cross-relaxation, it is possible to create two excited Tm^{3+} ions in the upper level of the energy diagram for each single absorbed photon. This practically results in a significant increase in efficiency. Values as much as 74% efficiency were reported for Tm^{3+}-doped fiber lasers pumped at 790 nm (Jackson 2004).

8.5 HIGH-POWER FIBER LASER EFFICIENCY

Fiber laser efficiency budget depends on emission and absorption cross-sections of the doping ions. These two parameters determine the signal saturation energy and power extraction efficiency. Other parameters include the quantum defect (pump/signal wavelength ratio), as well as, the absorption, and total fiber length.

Figure 8.7 illustrates a typical efficiency budget for Yb^{3+}-doped fiber laser (Zervas and Codemard 2014). The output power budget ranges from 60 to 80% of total power. The main sources of loss include quantum defect which is the largest with 8–12% followed by the excess pump and signal loss with 5–10% and non-optimized cavity loss with values between 0.5% and 1%.

All these losses can be reduced by proper choice of pumping and signal wavelengths as well as careful design of the cavity. A judicious choice of core and cladding material also plays an important role in minimizing the losses. The quantum defect is the ratio between pump and signal wavelength. Therefore, quantum defect-induced loss depends strongly on the choice of the two-wavelength and is strongly minimized in the case of in-band (tandem pumping), and the optical-to-optical efficiency is strongly increased in such a case. This gain in efficiency is, however, reduced by the pump conversion efficiency; hence, the overall efficiency remained close to previously reported values. Nevertheless, in-band pumping remains a powerful approach for heat management and power scaling-up of high-

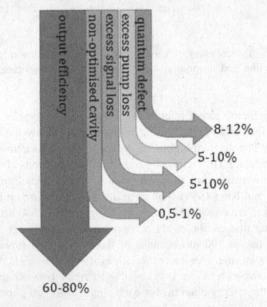

FIGURE 8.7 Typical efficiency budget for Yb^{3+}-doped fiber lasers.

power fiber lasers. Power-scaling up to 3 kW can be achieved by relying solely on in-band pumping.

8.6 BEAM QUALITY ANALYSIS

The beam quality is the measure of how well a laser beam can be focused. Beam quality is quantified by defining the beam quality factor M^2. Physically, this parameter tells how fast the beam diverges with respect to a diffraction-limited Gaussian beam with the same wavelength and waist diameter. This parameter is expressed as the ratio of the beam-parameter product BPP_Q of the beam of interest—which is the product of the beam waist radius and the far-field beam divergence angle—with the beam-parameter product of a diffraction-limited Gaussian beam BPP_G. The expression of beam quality factor can be written as:

$$M^2 = \frac{BPP_Q}{BPP_G} = \frac{\omega_Q \theta_Q}{\omega_G \theta_G} = \frac{\pi}{\lambda} \omega_Q \theta_Q \tag{8.4}$$

where ω_Q is the mode-filed radius and θ_Q is the far-field divergence of the beam. In the case of an ideal Gaussian beam $M^2 = 1$, while in practice $M^2 > 1$. It is worth mentioning that the beam quality of a fiber laser can be tailored by a careful fiber design. In Figure 8.8, the variation of beam quality factor as a function of V number (normalized frequency) for a step-index fiber for several transverse modes is illustrated.

From Figure 8.8, one can see that the beam quality factor of the LP_{01} FM is bigger than that for a V number near 1.5. This behaviour is attributed to the increasing evanescent field into the cladding, thus the field losing its Gaussian profile as explained by the Marcuse's formula (Marcuse 1977). The same behaviour can also be observed for higher-order modes at values of V number corresponding to their cut-off wavelength where their field starts to extend deeply into the cladding.

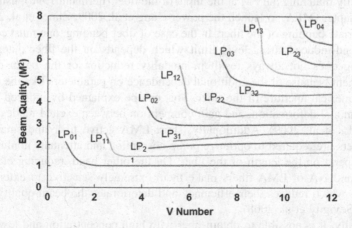

FIGURE 8.8 Beam quality as a function of V number for several modes of a step-index fiber.

In the case of an incoherent multimode beam, the beam quality factor is given by the weighted average of the beam quality factors of the participating modes. The weighted average $<M^2>$ of a multimode beam is approximated by:

$$\langle M^2 \rangle \approx \pi (2p_{max} + 1)/12 = \pi/12 + V/3 \qquad (8.5)$$

where $p_{max} = 2V/\pi$ and both polarizations are taken into account. Here one has to keep in mind that the weighted average $<M^2>$ is obtained when all supported modes are equally excited.

8.7   DOPED AND UNDOPED FIBERS

Increasing the output power of fiber lasers can be done through the development of large mode area (LMA) fibers. These fibers possess a large effective mode area, yet they only support a single or only a few transverse modes. Their effect is to reduce peak intensities and attenuate nonlinear effects. Because of their relatively big core section, these fibers end up being highly multimoded. In order to operate them on single-mode (SM) or low-mode (LM) regime, high order mode (HOM) differential losses must be introduced. However, while HOM differential loss is introduced, care must be taken to keep the loss of the FM as low as possible. Achieving this is very challenging in the context of big section fiber where the mode effective index differences decrease. Also, due to small perturbations resulting from fiber drawing or packaging-induced micro bending, small modal effective index difference promotes modal coupling between the FM and HOMs with a consequence of efficiency and beam quality badly affected.

8.7.1   RARE-EARTH DOPED FIBERS

Controlling the modality in MM fibers can be achieved by several techniques such as exactly matching the FM at the input of the fiber (Fermann 1998), using a mode-field adapter (MFA) to launch the power (Jung et al. 2009; Zimer et al. 2011) or by appropriate bending of the fiber. In the case of fiber bending, one must consider the safe bend-induced stress level limit which depends on the fiber outer diameter. Bending does not always result in modality reduction or the beam quality improvement because of the additional dependence on parameters like the core radius and numerical aperture of the fiber. This can be explained by induced mode coupling, modal deformation, and gain competition between excited modes (Schermer 2007; Li et al. 2009). Additionally, short LMA active fiber and small cladding diameters are required to optimize pump absorption and attenuate nonlinear effects that depend on the length of the fiber. On the other hand, reducing cladding diameter and NA of LMA fibers make them extremely sensitive to external perturbations which reduces their efficiency and deteriorates the beam quality (Fermann 1998; Sévigny et al. 2008).

Finally, it is possible to obtain fiber with high concentration and low numerical aperture using phosphate glasses (Wu et al. 2001). Phosphate glasses make it possible to improve dopant solubility and increase concentration without triggering luminosity quenching effects due to ion-ions interactions. However, the presence of phosphorus, in addition to increasing refractive index, also results in refractive index central dip as a consequence of the difficulty to control phosphorus evaporation during the MCVD fabrication technique. The consequence of this is the loss of the uniformity of the refractive index across the core area resulting in poor quality laser beam and fragile power stability of LMA fiber lasers. To alleviate this drawback, methods that put much less stringent requirements on the MCVD process have been proposed (Arismar Cerqueira 2010; Wadsworth et al. 2003; Canat et al. 2008). Another approach consists of adjusting the hole size and spacing of the airholes around the LMA doped core of a photonic crystal fiber (PCF) resulting in fiber single-mode operation for core diameter up to 100 μm (Teodoro et al. 2010). Other LMA fibers have been demonstrated in PCFs with the use of the "modal sieve" effect (Russell 2006). Further improvement of the LMA is the very large-mode-area (VLMA) fiber design. This design uses a large-pitch fiber (LPF). This type of fibers is PCF with a hole-to-hole spacing of approximately ~10–30 times the operating wavelength.

All the above designs show different degrees in terms of HOM leakage or delocalization into the cladding (Limpert et al. 2012; Dauliat et al. 2013). LPFs have less strict fabrication tolerances with respect to LMA PCFs; therefore, core diameters of 135 μm for a mode field diameter (MFD) of approximately 130 μm could be achieved in passive operation. These fibers must be kept straight during operation to avoid bend-induced MFD collapse. Also, at increased values of power, a change in waveguide properties of Yb^{3+}-doped LPF was observed. This change was attributed to the temperature and is responsible for a significant reduction of FM MFD increasing modality (Jansen et al. 2012).

Another LMA fiber design is based on the chirally coupled core (CCC) concept. This concept provides resonant filtering of HOMs and enables effective SM index-guiding. Using this technology, fibers with core diameters exceeding 50 μm have been produced (Chi-Hung et al. 2007). Finally, it is worth mentioning that single-mode operation of LMA MM fiber can be obtained by engineering the dopant distribution inside the core in such a way that the gain is predominantly attributed to the FM (Sudesh et al. 2008).

8.7.2 Undoped Fibers for High-Power Applications

The role played by LMA passive fiber consists essentially of delivering the generated power to the place of usage. Some applications require that power is delivered to distances up to 10–20 meters; therefore, LMA passive fibers of similar length are desirable. When designing this kind of fiber, special care must be taken to avoid excessive spectral broadening and apparition of temporal instabilities resulting from nonlinear effects, including stimulated Brillouin scattering (SBS), stimulated Raman scattering (SRS), self-phase modulation (SPM), and four-wave mixing (FWM). Similar care must also be taken to avoid beam quality degradation due to modal scrambling.

As mentioned before, the single-mode operation depends on the bending radius. For example, a standard single-mode fiber with 30 μm core diameter, an NA of 0.06, and effective areas of approximately 360 μm^2 can still be SM at a 5-cm bending radius. This is the lowest limit achievable in today's fabrication techniques (Li et al. 2009). However, standard single step-index fibers can show single-mode operations for diameters below 15 μm when kept straight (Richardson et al. 2010). To extend the FM area beyond the step-index limit, several designs of fibers have been proposed and experimentally verified. These designs use typical multimode fiber to which different techniques are applied to eliminate HOMs. An illustration of these techniques is the W-type fiber, in which a refractive-index dip is used around the core to convert the second-mode beyond cut-off into a leaking mode (Kawakami and Nishida 1974), or LMA segmented-cladding fibers, which convert all the HOMs into leaking modes (Rastogi and Chiang 2001).

A technique called leakage-channel fiber (LCF) involving the use of a small number of air holes inserted in the silica cladding to provide the required differential loss for the HOMs has been demonstrated. LCFs with core diameters in the range of 170 μm and effective areas over 10 000 μm^2 have been reported (Dong et al. 2009). The hollow micro-structured fibers have also been used for power delivery. Their key advantage is the capacity to effectively contain nonlinear effects. This is because the majority of the power (>99%) is guided in air.

8.8 DETRIMENTAL EFFECT AFFECTING HIGH-POWER OPERATION

The non-linear effects taking place in optical fibers are mostly χ^3 phenomena and therefore depend on intensity and length of the medium. The intensity dependence makes them more severe at high peak power in pulsed, kilowatt operation in continuous-wave regime. In silica, the non-linear effect is almost not observable due to its small nonlinear coefficient (in the range of 3.2×10^{-16} cm^2/W). At high-intensity values and lengths, the non-linearity enhancement factor (NEF) (Zervas and Codemard 2014) becomes large and non-linear effects are observable.

The values measured and reported, as it is shown in Figure 8.9, make optical fibers one of the most non-linear media.

In Figure 8.9, the NEF is plotted against the gain of the fiber laser and it can be seen that its value decreases significantly with gain. This is because the power distribution varies greatly along the fiber length for higher values of gain. Figure 8.9 shows that NEF also decreases with an increased value of V number because when the V number increases, the FM effective radius decreases. Quantitatively, when a value of V > 10 is reached, the NEF decreases by about one order of magnitude. The third observation is that NEF varies inversely proportional to the wavelength. Calculations show that moving the wavelength to values approaching 2 μm tends to half the non-linearity strength.

Non-linear effects are the most limiting factor in scaling up the power in fiber lasers. Their effect is fundamentally transferring energy to unwanted spectral regions resulting in laser operation disturbance. To reduce their deleterious effects in

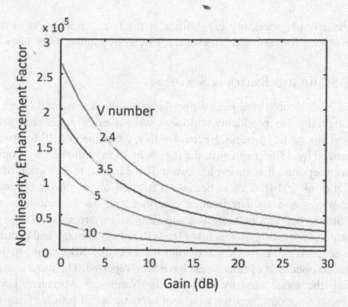

FIGURE 8.9 Nonlinear enhancement factor as a function of amplifier gain, for $L_0 = 15\ m$ and $NA = 0.1$.

fiber laser operations, proper fiber design and appropriate spectral filtering are recommended.

8.9 STIMULATED BRILLOUIN AND STIMULATED RAMAN SCATTERING

SRS and SBS are both inelastic non-linear processes. Both are related to the interaction of light waves with optical or acoustic-phonon waves, respectively (Singh et al. 2007).

8.9.1 STIMULATED RAMAN SCATTERING

Depending on the conditions, the effect of SRS and SBS can be beneficial or detrimental. Special fiber designs permit to minimize their influence or use then advantageously to enhance wavelength coverage of available fiber laser output spectrum. One way to suppress the influence of SRS is to use phosphosilicate fiber, allowing high power output for a relatively short length of the fiber. Another approach consists of using a filter element that filters out the stock component of the Raman scattering spectra and allows length independent nonlinearity threshold, this is particularly interesting in fiber beam delivery in material processing (Kim et al. 2006; Ma et al. 2011).

In the Raman scattering process, a photon of high energy is converted into one of lower energy. The lost energy is carried away by a phonon (a quantum of the lattice vibration). The lower energy corresponding to a longer wavelength, SRS results in a spectral broadening that can make the design of focusing optics difficult and badly affect the overall processing capabilities of the laser. In addition to this, the presence of strong forward and backward SRS can destabilize fiber laser cavities.

8.9.2 STIMULATED BRILLOUIN SCATTERING

SBS in a Yb^{3+} double-clad pulsed fiber laser has the detrimental effect of breaking up the original pulse, producing multipeak sub-pulses of higher peak power. SBS is a random (Stochastic) process. Because of this, the shape of SBS backward pulses is characterized by a sharp spike (in the ranges of 10 ns) followed by a quite long tail. It has been reported that strong backward SBS can lead to distortion of the forward pulse (Su et al. 2014). This is because a fraction of the power is transferred into acoustic waves and another fraction contributes to the generation of the forward second Stokes that can be seen as a superimposed sharp spike. This generation of the forward propagating Stokes often leads to optical damage and catastrophic fiber failure. Several techniques for reducing the effect of SBS while maintaining the single-mode operation of fiber lasers have been reported. The most common include increasing the mode area by reducing the Numerical Aperture (Taverner et al. 1997), using fiber designed for a tailored acoustic speed profile (Gray et al. 2009), increasing the effective linewidth via phase modulation (Li 2009; Khitrov et al. 2010) taking advantage of laser gain competition or simply using highly doped fibers to make sure that much of pump power is absorbed in a short length fiber. Factors that contribute to SBS threshold include self-heating and strong rate of change of temperature due to pump absorption (Jeong et al. 2007).

In the case of MM fibers, an additional SBS signal in the forward direction called forward SBS (FSBS) has been observed beside normal backwards propagating SBS. This process triggers power exchange between the forward propagating modes LP_{01} and LP_{11}. The presence of FSBS can be explained by the much longer lifetime of the generated phonon compared to the case of conventional SBS. However, unlike the case of normal SBS, FSBS does not depend significantly on the laser linewidth. It is one of the examples where increasing the optical mode area result in an enhanced actual power. Finally, FSBS is possible in both active and passive fibers and can result in power transfer from the FM into the high order modes, therefore, deteriorating the output beam quality.

8.10 SELF-PHASE MODULATION AND FOUR-WAVE MIXING

There is a dependency between the phase of an optical wave and its intensity, expressed via the Kerr effect (Buckingham and Pople 1997). This effect induces a spectral broadening which has a strong dependence with the pulse shape. Observation proved that the broadening is more pronounced for pulses with steep leading and trailing edges. SPM is a limiting factor in the quest of scaling up power especially in short pulse fiber laser using phase locking or coherent combination of multiple lasers.

FWM is a third-order nonlinear (χ^3) effect. In this process, the interaction of two photons results in the annihilation of both of them and the creation of Stoke and anti-Stoke photons. The frequency of the resulting photons is defined by the energy conservation principle. The efficiency of FWM process depends strongly on the exact phase matching between the signals involved because it is a coherent process. In high-power fiber lasers, the FWM pump is represented by the length-dependent exponentially growing signal. As a consequence, a substantial FWM phenomenon is generated despite the phase mismatching (Brooks and Teodoro 2005).

FWM also appears to be stronger in birefringent or multimode fibers. This is because phase matching is greatly promoted by the fact that the Stokes and anti-Stokes signals can propagate in different modes, hence, different group velocities (Limpert et al. 2002; Stolen and Bjorkholm 1982).

SRS and FWM are not always detrimental. In applications like efficient super-continuum generation, they can be beneficial. In other to efficiently build super-continuum source, a special design must be applied to the fibers. High-power peak powers trigger the spectral broadening via FWM and Raman shifting resulting in large bandwidth desirable for super-continuum sources (Fermann and Hartl 2009; Travers et al. 2008).

8.11 INFLUENCE OF OPTICAL DAMAGE

Optical damage is present in fibers at nanoseconds and sub-nanoseconds pulses. This destructive effect is associated with an electron avalanche effect (Smith et al. 2009). The effect is only reported for electron densities beyond the critical value of $2 \times 10^8 \ \mu m^{-3}$, where the plasma frequency is close enough to the optical frequency causing a strong absorption of the propagating light. In these conditions, the accumulated absorbed energy is high enough to cause the melting or the fracture of the silica glass.

Another type of optical damage is associated with the onset of strong SBS in pulsed fiber lasers. This later damage is the result of internal stresses induced by the acoustic waves generated by the SBS material/light interaction (Kashyap and Blow 1988).

8.12 INFLUENCE OF PHOTODARKENING

Optical power losses in a medium can grow when the medium is irradiated with light at certain wavelengths. This phenomenon known as photodarkening (PD) can lead to serious performance degradation of high-power fiber lasers. The optical loss in the event of PD is the result of colour centre formation in the glass matrix. The exact origin of the colour centre is still not clearly understood. Reported studies tend to prove that there is a connection between the PD rate and the Ytterbium inversion (Jetschke et al. 2007; Koponen et al. 2007). This order of inversion was quantified and reported to be in the range between 3.5 (Jetschke et al. 2007) and 7 (Koponen et al. 2007). Also, the PD-induced loss distribution is not uniform along the active fiber length. This loss variation is proportional to the square of the Ytterbium inversion. It has also been reported that the output power of the laser is inversely proportional to the operating temperature variation (Zervas et al. 2011). Finally, the PD entirely depends on the core material. By carefully choosing the

Ytterbium concentration and other dopants such as Phosphorous (Engholm and Norin 2008), Aluminium (Jetschke et al. 2012; Morasse et al. 2007), or Cerium (Engholm et al. 2009), the PD can be critically reduced or even eliminated. PD has also been reported in Thulium-doped fibers (Broer et al. 1993; Brocklesby et al. 1993).

8.13 INFLUENCE OF TMI

TMI in the LMA fiber amplifier is observed as a sudden output beam quality deterioration when output power exceeds a given threshold. The threshold is particularly low when large core diameter fibers are used. The origin of TMI, its nature, and its characteristics have been theoretically investigated using several numerical models (Smith and Smith 2011; 2013a; 2013b; Jauregui et al. 2012; Hansen et al. 2013; Dong 2013; Hu et al. 2013; Naderi et al. 2013; Ward 2013). It is strongly agreed that TMI is the consequence of forward mode coupling caused by thermo-optically induced gratings formation. The modes interact through either Stimulated Thermal Rayleigh Scattering (STRS) (Smith and Smith 2011; 2013b), or static or dynamic mode interactions (Naderi et al. 2013; Ward 2013).

TMI effects have been observed and reported in many fibers including PCF, LPF high power Yb^{3+}-doped fiber amplifiers (Eidam et al. 2011a; 2011b; Otto et al. 2012; Laurila et al. 2012; Haarlammert et al. 2012; Karow et al. 2012; Ward et al. 2012; Jansen et al. 2012) and solid-core LMA fiber amplifiers (Engin et al. 2013; Brar et al. 2014). It was reported that the highest values of TMI threshold were found with step-index single-mode actively cooled fibers, with an optimum diameter of 22 and 30 μm (Wirth et al. 2010). TMI threshold depends on the type of fibers, even if its value always decreases directly proportional to the core diameters irrespective of the fiber used. However, observation has demonstrated that TMI threshold is affected by several amplifier characteristics. In this regard, it was observed that HOM excess leakage doubles the TMI threshold of DMF (Laurila et al. 2012), gain tailoring results in significant TMI threshold increase in SI-GT fibers, and does not affect the PCF-GT fibers in the same proportions. Fiber cooling greatly increases the TMI threshold of PCF-SAT fibers (Ward et al. 2012). Finally, in the case of PM25 polarization-maintaining fiber, the TMI threshold was only increased by pumping the amplifier outside of the absorption peak and increasing the input power. Under the same conditions, a non-PM fiber has no onset of TMI up to power as high as 1 kW (Brar et al. 2014). It should also be noted that no TMI was reported in fiber with diameters of less than 20 μm even for output powers higher than 2 kW (Huang et al. 2014).

8.14 WORKING REGIMES OF HIGH-POWER FIBER LASERS

As seen in the previous sections of this chapter, scaling up the power of fiber lasers relies on tailoring the core and the cladding of the fiber to allow full control of the beam modality, and non-linear effects.

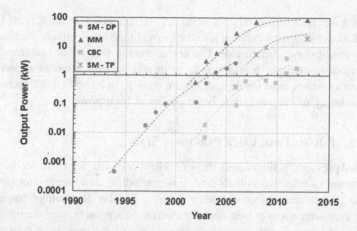

FIGURE 8.10 Power evolution with time in high-power Yb^{3+}-doped cladding pumped lasers. [Multimode (MM), single-mode diode-pumped (SM-DP), single-mode tandem pumped (SM-TP), and coherent beam combination (CBC)].

8.14.1 SINGLE-FIBER, SINGLE-MODE CW OUTPUT POWER

Another factor that contributed to power evolution in fiber lasers is the maturity in pumping technology. Starting from the early low brightness, the technology evolves to combined high brightness diode modules and in-band tandem pumping (Connor and Shiner 2011).

Figure 8.10 provides a map of power evolution of a single-mode Ytterbium-doped fiber laser with different pumping schemes, namely diode-pumped and in-band tandem pumped (SM-TP).

Since the introduction of high brightness diode modules, single-mode output power has been growing steadily. At this rate, an output power of about 36 kW of the diffraction-limited beam could be reached with an arbitrary increase of MFD. However, in practice, an arbitrary increase of MFD cannot be achieved because of thermal and SRS effects. Also, modal instability effects can reduce critically output power (Dawson et al. 2008). Requirements to scale up the single-mode fiber laser output beyond the value of 3 kW include in-band (or tandem) pumping to reduce the thermal load on the fiber amplifier. Applying this technique has resulted in single-mode output in the order of 20 kW (Shiner 2013). In theory, using inbound pumping can help reach about 70 kW of single-mode operation (Zhu et al. 2011). To reach such order of powers in SM, very large diameter fiber will be needed. As already mentioned, achieving single-mode operation for such a diameter is not possible. Therefore, robust fiber lasers can reasonably be operated at about 25 kW quasi-diffraction-limited single-mode output which is rather weak compared to the MM output of 100 kW obtained using a geometric incoherent combination (Figure 8.10). With such powers, the future of fiber lasers into industrial and directed energy applications is very promising.

The best option to increase the output power of fiber lasers remains the coherent combination of several SM fiber lasers. With this technique, multi-kW quasi-Gaussian outputs have been achieved (Yu et al. 2011; Redmond et al. 2012). In coherent combination, the optical gain but also the thermal load is effectively distributed among several contributing fiber strands. The key advantage is to break free from the vicious circle where increasing power results in increasing heat generation and increasing nonlinearity which in terms degrades beam quality. In a nutshell, the technique offers an opportunity to keep excellent beam quality at high power.

8.14.2 PULSED FIBER LASER PARAMETER SPACE

As stated previously, fiber lasers show excellent response to power scaling. However, they are not as effective in their peak power handling and energy storage. The good news is that in this aspect, the huge progress in fiber technology has enabled the drastic improvement of pulsed fiber laser performance, increasing the market share of this type of lasers in industrial applications. One of the most used configurations of pulsed fiber lasers is the Q-switched fiber laser. This type of lasers is excellent because of its simplicity and ability to deliver high-energy pulses (Wang and Xu 2007; Renaud et al. 2001; Chen et al. 2004; Limpert et al. 2005; Wei et al. 2013; Zheng et al. 2012). High-power Q-switched fiber lasers with pulse durations below 10 ns have been reported. The laser used a short-length Ytterbium-doped rod-type PCF as the gain medium. The reported pulse energy was above 0.5 mJ for excess power of more than 30 W, in single-mode and 100 kHz repetition rate (Limpert et al. 2005). The repetition rate is a limiting factor because, at repetition rates, ASE signal has enough time to depopulate the metastable level, reduce the inversion, and limit the pulse energy (Wei et al. 2013; Zheng et al. 2012).

An excellent way of controlling the shape of the pulse is to use a MOPA structure amplifying a signal from a diode seed. In this way, a large range of pulse duration and repetition rates can be controlled. This is possible because of the combination of the fast turn on and off characteristic of modern semiconductor lasers and the high gain and high power scalability of fiber amplifiers. This type of pump modulated pulsed fiber lasers result in high-performance sources with excellent beam quality for many applications such as marking and material microprocessing (Zervas et al. 2006; Horley et al. 2007).

8.15 OTHER FIBER LASERS

The possibility offered by fiber laser is immense in terms of covered wavelength. This is due to the variety of rare-earth ions used ranging from Er^{3+}, Yb^{3+}, Nd^{3+} to name a few. However, the development of high-power fiber lasers in other materials has been hampered by the technical limitation like small quantum defects as well as unattainable thermal management requirements. One of the most promising types of fiber lasers is the Tm^{3+}-doped fiber laser which emits in the eye-safe 2-μm region required for certain type of material processing. Scaling-up lasers in this range can be very interesting in some applications. A Holmium fiber was developed producing output power higher than 400 W in the 2.05–2.15 μm wavelength region (Hemming et al. 2014). Using non-linear processes such as SRS combined with properly designed and optimized fiber can produce a significant amount of output power in the

1–2.5 µm range. Another interesting application of fiber laser is using them as a seed to produce supercontinuum. A CW supercontinuum source extending in the visible range spectra has been demonstrated using a PCF pumped at 1070 nm with a 400 W single-mode CW Ytterbium-doped fiber laser. The supercontinuum cover 1300 nm with average power up to 50 W and spectral power over 50 mW/nm (Travers et al. 2008). In another experiment, high-energy pulsed supercontinuum with a span region of 450–1750 nm was also demonstrated. The energy spectra density was 1 nJ/nm in the visible region only. The potential application for such a source is in STED microscopy (Almeida et al. 2009).

8.16 CONCLUSION

Fundamentals of high-power fiber lasers were reviewed in this chapter. High-power fiber lasers are a subject of intense research in modern photonics because of their applications in material processing. Throughout the chapter, physical concepts and the technological aspect of high-power fiber lasers were presented in a simplified manner. It was shown that increasing the output power of the fiber laser often corresponds to designing a fiber capable of absorbing pump power and limit the effect of deleterious effects such as non-linear phenomena. This is done by carefully choosing appropriate shapes and dimensions for the core of the doped fibers. Several types of fiber were proposed such as LMA fiber as well as various types of PCFs. Pumping is also important in scaling-up the output power and several pumping schemes were proposed to increase the output power while preserving the beam quality. Finally, among all the rare-earth ions, the most promising for high-power fiber lasers are Ytterbium and Thulium.

The topic of high-power fiber laser is vast and can be the subject of an entire book in itself; therefore, what was presented here was a comprehensive review to complement what has been the subject of discussion of the book.

REFERENCES

Alam, S.U., A.T. Harker, R.J. Horley, F. Ghiringhelli, M.P. Varnham, P.W. Turner, M.N. Zervas, and S.R. Norman. 2008. "All-Fiber, High Power, Cladding-Pumped 1565nm MOPA Pumped by High Brightness 1535nm Pump Sources." In *2008 Conference on Lasers and Electro-Optics and 2008 Conference on Quantum Electronics and Laser Science*, 1–2. doi:10.1109/CLEO.2008.4551873.

Almeida, P.J., P. Dupriez, J. Clowes, E. Bricchi, M. Rusu, and A.B. Grudinin. 2009. "Ultrafast Fiber Lasers and Nonlinear Generation of Light." In *2009 11th International Conference on Transparent Optical Networks*, 1–4. doi:10.1109/ICTON.2009.5185253.

Arismar Cerqueira, S. 2010. "Recent Progress and Novel Applications of Photonic Crystal Fibers." *Reports on Progress in Physics* 73 (2): 24401. doi:10.1088/0034-4885/73/2/024401.

Åslund, M., S.D. Jackson, J. Canning, A. Teixeira, and K. Lyytikäinen-Digweed. 2006. "The Influence of Skew Rays on Angular Losses in Air-Clad Fibers." *Optics Communications* 262 (1): 77–81. doi:10.1016/j.optcom.2005.12.050.

Barua, P., E.H. Sekiya, K. Saito, and A.J. Ikushima. 2008. "Influences of Yb3+ Ion Concentration on the Spectroscopic Properties of Silica Glass." *Journal of Non-Crystalline Solids* 354 (42): 4760–4764. doi:10.1016/j.jnoncrysol.2008.04.020.

Böhme, S., S. Fabian, T. Schreiber, R. Eberhardt, and A. Tünnermann. 2012. "End Cap Splicing of ´ Photonic Crystal Fibers with Outstanding Quality for High-Power Applications." In laser-based Micro and Nanopackaging and Assembly VI vol. 8244, p. 824406(1-9). International Society for Optics and Photonics.

Brar, K., M. Savage-Leuchs, J. Henrie, S. Courtney, C. Dilley, R. Afzal, and E. Honea. 2014. "Threshold Power and Fiber Degradation Induced Modal Instabilities in High-Power Fiber Amplifiers Based on Large Mode Area Fibers." In fiber laser XI: Technology, Syatems, an Applications (vol. 8961, p. 8911R). International Society for Optics and Photonics.

Brocklesby, W.S., A. Mathieu, R.S. Brown, and J.R. Lincoln. 1993. "Defect Production in Silica Fibers Doped with Tm3+." Optics Letters 18 (24): 2105–2107. doi:10.1364/OL.18.002105.

Broer, M.M., D.M. Krol, and D.J. DiGiovanni. 1993. "Highly Nonlinear Near-Resonant Photodarkening in a Thulium-Doped Aluminosilicate Glass Fiber." Optics Letters 18 (10): 799–801. doi:10.1364/OL.18.000799.

Brooks, C.D. and F.D. Teodoro. 2005. "1-MJ Energy, 1-MW Peak-Power, 10-W Average-Power, Spectrally Narrow, Diffraction-Limited Pulses from a Photonic-Crystal Fiber Amplifier." Optics Express 13 (22): 8999–9002. doi:10.1364/OPEX.13.008999.

Buckingham, A.D. and J.A. Pople. 1997. "Theoretical Studies of the Kerr Effect I: Deviations from a Linear Polarization Law." In Optical, Electric and Magnetic Properties of Molecules, 49–53. Elsevier.

Canat, G., S. Jetschke, S. Unger, L. Lombard, P. Bourdon, J. Kirchhof, V. Jolivet, A. Dolfi, and O. Vasseur. 2008. "Multifilament-Core Fibers for High Energy Pulse Amplification at 1.5 Mm with Excellent Beam Quality." Optics Letters 33 (22): 2701–2703. doi:10.1364/OL.33.002701.

Chen, M.-Y., Y.-C. Chang, A. Galvanauskas, P. Mamidipudi, R. Changkakoti, and P. Gatchell. 2004. "27-MJ Nanosecond Pulses in M2 = 6.5 Beam from a Coiled Highly Multimode Yb-Doped Fiber Amplifier." In Conference on Lasers and Electro-Optics/International Quantum Electronics Conference and Photonic Applications Systems Technologies, CTuS4. Technical Digest (CD), San Francisco, California, Optical Society of America.

Cheng, X.P., P. Shum, J. Zhang, and M. Tang. 2006. "Analysis of Nonlinear Effective Absorption Coefficient of Double Cladding Fiber." IEEE Region 10 Annual International Conference, Proceedings/TENCON, January. 10.1109/TENCON.2006.344167.

Chi-Hung, L., G. Chang, N. Litchinitser, D. Guertin, N. Jacobsen, K. Tankala, and A. Galvanauskas. 2007. Chirally Coupled Core Fibers at 1550-Nm and 1064-Nm for Effectively Single-Mode Core Size Scaling. doi:10.1109/CLEO.2007.4452926.

Dauliat, R., D. Gaponov, A. Benoit, F. Salin, K. Schuster, R. Jamier, and P. Roy. 2013. "Inner Cladding Microstructuration Based on Symmetry Reduction for Improvement of Singlemode Robustness in VLMA Fiber." Optics Express 21 (16): 18927–18936. doi:10.1364/OE.21.018927.

Dawson, J.W., M.J. Messerly, R.J. Beach, M.Y. Shverdin, E.A. Stappaerts, A.K. Sridharan, P.H. Pax, J.E. Heebner, C.W. Siders, and C.P.J. Barty. 2008. "Analysis of the Scalability of Diffraction-Limited Fiber Lasers and Amplifiers to High Average Power." Optics Express 16 (17): 13240–13266. doi:10.1364/OE.16.013240.

Dong, L. 2013. "Stimulated Thermal Rayleigh Scattering in Optical Fibers." Optics Express 21 (3): 2642–2656.

Dong, L., T. Wu, H.A. McKay, L. Fu, J. Li, and H.G. Winful. 2009. "All-Glass Large-Core Leakage Channel Fibers." IEEE Journal of Selected Topics in Quantum Electronics 15 (1): 47–53. doi:10.1109/JSTQE.2008.2010238.

Eidam, T., S. Hädrich, F. Jansen, F. Stutzki, J. Rothhardt, H. Carstens, C. Jauregui, J. Limpert, and A. Tünnermann. 2011a. "Preferential Gain Photonic-Crystal Fiber for

Mode Stabilization at High Average Powers." *Optics Express* 19 (9): 8656–8661. doi: 10.1364/OE.19.008656.

Eidam, T., C. Wirth, C. Jauregui, F. Stutzki, F. Jansen, H.-J. Otto, O. Schmidt, T. Schreiber, J. Limpert, and A. Tünnermann. 2011b. "Experimental Observations of the Threshold-Like Onset of Mode Instabilities in High Power Fiber Amplifiers." *Optics Express* 19 (14): 13218–13224. doi:10.1364/OE.19.013218.

Engholm, M., P. Jelger, F. Laurell, and L. Norin. 2009. "Improved Photodarkening Resistivity in Ytterbium-Doped Fiber Lasers by Cerium Codoping." *Optics Letters* 34 (8): 1285–1287. doi:10.1364/OL.34.001285.

Engholm, M. and L. Norin. 2008. "Preventing Photodarkening in Ytterbium-Doped High Power Fiber Lasers; Correlation to the UV-Transparency of the Core Glass." *Optics Express* 16 (2): 1260. doi:10.1364/oe.16.001260.

Engin, D., W. Lu, H. Verdun, and S. Gupta. 2013. *High Power Modal Instability Measurements of Very Large Mode Area (VLMA) Step Index Fibers.* Vol. 8733. 10.1117/12.2016106.

Fermann, M.E. 1998. "Single-Mode Excitation of Multimode Fibers with Ultrashort Pulses." *Optics Letters* 23 (1): 52–54. doi:10.1364/OL.23.000052.

Fermann, M.E. and I. Hartl. 2009. "Fiber Laser Based Hyperspectral Sources." *Laser Physics Letters* 6 (1): 11–21. doi:10.1002/lapl.200810090.

Filippov, V., Y. Chamorovskii, J. Kerttula, K. Golant, M. Pessa, and O.G. Okhotnikov. 2008. "Double Clad Tapered Fiber for High Power Applications." *Optics Express* 16 (3): 1929–1944. doi:10.1364/OE.16.001929.

Gapontsev, V., D. Gapontsev, N. Platonov, O. Shkurikhin, V. Fomin, A. Mashkin, M. Abramov, and S. Ferin. 2005. "2 KW CW Ytterbium Fiber Laser with Record Diffraction-Limited Brightness." In *CLEO/Europe. 2005 Conference on Lasers and Electro-Optics Europe, 2005*, 508. doi:10.1109/CLEOE.2005.1568286.

Gapontsev, V. and I. Samartsev. 1999. "Coupling Arrangement Between a Multimode Light Source and an Optical Fiber Through an Intermediate Optical Fiber Length." *US Patent 5,999,673.* Google Patents.

Goldberg, L., D.J. Ripin, E. Snitzer, and B. Cole. 1996. "V-Groove Side-Pumped 1.5-µm Fiber Amplifier." In *Conference on Lasers and Electro-Optics*, 9:CTuU1. OSA Technical Digest. Anaheim, California, Optical Society of America.

Gray, S., D.T. Walton, X. Chen, J. Wang, M. Li, A. Liu, A.B. Ruffin, J.A. Demeritt, and L.A. Zenteno. 2009. "Optical Fibers with Tailored Acoustic Speed Profiles for Suppressing Stimulated Brillouin Scattering in High-Power, Single-Frequency Sources." *IEEE Journal of Selected Topics in Quantum Electronics* 15 (1): 37–46. doi:10.1109/JSTQE.2008.2010240.

Grudiain, A.B., J. Nilsson, P.W. Turner, C.C. Renaud, W.A. Clarkson, and D.N. Payne. 1999. "Single Clad Coiled Optical Fiber for High Power Lasers and Amplifiers." In *Technical Digest. Summaries of Papers Presented at the Conference on Lasers and Electro-Optics. Postconference Edition. CLEO '99. Conference on Lasers and Electro-Optics (IEEE Cat. No. 99CH37013)*, CPD26/1–CPD26/2. doi:10.1109/CLEO.1999.834627.

Grudinin, A.B., D.N. Payne, P.W. Turner, J. Nilson, M.N. Zervas, M. Ibsen, and M.K. Durkin. 2000. "Multi-Fiber Arrangements for High Power Lasers and Amplifiers." *U.S. Patent 6826335.*

Haarlammert, N., O. de Vries, A. Liem, A. Kliner, T. Peschel, T. Schreiber, R. Eberhardt, and A. Tünnermann. 2012. "Build up and Decay of Mode Instability in a High Power Fiber Amplifier." *Optics Express* 20 (12): 13274–13283. doi:10.1364/OE.20.013274.

Hansen, K. R., T. T. Alkeskjold, J. Broeng, and J. Lægsgaard. 2013. "Theoretical Analysis of Mode Instability in High-Power Fiber Amplifiers." *Optics Express* 21 (2): 1944–1971.

Hemming, A., N. Simakov, J. Haub, and A. Carter. 2014. "High Power Resonantly Pumped Holmium-Doped Fiber Sources." *Proceedings of SPIE - The International Society for Optical Engineering* 8982 (February). doi:10.1117/12.2045489.

Horley, R., S. Norman, and M.N. Zervas. 2007. "Progress and Development in Fiber Laser Technology." In *Proceedings of SPIE* 6738. doi:10.1117/12.753171.

Hu, I.-N., C. Zhu, C. Zhang, A. Thomas, and A. Galvanauskas. 2013. "Analytical Time-Dependent Theory of Thermally Induced Modal Instabilities in High Power Fiber Amplifiers." *Fiber Lasers X: Technology, Systems, and Applications* 8601: 860109. International Society for Optics and Photonics.

Huang, L., W. Wang, J. Leng, S. Guo, X. Xu, and X. Cheng. 2014. "Experimental Investigation on Evolution of the Beam Quality in a 2-KW High Power Fiber Amplifier." *IEEE Photonics Technology Letters* 26 (1): 33–36. doi:10.1109/LPT.2013.2287195.

Jackson, S.D. 2004. "Cross Relaxation and Energy Transfer Upconversion Processes Relevant to the Functioning of 2 Mm Tm^{3+}-Doped Silica Fiber Lasers." *Optics Communications* 230 (1–3): 197–203. doi:10.1016/j.optcom.2003.11.045.

Jackson, S.D. and T.A. King. 1999. "Theoretical Modeling of Tm-Doped Silica Fiber Lasers." *Journal of Lightwave Technology* 17 (5): 948–956. doi:10.1109/50.762916.

Jansen, F., F. Stutzki, H.-J. Otto, T. Eidam, A. Liem, C. Jauregui, J. Limpert, and A. Tünnermann. 2012. "Thermally Induced Waveguide Changes in Active Fibers." *Optics Express* 20 (4): 3997–4008. doi:10.1364/OE.20.003997.

Jauregui, C., T. Eidam, H.-J. Otto, F. Stutzki, F. Jansen, J. Limpert, and A. Tünnermann. 2012. "Physical Origin of Mode Instabilities in High-Power Fiber Laser Systems." *Optics Express* 20 (12): 12912–12925.

Jeong, Y., J.K. Sahu, D.N. Payne, and J. Nilsson. 2004. "Ytterbium-Doped Large-Core Fiber Laser with 1.36 KW Continuous-Wave Output Power." *Optics Express* 12 (25): 6088. doi:10.1364/OPEX.12.006088.

Jeong, Y., J. Nilsson, J.K. Sahu, D.N. Payne, R. Horley, L.M.B. Hickey, and P.W. Turner. 2007. "Power Scaling of Single-Frequency Ytterbium-Doped Fiber Master-Oscillator Power-Amplifier Sources up to 500 W." *IEEE Journal of Selected Topics in Quantum Electronics* 13 (3): 546–551. doi:10.1109/JSTQE.2007.896639.

Jetschke, S., S. Unger, M. Leich, and J. Kirchhof. 2012. "Photodarkening Kinetics as a Function of Yb Concentration and the Role of Al Codoping." *Applied Optics* 51 (32): 7758–7764. doi:10.1364/AO.51.007758.

Jetschke, S., S. Unger, U. Röpke, and J. Kirchhof. 2007. "Photodarkening in Yb Doped Fibers: Experimental Evidence of Equilibrium States Depending on the Pump Power." *Optics Express* 15 (22): 14838–14843. doi:10.1364/OE.15.014838.

Jung, Y., Y. Jeong, G. Brambilla, and D.J. Richardson. 2009. "Adiabatically Tapered Splice for Selective Excitation of the Fundamental Mode in a Multimode Fiber." *Optics Letters* 34 (15): 2369–2371. doi:10.1364/OL.34.002369.

Karlsen, S.R., R.K. Price, M. Reynolds, A. Brown, R. Mehl, S. Patterson, and Robert J. Martinsen. 2009. "100-W 105-Um 0.15NA Fiber Coupled Laser Diode Module." In *Proceedings of SPIE* 7198. doi:10.1117/12.809710.

Karow, M., H. Tünnermann, J. Neumann, D. Kracht, and P. Weßels. 2012. "Beam Quality Degradation of a Single-Frequency Yb-Doped Photonic Crystal Fiber Amplifier with Low Mode Instability Threshold Power." *Optics Letters* 37 (20): 4242–4244. doi:10.1364/OL.37.004242.

Kashyap, R. and K.J. Blow. 1988. "Observation of Catastrophic Self-Propelled Self-Focusing in Optical Fibers." *Electronics Letters* 24 (1): 47–49. doi:10.1049/el:19880032.

Kawakami, S. and S. Nishida. 1974. "Characteristics of a Doubly Clad Optical Fiber with a Low-Index Inner Cladding." *IEEE Journal of Quantum Electronics* 10 (12): 879–887. doi:10.1109/JQE.1974.1068118.

Khitrov, V., K. Farley, R. Leveille, J. Galipeau, I. Majid, S. Christensen, B. Samson, and K. Tankala. 2010. "KW Level Narrow Linewidth Yb Fiber Amplifiers for Beam Combining." *Proceedings of SPIE* 7686 (April). doi: 10.1117/12.862648.

Kim, J., P. Dupriez, C. Codemard, J. Nilsson, and J.K. Sahu. 2006. "Suppression of Stimulated Raman Scattering in a High Power Yb-Doped Fiber Amplifier Using a W-Type Core with Fundamental Mode Cut-Off." *Optics Express* 14 (12): 5103–5113. doi: 10.1364/OE.14.005103.

Kliner, D.A.V., K. Chong, J. Franke, T. Gordon, J. Gregg, W. Gries, H. Hu, et al. 2011. "4-KW Fiber Laser for Metal Cutting and Welding." In Fiber Lasers VIII: Technology, Systems, and Applications (Vol. 7914, p. 791418). International Society for Optics and Photonics.

Kliner, D., J. Koplow, L. Goldberg, A. Carter, and J.A. Digweed. 2001. "Polarization-Maintaining Amplifier Employing Double-Clad Bow-Tie Fiber." *Optics Letters* 26 (February): 184–186. doi: 10.1364/OL.26.000184.

Koponen, J., M. Söderlund, H.J. Hoffman, D. Kliner, and J. Koplow. 2007. "Photodarkening Measurements in Large-Mode-Area Fibers - Art. No. 64531E." *Proceedings of SPIE - The International Society for Optical Engineering* 6453 (January). doi: 10.1117/12.712545.

Kosterin, A., V. Temyanko, M. Fallahi, and M. Mansuripur. 2004. "Tapered Fiber Bundles for Combining High-Power Diode Lasers." *Applied Optics* 43 (19): 3893–3900. doi: 10.1364/AO.43.003893.

Laurila, M., M.M. Jørgensen, K.R. Hansen, T.T. Alkeskjold, J. Broeng, and J. Lægsgaard. 2012. "Distributed Mode Filtering Rod Fiber Amplifier Delivering 292W with Improved Mode Stability." *Optics Express* 20 (5): 5742–5753. doi: 10.1364/OE.20.005742.

Leger, J.R. and W.C. Goltsos. 1992. "Geometrical Transformation of Linear Diode-Laser Arrays for Longitudinal Pumping of Solid-State Lasers." *IEEE Journal of Quantum Electronics* 28 (4): 1088–1100. doi: 10.1109/3.135232.

Li, M.-J., X. Chen, A. Liu, S. Gray, J. Wang, D.T. Walton, and L.A. Zenteno. 2009. "Limit of Effective Area for Single-Mode Operation in Step-Index Large Mode Area Laser Fibers." *Journal of Lightwave Technology* 27 (15): 3010–3016.

Li, Q. 2009. "Research on Stimulated Brillouin Scattering Suppression Based on Multi-Frequency Phase Modulation." *Chinese Optics Letters - CHIN OPT LETT* 7 (January): 29–31. doi: 10.3788/COL20090701.0029.

Li, Z., J. Zhou, W. Wang, B. He, Y. Xue, and Q. Lou. 2009. "Limitations of Coiling Technique for Mode Controlling of Multimode Fiber Lasers." In *2009 Conference on Lasers & Electro-Optics & the Pacific Rim Conference on Lasers and Electro-Optics*, 1–2. doi: 10.1109/CLEOPR.2009.5292572.

Limpert, J., N. Deguil-Robin, S. Petit, I. Manek-Hönninger, F. Salin, P. Rigail, C. Hönninger, and E. Mottay. 2005. "High Power Q-switched Yb-Doped Photonic Crystal Fiber Laser Producing Sub-10 Ns Pulses." *Applied Physics B* 81 (1): 19–21. doi: 10.1007/s00340-005-1820-7.

Limpert, J., S. Höfer, A. Liem, H. Zellmer, A. Tünnermann, S. Knoke, and H. Voelckel. 2002. "100-W Average-Power, High-Energy Nanosecond Fiber Amplifier." *Applied Physics B* 75 (January): 477–479. doi: 10.1007/s00340-002-1018-1.

Limpert, J., F. Stutzki, F. Jansen, H.-J. Otto, T. Eidam, C. Jauregui, and A. Tünnermann. 2012. "Yb-Doped Large-Pitch Fibers: Effective Single-Mode Operation Based on Higher-Order Mode Delocalisation." *Light: Science & Applications* 1 (April): e8.

Liu, A. and K.-i. Ueda. 1996. "Propagation Losses of Pump Light in Rectangular Double-Clad Fibers." *Optical Engineering* 35 (11): 3130–3134.

Liu, C., B. Ehlers, F. Doerfel, S. Heinemann, A. Carter, K. Tankala, J. Farroni, and A. Galvanauskas. 2004. "810 W Continuous-Wave and Single-Transverse-Mode Fiber

Laser Using 20 /Spl Mu/m Core Yb-Doped Double-Clad Fiber." *Electronics Letters* 40 (23): 1471–1472. doi:10.1049/el:20046464.

Ma, X., I.-N. Hu, and A. Galvanauskas. 2011. "Propagation-Length Independent SRS Threshold in Chirally-Coupled-Core Fibers." *Optics Express* 19 (23): 22575–22581. doi:10.1364/OE.19.022575.

Marcuse, D. 1977. "Loss Analysis of Single-Mode Fiber Splices." *The Bell System Technical Journal* 56 (5): 703–718. doi:10.1002/j.1538-7305.1977.tb00534.x.

Minelly, J., L. Spinelli, R. Tumminelli, S. Govorkov, D. Anthon, E. Pooler, R. Pathak, et al. 2011. "All-Glass KW Fiber Laser End-Pumped by MCCP-Cooled Diode Stacks." In *CLEO/Europe and EQEC 2011 Conference Digest*, CJ_P14. OSA Technical Digest (CD). Munich, Optical Society of America.

Morasse, B., S. Chatigny, E. Gagnon, C. Hovington, J.-P. Martin, and J.-P. de Sandro. 2007. "Low Photodarkening Single Cladding Yttetbium Fiber Amplifier - Art. No. 64530H." *Proceedings of SPIE - The International Society for Optical Engineering* 6453 (February). doi:10.1117/12.700529.

Mortensen, N.A. 2007. "Air-Clad Fibers: Pump Absorption Assisted by Chaotic Wave Dynamics?" *Optics Express* 15 (14): 8988. doi:10.1364/oe.15.008988.

Naderi, S., I. Dajani, T. Madden, and C. Robin. 2013. "Investigations of Modal Instabilities in Fiber Amplifiers through Detailed Numerical Simulations." *Optics Express* 21 (13): 16111–16129.

Nilsson, J., S. Alam, J.A. Alvarez-Chavez, P.W. Turner, W.A. Clarkson, and A.B. Grudinin. 2003. "High-Power and Tunable Operation of Erbium-Ytterbium Co-Doped Cladding-Pumped Fiber Lasers." *IEEE Journal of Quantum Electronics* 39 (8): 987–994. doi: 10.1109/JQE.2003.814373.

Nilsson, J., W.A. Clarkson, R. Selvas, J.K. Sahu, P.W. Turner, S.-U. Alam, and A.B. Grudinin. 2004. "High-Power Wavelength-Tunable Cladding-Pumped Rare-Earth-Doped Silica Fiber Lasers." *Optical Fiber Technology* 10 (1): 5–30. doi:10.1016/j.yofte.2003.07.001.

Norman, S., M. Zervas, A. Appleyard, P. Skull, D. Walker, P. Turner, and I. Crowe. 2006. "Power Scaling of High-Power Fiber Lasers for Micromachining and Materials Processing Applications." In Fiber Lasers III: Technology, Systems, and Applications (Vol. 6102, p. 61021P). International Society for Optics and Photonics.

Norman, S. and M.N. Zervas. 2007. "Fiber Lasers Prove Attractive Industrial Applications." *Laser Focus World* 43 (8): 93–96.

O'Connor, M. and B. Shiner. 2011. "High Power Fiber Lasers for Industry and Defense." In *High Power Laser Handbook*. vol. 517, 532.

Otto, H.-J., F. Stutzki, F. Jansen, T. Eidam, C. Jauregui, J. Limpert, and A. Tünnermann. 2012. "Temporal Dynamics of Mode Instabilities in High-Power Fiber Lasers and Amplifiers." *Optics Express* 20 (14): 15710–15722. doi:10.1364/OE.20.015710.

Ou, P., P. Yan, M. Gong, W. Wei, and Y. Yuan. 2004. "Studies of Pump Light Leakage out of Couplers for Multi-Coupler Side-Pumped Yb-Doped Double-Clad Fiber Lasers." *Optics Communications* 239 (4): 421–428. doi:10.1016/j.optcom.2004.05.055.

Pare, C. 2003. "Influence of Inner Cladding Shape and Stress-Applying Parts on the Pump Absorption of a Double-Clad Fiber Amplifier."In Applications of Photonic Technology 6 (Vol. 5260, pp. 272-277). International Society for Optics and Photonics.

Pask, H.M., D.C. Hanna, A.C. Tropper, C.J. Mackechnie, P.R. Barber, J.M. Dawes, and R.J. Carman. 1995. "Ytterbium-Doped Silica Fiber Lasers: Versatile Sources for the 1–1.2 Mm Region." *IEEE Journal of Selected Topics in Quantum Electronics* 1 (1): 2–13. doi:10.1109/2944.468377.

Po, H., E. Snitzer, R. Tumminelli, L. Zenteno, F. Hakimi, N.M. Cho, and T. Haw. 1989. "Double Clad High Brightness Nd Fiber Laser Pumped by Gaalas Phased Array." In

Optical Fiber Communication Conference, 5:PD7. 1989 OSA Technical Digest Series. Houston, Texas, Optical Society of America. doi:10.1364/OFC.1989.PD7.

Polynkin, P., V. Temyanko, M. Mansuripur, and N. Peyghambarian. 2004. "Efficient and Scalable Side Pumping Scheme for Short High-Power Optical Fiber Lasers and Amplifiers." *IEEE Photonics Technology Letters* 16 (9): 2024–2026. doi:10.1109/LPT.2004.831977.

Rastogi, V. and K.S. Chiang. 2001. "Propagation Characteristics of a Segmented Cladding Fiber." *Optics Letters* 26 (8): 491–493. doi:10.1364/OL.26.000491.

Rath, W. 2012. "Lasers For Industrial Production Processing: Tailored Tools With Increasing Flexibility," February, 4. doi:10.1117/12.906643.

Redmond, S.M., D.J. Ripin, C.X. Yu, S.J. Augst, T.Y. Fan, P.A. Thielen, J.E. Rothenberg, and G.D. Goodno. 2012. "Diffractive Coherent Combining of a 2.5 KW Fiber Laser Array into a 1.9 KW Gaussian Beam." *Optics Letters* 37 (14): 2832–2834. doi:10.1364/OL.37.002832.

Renaud, C.C., H.L. Offerhaus, J.A. Alvarez-Chavez, J. Nilsson, W.A. Clarkson, P.W. Turner, D.J. Richardson, and A.B. Grudinin. 2001. "Characteristics of Q-switched Cladding-Pumped Ytterbium-Doped Fiber Lasers with Different High-Energy Fiber Designs." *IEEE Journal of Quantum Electronics* 37 (2): 199–206. doi:10.1109/3.903069.

Richardson, D.J., J. Nilsson, and W.A. Clarkson. 2010. "High Power Fiber Lasers: Current Status and Future Perspectives [Invited]." *Journal of the Optical Society of America B* 27 (11): B63–B92. doi:10.1364/JOSAB.27.000B63.

Russell, P.J. 2006. "Photonic-Crystal Fibers." *Journal of Lightwave Technology* 24 (12): 4729–4749.

Sahu, J., C. Renaud, K. Furusawa, R. Selvas, J. Alvarez-Chavez, D.J. Richardson, and J. Nilsson. 2001. "Jacketed Air-Clad Cladding Pumped Ytterbium-Doped Fiber Laser with Wide Tuning Range." *Electronics Letters* 37 (September): 1116–1117. doi:10.1049/el:20010753.

Schermer, R.T. 2007. "Mode Scalability in Bent Optical Fibers." *Optics Express* 15 (24): 15674–15701. doi:10.1364/OE.15.015674.

Selvas, R., J.K. Sahu, L.B. Fu, J.N. Jang, J. Nilsson, A.B. Grudinin, K.H. Ylä-Jarkko, S.A. Alam, P.W. Turner, and J. Moore. 2003. "High-Power, Low-Noise, Yb-Doped, Cladding-Pumped, Three-Level Fiber Sources at 980nm." *Optics Letters* 28 (13): 1093–1095. doi:10.1364/OL.28.001093.

Seo, H.-S., J.T. Ahn, B.J. Park, J.-H. Song, and W. Chung. 2011. "Efficient Pump Beam Multiplexer Based on Single-Mode Fibers." *Japanese Journal of Applied Physics* 51 (1): 10203. doi:10.1143/jjap.51.010203.

Sévigny, B., X. Zhang, M. Garneau, M. Faucher, Y.K. Lizé, and N. Holehouse. 2008. "Modal Sensitivity Analysis for Single Mode Operation in Large Mode Area Fiber." *Proceedings of SPIE* 6873. doi:10.1117/12.775101.

Shiner, B. 2013. "The Impact of Fiber Laser Technology on the World Wide Material Processing Market." In *CLEO: 2013*, AF2J.1. OSA Technical Digest (Online). San Jose, CA, Optical Society of America. doi:10.1364/CLEO_AT.2013.AF2J.1.

Singh, S., R. Gangwar, and N. Singh. 2007. "Nonlinear Scattering Effects in Optical Fibers." *Progress in Electromagnetics Research-Pier - PROG ELECTROMAGN RES* 74 (January): 379–405. doi:10.2528/PIER07051102.

Smith, A.V., B.T. Do, G.R. Hadley, and R.L. Farrow. 2009. "Optical Damage Limits to Pulse Energy From Fibers." *IEEE Journal of Selected Topics in Quantum Electronics* 15 (1): 153–158. doi:10.1109/JSTQE.2008.2010331.

Smith, A.V. and J.J. Smith. 2011. "Mode Instability in High Power Fiber Amplifiers." *Optics Express* 19 (11): 10180–10192.

Smith, A.V. and J.J. Smith. 2013a. "Increasing Mode Instability Thresholds of Fiber Amplifiers by Gain Saturation." *Optics Express* 21 (13): 15168–15182.

Smith, A.V. and J.J. Smith. 2013b. "Steady-Periodic Method for Modeling Mode Instability in Fiber Amplifiers." *Optics Express* 21 (3): 2606–2623.

Snitzer, E., H. po, F. Hakimi, R. Tumminelli, and B. Mccollum. 1988. *Double-Clad, Offset Core Nd Fiber Laser.* doi:10.1364/OFS.1988.PD5.

Stolen, R. and J. Bjorkholm. 1982. "Parametric Amplification and Frequency Conversion in Optical Fibers." *IEEE Journal of Quantum Electronics* 18 (7): 1062–1072. doi:10.1109/JQE.1982.1071660.

Su, R., P. Zhou, H. Lü, X. Wang, C. Luo, and X. Xu. 2014. "Numerical Analysis on Impact of Temporal Characteristics on Stimulated Brillouin Scattering Threshold for Nanosecond Laser in an Optical Fiber." *Optics Communications* 316: 86–90. doi:10.1016/j.optcom.2013.10.073.

Sudesh, V., T. Mccomb, Y. Chen, M. Bass, M. Richardson, J. Ballato, and A.E. Siegman. 2008. "Diode-Pumped 200 Mm Diameter Core, Gain-Guided, Index-Antiguided Single Mode Fiber Laser." *Applied Physics B: Lasers and Optics* 90 (March): 369–372. doi:10.1007/s00340-008-2947-0.

Taverner, D., D.J. Richardson, L. Dong, J.E. Caplen, K. Williams, and R.V. Penty. 1997. "158-MJ Pulses from a Single-Transverse-Mode, Large-Mode-Area Erbium-Doped Fiber Amplifier." *Optics Letters* 22 (6): 378–380. doi:10.1364/OL.22.000378.

Teodoro, F.D., M.K. Hemmat, J. Morais, and E.C. Cheung. 2010. "High Peak Power Operation of a 100μm-Core Yb-Doped Rod-Type Photonic Crystal Fiber Amplifier." In Fiber Lasers VII: Technology, Systems, and Applications (Vol. 7580, p. 758006). International Society for Optics and Photonics.

Tobergte, D.R. and S. Curtis. 2013. "Tapered Fiber Bundles for Coupling Light into and out of Cladding-Pumped Fiber Devices." *Journal of Chemical Information and Modeling*. Google Patents. doi:10.1017/CBO9781107415324.004.

Travers, J.C., A.B. Rulkov, B.A. Cumberland, S.V. Popov, and J.R. Taylor. 2008. "Visible Supercontinuum Generation in Photonic Crystal Fibers with a 400W Continuous Wave Fiber Laser." *Optics Express* 16 (19): 14435–14447. doi:10.1364/OE.16.014435.

Tünnermann, A., T. Schreiber, and J. Limpert. 2010. "Fiber Lasers and Amplifiers: An Ultrafast Performance Evolution." *Applied Optics* 49 (25): F71–F78. doi:10.1364/AO.49.000F71.

Wadsworth, W.J., R.M. Percival, G. Bouwmans, J.C. Knight, T.A. Birks, T.D. Hedley, and P.S.J. Russell. 2004. "Very High Numerical Aperture Fibers." *IEEE Photonics Technology Letters* 16 (3): 843–845. doi:10.1109/LPT.2004.823689.

Wadsworth, W.J., R.M. Percival, G. Bouwmans, J.C. Knight, and P.J. Russell. 2003. "High Power Air-Clad Photonic Crystal Fiber Laser." *Optics Express* 11 (1): 48–53. doi:10.1364/OE.11.000048.

Wang, Y. 2004. "Heat Dissipation in Kilowatt Fiber Power Amplifiers." *IEEE Journal of Quantum Electronics* 40 (6): 731–740. doi:10.1109/JQE.2004.828271.

Wang, Y. and C.Q. Xu. 2007. "Actively Q-switched Fiber Lasers: Switching Dynamics and Nonlinear Processes." *Progress in Quantum Electronics* 31 (3–5): 131–216. doi:10.1016/j.pquantelec.2007.06.001.

Ward, B., C. Robin, and I. Dajani. 2012. "Origin of Thermal Modal Instabilities in Large Mode Area Fiber Amplifiers." *Optics Express* 20 (10): 11407–11422. doi:10.1364/OE.20.011407.

Ward, B.G. 2013. "Modeling of Transient Modal Instability in Fiber Amplifiers." *Optics Express* 21 (10): 12053–12067.

Wei, T., Z. Tan, J. Li, and J. Zhu. 2013. "Theoretical and Experimental Study of the Pump Pulse Width Optimization of the Yb-Doped Fiber Amplifier." *Optik* 124 (16): 2459–2462. doi:10.1016/j.ijleo.2012.08.010.

Wetter, A., M. Faucher, M. Lovelady, and F. Séguin. 2007. "Tapered Fused-Bundle Splitter Capable of 1kW CW Operation – Art. No. 64530I." *Proceedings of SPIE – The International Society for Optical Engineering* 6453 (February). doi:10.1117/12.700466.

Wirth, C., T. Schreiber, M. Rekas, I. Tsybin, T. Peschel, R. Eberhardt, and A. Tünnermann. 2010. "High-Power Linear-Polarized Narrow Linewidth Photonic Crystal Fiber Amplifier." In *Fiber Lasers VII: Technology, Systems, and Applications*, 7580:75801H. International Society for Optics and Photonics.

Wu, R., J.D. Myers, and M. Myers. 2001. "High Power Rare-Earth-Doped Phosphate Glass Fiber and Fiber Laser." *Proceedings of SPIE – The International Society for Optical Engineering* 4267 (January). doi:10.1364/ASSL.2001.MB2.

Xiao, Q.-r., P. Yan, S. Yin, J. Hao, and M. Gong. 2010. "100 W Ytterbium-Doped Monolithic Fiber Laser with Fused Angle-Polished Side-Pumping Configuration." *Laser Physics Letters* 8 (2): 125.

Xiao, Q., P. Yan, J. He, Y. Wang, X. Zhang, and M. Gong. 2011. "Tapered Fused Fiber Bundle Coupler Capable of 1 KW Laser Combining and 300 W Laser Splitting." *Laser Physics - LASER PHYS* 21 (August): 1415–1419. doi:10.1134/S1054660X11150308.

Xiao, Y., F. Brunet, M. Kanskar, M. Faucher, A. Wetter, and N. Holehouse. 2012. "1-Kilowatt CW All-Fiber Laser Oscillator Pumped with Wavelength-Beam-Combined Diode Stacks." *Optics Express* 20 (3): 3296–3301. doi:10.1364/OE.20.003296.

Yu, C.X., S.J. Augst, S.M. Redmond, K.C. Goldizen, D.V. Murphy, A. Sanchez, and T.Y. Fan. 2011. "Coherent Combining of a 4 KW, Eight-Element Fiber Amplifier Array." *Optics Letters* 36 (14): 2686–2688. doi:10.1364/OL.36.002686.

Zervas, M., M. Durkin, F. Ghiringhelli, K. Vysniauskas, L. Hickey, A. Gillooly, P.W. Turner, and B. Kao. 2006. "High Peak Power, High Rep-Rate Pulsed Fiber Laser for Marking Applications - Art. No. 61020Q." *Proceedings of SPIE - The International Society for Optical Engineering* (February). doi:10.1117/12.645553.

Zervas, M.N. 2010. "High Power Fiber Lasers: From Lab Experiments to Real World Applications." *AIP Conference Proceedings* 1288 (1): 63–66. doi:10.1063/1.3521373.

Zervas, M.N. and C.A. Codemard. 2014. "High Power Fiber Lasers: A Review." *IEEE Journal on Selected Topics in Quantum Electronics* 20 (5). doi:10.1109/JSTQE.2014.2321279.

Zervas, M.N., F. Ghiringhelli, M.K. Durkin, and I. Crowe. 2011. "Distribution of Photodarkening-Induced Loss in Yb-Doped Fiber Amplifiers."In Fiber Lasers VIII: Technology, Systems, and Applications (Vol. 7914, p. 79140L). International Society for Optics and Photonics.

Zheng, C., H.T. Zhang, W.Y. Cheng, M. Liu, P. Yan, and M.L. Gong. 2012. "11-MJ Pulse Energy Wideband Yb-Doped Fiber Laser." *Optics Communications* 285 (17): 3623–3626. doi:10.1016/j.optcom.2012.04.033.

Zhu, J., P. Zhou, Y. Ma, X. Xu, and Z. Liu. 2011. "Power Scaling Analysis of Tandem-Pumped Yb-Doped Fiber Lasers and Amplifiers." *Optics Express* 19 (19): 18645–18654. doi:10.1364/OE.19.018645.

Zimer, H., M. Kozak, A. Liem, F. Flohrer, F. Doerfel, P. Riedel, S. Linke, et al. 2011. "Fibers and Fiber-Optic Components for High-Power Fiber Lasers."In Fiber Lasers VIII: Technology, Systems, and Applications (Vol. 7914, p. 791414). International Society for Optics and Photonics.

Index

Printed in the United States
by Baker & Taylor Publisher Services

Printed in the United States
by Baker & Taylor Publisher Services